Industrial Design with Microcomputers

Industrial Design with Microcomputers

STEVEN K. ROBERTS

PRENTICE-HALL, INC.

Englewood Cliffs, New Jersey 07632

Library of Congress Cataloging in Publication Data

Roberts, Steven K.
  Industrial design with microcomputers.

  Includes index.
  1. Process control—Data processing.
  2. Manufacturing processes—Data processing.
  3. Microcomputers. I. Title.
  TS156.8.R6   620'.00425'02854   81-18265
  ISBN   0-13-459461-4

Editorial/production supervision by *Lori Opre*
Cover design by *David Caudill*
Manufacturing buyer: *Gordon Osbourne*

Printed in the United States of America

10   9   8   7   6   5   4   3   2   1

ISBN   0-13-459461-4

Prentice-Hall International, Inc., *London*
Prentice-Hall of Australia Pty. Limited, *Sidney*
Prentice-Hall of Canada, Ltd., *Toronto*
Prentice-Hall of India Private Limited, *New Delhi*
Prentice-Hall of Japan, Inc., *Tokyo*
Prentice-Hall of Southeast Asia Pte. Ltd., *Singapore*
Whitehall Books Limited, *Wellington, New Zealand*

to
LORI

Contents

2 THE INTELLIGENT BLACK BOX   17

3 INSIDE THE BOX   30

4 HARDWARE TOOLS   52

PART II

7 TO TEACH A MACHINE   155

8 SOFTWARE TOOLS   170

9 ARTIFICIAL INTELLIGENCE   222

10 APPROACHING A DESIGN PROJECT   244

11   SOFTWARE TACTICS   267

12   SYSTEM INTEGRATION   286

PART III

13   A DISTRIBUTED INDUSTRIAL SYSTEM   301

14   THE DATA COLLECTION TERMINALS   327

15   A POTPOURRI OF APPLICATIONS   360

Preface

There was a time, not so long ago, when it was possible to speak of "the mighty microprocessor" as a relatively isolated phenomenon. Although the devices were supremely flexible replacements for random logic, they had not quite changed the world. Not yet, anyway.

But it was inevitable. Such concentrated and cheap processing horsepower couldn't help but endow everything it touched with new energy, and the infant technology became, within a few years, an essential part of the environment in which every other industry functions.

This caused as many problems as it solved. Not the least of these was the supreme difficulty of learning what it was all about: where most other technologies have a rich heritage on which to draw, this one had only the obscurest of roots in the vast body of engineering literature. Device data was readily available from the manufacturers, of course, but there was little or no established art that could serve as the basis for the creative efforts of individual system designers.

The most frequent question I heard during eight years of consulting, designing industrial systems, and selling small computers was, "How can I learn to design with microprocessors? What book do you recommend?"

I always had a hard time coming up with an answer, even when, in the late 1970s, the market became flooded with literature. There were books on languages, the micros themselves, interfacing, logic design, communications, and even engineering management—but none that adequately integrated the whole field of microcomputer-based design into a cohesive whole that could serve as the basis for detailed and task-specific information.

It was with awareness of this void that this book was written. With a real-world emphasis gleaned from (often painful) experience, I have attempted to bring together the widely diverse ideas that must be assimilated before creative system design can be effectively undertaken. Far from being another "how-to" text that starts with binary numbers and the author's favorite instruction set, this book is designed to provide the philosophical underpinnings for the wealth of information that can be found elsewhere.

Part I opens with an exploration of industrial system design requirements, and then sets the stage for an overview of microcomputer technology with a real application context. Next, avoiding product-specific data as much as possible, I sketch both the chip- and board-level tools that will be used as "black boxes" throughout the text. Our discussion of reality begins with the consideration of a wide range of interfacing techniques, after which the bad news is revealed: the ideal textbook world is not what is really out there.

Part II brings a sharp shift of emphasis with an introduction to the concepts of software design and a discussion of the various tools and methods that have evolved to make programming not only possible but productive as well. In this context, I take the time to consider industrial applications of artificial intelligence techniques, as well as the more prosaic but eminently worthwhile principles of structure, hardware/software balance, and system design philosophy. Part II ends with a close look at the development environment, including emulation and debugging techniques.

Part III is designed to apply everything covered so far by tracing the development and operation of a distributed industrial control system. This leads us into network communications, application of real-world design ideas, interfacing, human factors, software engineering, and a concrete system integration example. The book closes with a look at digital signal processing, a few relatively esoteric design techniques, and a brief potpourri of applications.

But as important as the specific subjects discussed is the attitude of respect for the reader's creativity and awareness that permeates the entire book. I have made the text casual—not in a trite or "underground" sort of way, but in the same fashion that would characterize correspondence with an intelligent friend. Much of the true meaning is between the lines, where ideas too delicate for blatant exposition are gently explored. While the classic textbook places a formal distance between the author and the student, this one invites readers to relax, kick off their shoes, and share understanding.

S.K.R.

Acknowledgments

A book, like any human creation, springs not from the author alone. This one is no exception: many people have touched my life in ways which have had a profound effect (directly or indirectly) upon these words. Although an exhaustive list is impossible, I would like to thank certain friends, associates, mentors, authors, and machines.

First, I am deeply indebted to Bob Phare, who not only devoted hundreds of evening and weekend hours to the creation of the illustrations, but also participated strongly in the refinement of the text. His penciled notes in the manuscript more than once spotlighted clumsy structure and logical gaps, and the text is scattered with the echoes of countless midnight tea-sodden brainstorming sessions. (Bob, in turn, thanks Rita, who somehow put up with all this.)

This book has also profited immeasurably from the close involvement of Lori Opre (editor par excellence), whose loving touch guided it through the production process, and held the enemies of quality at bay.

I would like to thank Steve Orr for over a decade of close and stimulating friendship, general consultation, insights into digital signal processing, and the PACK macro. Steve is very much a part of this book.

My parents, Phyllis and Ed Roberts, are also represented here: by nurturing my childhood fascination with all things electronic, by representing an attitude of quality in workmanship and expression, and by supporting me even through times of disagreement, they have shaped my work in ways for which I shall be forever grateful.

At this point, I quickly encounter the problem of implied prioritization embedded in anything other than arbitrary sequencing. That would be intolerable, so I now alphabetically thank the following:

Doug Bucheit of Robinson-Nugent Company, for the original concept behind the insertion-withdrawal force system.

David Caudill, for his inspired participation in this project through creation of the cover art.

Jim Dudick, for general consultation and the PL/I word count program.

David Heiserman, for introducing me to Prentice-Hall.

Doug Hofstadter, for his magnificent book, *Gödel, Escher, Bach*—an endlessly stimulating and thought-provoking volume if ever there was one.

Jim Jenal, for assistance in understanding UNIX, insights into the AI field, and off-site file backup facilities.

Jency Kelly, for CROMIX assistance and a long-desired system renovation.

Dan Landon, for his excellent work on the data collection system and for many hours of enthusiastic home-computer interaction during those exciting pre-kit years.

Al Messing, for his major creative role in the IDAC system and his subsequent efforts in helping me market it.

Susan Mowers, my meta-assistant, for dedicated proofreading, broad-spectrum support, and boundless enthusiasm.

Sylvia Nelson and Ron Pearson, for passing along the spark.

Vince Phillips of the Alkon Corporation, for contributing strongly to my understanding of structured design and software engineering.

Debra Roberts, for proofreading, frequent encouragement, superhuman tolerance, and much more.

David Wright, for creative assistance in a variety of microcomputer-based projects which helped shape this book.

I also thank the following for a variety of reasons far too diverse and complex for elaboration here:

John Barnard, Jim Berkey, David Boelio, Judy Borsuk, S. Jerome Clarke, Sam Cooke, Jim Dome, Glenn Glassner, Bill Godbout, Jim Helton, Roy Humphress, Ernest Kent, Brad Lizotte, Dick Lorimor, David Martin, Edgar Marven, Dan Meinerz, Jyll Matheny, Cheryl Miller, Chris Morgan, Phil Morgan, Alan O'Neill, John Raffauf, Tom Rouse, Frank Sharp, George Stege, John Stork, and the Dublin Ohio Police Department.

Last and most certainly not least, I want to thank BEHEMOTH, my reliable and 100% dedicated computer system, for tirelessly and heroically manipulating the text of this book and for potentiating a writing career by obsoleting the typewriter.

STEVEN K. ROBERTS

Dublin, Ohio

Introduction

0.0 A SHORT HISTORY

Somewhere in the midst of the 1970s, the "microelectronics revolution" that was already well under way escalated into a meta-revolution. First tentatively, and then with unreserved gusto, the major IC manufacturers turned their collective resources to the development and marketing of microprocessors.

The effect on the engineering community was immediate and profound. Some engineers, eyes bright with excitement, plunged headlong into this strange new form of system design which promised liberation from random logic. Others, more set in their ways, flipped uncomfortably past the growing profusion of magazine articles about micros, preferring to maintain complacent confidence in their existing specialties rather than become awkward beginners in this bizarre technology. Thus the "old dog" syndrome surfaced with a vengeance, and companies faced with competitive market realities began searching far and wide for people who could take the magical single-chip processors and make them dance. There were a few safe niches wherein the meek could take shelter from the onrush of alien concepts. But not for long.

Bit by bit, microprocessors began chipping away, so to speak, at even such traditionally nondigital edifices as TV circuit design, automotive electronics, large appliances, and communications systems. Manufacturers grew more and more frantic, some desperate West Coast operations even paying their employees healthy bounties to bring in new talent. It became clear to even the stodgiest of anti-micro diehards that this incomprehensible new technology was here to stay. Some became managers.

1

There was good reason for this meta-revolution. Aside from the sublime pleasures to be had in the world of microprocessor system design, there were some compelling economic advantages. In a nutshell, the need for new and dedicated hardware configurations for each and every product or custom controller could be eliminated and replaced with an elegant black box, appropriately interfaced with the rest of the world and coaxed into operation through the creation of "software."

This new approach to design was not, of course, free from the tyranny of the trade-off. The apparent economic freebie gained with the elimination of all that custom hardware had to be weighed against the new and unfamiliar costs of software design ... and interface design ... and the learning curves associated with a host of foreign concepts.

But "the market" was forceful in its demands. Brutal reality tipped the trade-off more and more in favor of micros—in nearly any application characterized by the need for a controller. Now, years later, the question is not whether to use a microprocessor, but how.

0.1 THE DOCUMENTATION OF MAGIC

This book is a concession to reality. Despite the maturity of the technology (or perhaps because of it), it is exceedingly difficult for a design engineer to meaningfully approach his or her first micro development projects armed only with a manufacturer's data book. Despite the ready availability of "complete documentation," there is something missing: a set of heuristics which integrate all that new product-specific data with the broad design concepts already possessed by the engineer. What are the boundaries between the processor and the rest of the world? How do you handle asynchronous events? How can you develop hardware and software independently, then put these two unknowns together and make them fit? What about noise and other manifestations of unavoidable universal truth?

How smart does this box have to be, anyway?

Tough ones. A look in the data book provides the number of nanoseconds between the start of a memory write pulse and the end of valid bus data; it even provides details of the instruction set and some suggested circuit configurations.

But between that and an industrial control system that must, glitchless, keep a production batcher going 24 hours a day, there is a great void. One of the ultimate objectives of this book is to fill that void with light and substance.

Prior to our excursion into the world of microprocessor system design, it is worthwhile to spend a moment considering just how we presume to accomplish such a feat.

This is not a normal textbook.

The classical approach to education seems, alas, to possess a fundamental flaw. It is well typified by the experience of engineering school: you walk in as a freshman and are immediately assailed by a wide variety of essentially random bits of data, ranging from Boyle's Law to the formulas for X-ray diffraction in a crystal; from limits and differentials to DO loops. It's all great and useful stuff, of course, but unless you already have a very sophisticated conceptual framework into which it can be fit, it may as well be a sequence of pseudorandom numbers.

An exaggeration, perhaps, but consider the thought processes that are taking place as these seemingly unconnected facts and concepts are presented in relentless staccato bursts throughout those first few bewildering semesters. Your mind, architecturally defined at every level as an associative processor, optimized for storage of information within the context of information already stored, is suddenly being asked to behave just like a computer. "Remember this, and this, and this. Don't worry, it will all fit together someday."

An associative system like the brain is forced into an unnatural and uncomfortable mode when required to retain random data for later assimilation. In most cases, by the time the long-heralded "fitting together" rolls around in the engineering curriculum, much of the component data that was forcibly injected in dozens of classrooms is either associated inappropriately or lost entirely.

This is a great tragedy, and has been responsible for more than a few incidents of attrition. It is not uncommon for highly creative and intelligent people to feel so uncomfortable in this backward educational structure that they recoil in dismay and disappointment from engineering, incorrectly judging it to be a dry and colorless pursuit.

Well, all is not lost. The solution is completely obvious, and is, in fact, likely to be found in almost any situation in which one person explains—to someone he knows—how something works. The key here is "someone he knows." Hmmmm.

Human communication (bear with me—we'll get to microprocessors in a moment) is a phenomenon of much wider information bandwidth than is detectable in the linguistic content of the utterances involved. When you describe a concept to a friend, your words carry substantially more meaning than their dictionary definitions, for their primary function is the gradual modification of some aspect of your friend's internal model of the world—a model that you share to some degree. It is thus unnecessary (and boring) to reduce your explanation of the concept to its absolute, explicit form; it's much more efficient to take advantage of established contexts and mutually meaningful symbols.

A good professor recognizes this intuitively, and treats interaction with a class more as a continuous process of synchronous model building than its opposite: a series of discrete, open-loop lectures which assume nothing but a set of prerequisites defined by a curriculum. The problem is with books. Textbooks.

When someone, either in academia or industry, seats himself before a typewriter (or, hopefully, a word processing system) and writes something called a textbook, a strange phenomenon seems to take place. All contextual frameworks that contain the material are stripped away, leaving just the facts. However expertly the writer develops these facts into concepts and usable tools, there is still, too often, a sharp boundary between the textbook and the real world. This gives textbooks a bad name.

That's depressing. Despite the obvious necessity of interacting with the real world to gain a deeper and more useful understanding than can be had by just reading about it, it would seem that a textbook on some subject could be made to fit the optimal learning modes of the human brain a bit more than most of them do. The reason most of them don't may be that those shared internal models of the world I rhapsodized about a moment ago simply do not exist between the lone writer at his machine and the anonymous reader, somewhere ... out there.

Well. Let's consider that problem.

I have, in all likelihood, never met you, and probably never will. I have no way of knowing whether you are reading this book for a class at an engineering school, trying to enhance your value to your employer as a system designer, adding to your understanding of microprocessors simply because they are fun, or just browsing in a library. So what am I supposed to do, abandon all ideas of shared context just because you could be anybody in the world? Start conventionally with binary numbers, two's-complement arithmetic, Boolean algebra, and all that other tedious stuff? Provide a list of prerequisite readings? All of the above? None of the above?

Yup, you got it—the answer is (E). Here are my assumptions, and buried within them is our shared context: You already have enough general knowledge about electronics and digital logic to recognize gates and deal with circuitry. You already know that the ideal "textbook world" (that word certainly does have a bad name!) of zero-rise-time pulses, noise-free circuits, and lumped constants is not exactly what actually exists out there (and if you don't know that, don't panic—it becomes obvious very quickly). You are at least intelligent enough to deal with engineering at the philosophical as well as the "crank-turning" level, and your healthy childhood curiosity somehow survived the schooling to which you have already been subjected. (It doesn't matter whether you are reading this as course material or as "professional upgrade" material—our purposes are identical.)

That doesn't seem like much of a shared context, does it? We're going to build 400-odd pages of information about microprocessors on *that*?

Sure, because we're going to deliberately develop it as we go along. If, at certain points, some background knowledge is required outside the domain of this book, I'll throw in references that will let you make a subroutine call without crashing your stack.

We have already abused the prefix "meta-" once herein; why not go for broke and cockily call this a meta-textbook? Let's do it.

Part I

1

The Rationale

1.0 PRIME FOCUS

Microcomputers can be viewed on a variety of different levels, ranging from blender controllers to human intelligence amplifiers.

Their technology has become a part of the environment in which every other industry functions. It thus follows that in any applications-oriented discussion of micros, the principles of the technology itself must coexist with the varied objectives of the industries to which it has been applied. This could have the effect of scattering our philosophical cohesion to the winds, so let's first talk about what we are *not* going to talk about.

One vast enterprise we'll overlook herein involves the myriad applications for microprocessors in consumer goods of various sorts. Many of the same principles apply, of course (so this will still be useful if you yearn to computerize can openers), but the economics are so thoroughly different that it may as well be a different world. In the mass-production business, you can be a big hit with management if you knock a dime off the cost of each unit; in the industrial design business, yanking out an $80 linear power supply and replacing it with a $200 switcher might solve a $10,000/hour problem. Both of these orthogonal economic viewpoints call for surpassing cleverness, but in the industrial world, there is the added satisfaction of being able to justify quality.

Another robust industry based on the microprocessor is the vast personal computer phenomenon, within which we might as well include small business computers and development systems—we're not going to talk about them anyway. We will, however, use them as tools.

As micros have become more and more powerful, their usefulness has extended far beyond the areas their designers originally visualized. Even "big systems," once proud and aloof in the presence of silly chips presumptious enough to call themselves "computers," have been invaded by great phalanxes of the devices. They are I/O and disk controllers, diagnostic systems, comm supervisors, smart terminals ... and they're even crowding the stuffy old CPUs themselves, now that high-speed 32-bit chips and multiprocessor architectures exist.

But we're not going to talk much about them, either.

There is hardly ever such a thing as an absolutely clear distinction. Somewhere in here, we have to accept the existence of blurred, overlapping boundaries between these neat categories. If we identify the classic form of consumer product as one engineering extreme and a system whose development cost is no object as representative of another, then we find, between the clustered categories at each end, a rather uniform continuum. If we draw any line at all in this region, it is arbitrary. Smart instruments, communications controllers, medical systems, and a dizzying variety of other microcomputer applications fall somewhere in this range. Although our prime focus is within the nonconsumer end of this spectrum, our depth of field extends well into its midst.

All that we have really managed to say here, of course, is that we're more interested in industrial control systems than any other specific application category, but the techniques we will be discussing in that context are widely applicable to others—within reason.

1.1 INDUSTRIAL REQUIREMENTS

The world of industrial system design is a brutal one, fraught with a variety of engineering requirements rarely found in the more humane arenas of microcomputer application. The reasons for this are at least sixfold: the astronomical costs of interrupted production, a magnificent panoply of environmental abuses ranging from thousand-volt noise spikes to vandalism, a relentless need for operational and design flexibility, the possibility of unexpected inputs and strange machine conditions, plant management's ongoing thirst for statistical information about its processes, and an urgent, usually neglected requirement for serviceability. To set the stage for our discussion of industrial microcomputers, let's look at these a little more closely.

1.1.1 Downtime

Nothing makes a plant manager more miserable than production that has ground to a halt. The economics can be staggering: in a typical case, 70 employees at an average cost of $9.30 per hour, plus a ceramic furnace with energy requirements of $206 per minute, plus miscellaneous plant overhead, equals over $20,000 per hour of cash down the drain when the line stops. Depending on the type of factory, this figure may be dwarfed by the costs associated with bringing the line back up.

All too often, such a stoppage can be traced to some innocuous little component, itself worth perhaps $5. This can be very embarrassing if you happen to be the design engineer behind the control system that contains that component, especially if the reason for the failure is something stupid like a 12-volt electrolytic in the filter for the 12-volt supply. Such a thing is never excusable, of course, but in an industrial system it can do more than just annoy somebody.

Here is another example. One of the systems we discuss later in the book is a real-time assembly machine controller involved in the continuous manufacture of electromechanical components (we'll call 'em widgets—it's a proprietary process). The machine, including setup time and breaks, produces somewhere around 500 widgets a minute or about 720,000 per day for all three shifts. If each one represents a gross profit of, say, 10 cents, this single machine produces a daily cash flow of $72,000—over $2 million per month. But that's not all: widgets don't just tumble out of the machine into customers' waiting hands; they are inputs to other assembly processes that integrate them into finished products (we'll call 'em gadgets). If the controller dies, then not only does the widget machine lay idle, but the gadget machines do as well—as do the people who run them (Figure 1-1).

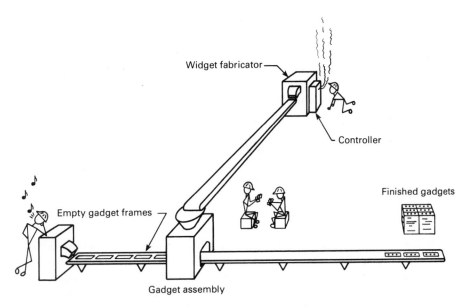

Figure 1-1   The failure of a "widget-machine" controller has resulted in production stoppage on the "gadget" assembly line, demonstrating that the scope of a failure is not necessarily local.

This all points to the need for very high reliability in any kind of industrial control system, a requirement that affects every stage of design and determines the necessary level of field support. A system whose mean time between failures would be impressive in an office environment may be viciously denounced as "garbage" by an industrial user.

From the standpoint of engineering aesthetics, such stringent requirements can be viewed as a very good thing. Where else could you freely use the best of everything, being more concerned with quality than with cost savings? Most industrial customers, once they have swallowed the bitter pill of bargain electronics a time or two, are more than happy to pay well for insurance against future disasters.

1.1.2 Environmental Assault

The industrial environment is such an awesome subject that it will not only have its very own chapter (Chapter 6), but will also appear in various forms to alternately haunt and entertain us as we progress through other matters.

Consider a foundry from a microcomputer's viewpoint (Figure 1-2).

Every few minutes, a 2800-volt spike bursts through the back door via the power line, and somebody is welding on a girder overhead—causing the ac to drop periodically to about 94 volts.

The communications channel from the plant computer is currently sending a string of command parameters, but we just missed a byte or two in a burst of electromagnetic interference from an unknown source.

The sun is shining on the side of the cabinet, the ambient is 32.8 degrees Celsius, and the temperature in the card cage is just over twice that because the fan's filter is clogged with a thick accumulation of airborne crud. A thin layer of that crud is all over the circuit boards, since the maintenance electrician lets the system run open while he cleans the filters. Unfortunately, it is a mixture of fine metallic particles and silicates coated with a sticky acid.

A solenoid driver is on the verge of going over the hill because the designer forgot to put in a protection diode. The negative inductive spikes are abusing the driver transistor.

The machine on which the control system is mounted vibrates continuously at about 16 hertz, with an excursion of 0.7 millimeter. One of the edge connectors is working loose because the last technician who serviced the unit hates those little PC board retaining clamps and leaves them lying in the bottom of the cabinet.

A connector of another variety has problems, too: a cost-saving octal socket on the side carrying some opto sense lines has three corroded pins—somebody spilled a Coke. A front panel switch is broken because a disgruntled employee saw fit to actuate it with a hammer. There is an ugly, rusted dent in the side from a fork-lift tine, and somebody has scrawled an obscene word across the Plexiglas display plate with an icepick.

We could continue, but it's too depressing: it isn't fiction. Scenes like this are common.

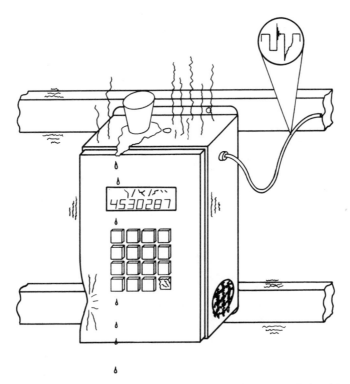

Figure 1-2   The wide range of environmental abuses found in an industrial setting threaten the very survival of a microcomputer system. Such machines must
commonly contend with heat, noise, vibration, human abuse, contaminants, and
poor maintenance.

Such environmental brutality calls not only for a mechanically rugged system, but also for a style of circuit and system design that is more sophisticated
than the old reliable, "fiddle with it till it works" approach. Noise must be battled, communications links must incorporate some redundancy, and margins of
safety that would be ridiculous anywhere else must exist between device specifications and reality. You can't coddle the box on the bench and ship it when it works
well under ideal conditions; you have to grit your teeth and try to destroy the
thing you just spent months creating. A drink helps.

1.1.3 Operational Flexibility

Once a controller is doing its job in an industrial setting, it is easy to assume that
the "design" part of the job is done and all that remains is the ongoing effort to
keep it working despite the environmental abuses noted above.
    Unfortunately, this is not always the case. One of the classic pains in the

control system world involves the changes that must be made when a controlled process is redefined or updated. New materials, more efficient methods of production, changes in the product line, and even variations in packaging brought about by market competition have each been responsible for their share of headaches.

In the Old Days, such production changes would bring the control system designer, slide rule at his belt and hard hat on his head, who would implement whatever magic circuit modifications were required. Perhaps a few relays were rewired, a couple of potentiometers adjusted, maybe the pins on a patch panel reconfigured. But if the changes were major, the line might be down for days or even weeks while the controller was rebuilt from the ground up.

Grim. Even if such a stoppage is planned, it's not cheap.

In the somewhat-less-old days, when logic designers were snickering at relays but had not yet heard of microcomputers, things actually got worse. Control systems were being built on PC boards densely packed with small-scale integration—RTL, DTL, and low-numbered 7400 series logic. It was a revolution in system design, of course, but it gave industrial people fits: for the first time, plant electricians were unable to reach in, burnish a contact or two, and wiggle all the relays to make certain that they were well seated. An operational change in the production process might have left factory personnel wringing their hands, for the logic was black magic and the local custom engineering firm consisting of one brilliant wierdo had evaporated in a puff of Chapter XI shortly after it had designed and built the box. The third-generation thermal photocopies of the logic circuitry gave no clue as to the reasoning behind any of those incomprehensible jumbles of gates, and mysterious undocumented patches made with X-Acto knives and No. 30 Kynar on the backs of the boards merely added to the confusion.

The mid-1970s introduction of the microprocessor promised to change all that. It didn't, of course, but at least the means became available for the designer, if so inclined, to build some flexibility into a system. Algorithms implemented by programs could take all their parameters from data tables rather than from a "hard-coded" sequence of instructions which are about as incomprehensible to the uninitiated as a board packed with gates and flip-flops. Those data tables could even be made accessible to the user via a clever setup program that translates his requirements into appropriate data without requiring too much sophistication outside his areas of expertise. Beautiful?

It doesn't happen very often. When it does, the system can mature right along with the process it serves; when it doesn't, the box is even less comprehensible than the random logic equivalent we considered a moment ago. A circuit built from TTL can at least be traced if one is desperate enough—but it is hard to make much sense out of a handful of 4K × 8 ROMs.

All this assumes that the industrial customer is going to be left, at some point, without a hand to hold. In the best of all possible worlds, the designer of a system would always be there to support it, but heck, we have our own lives to live. Right?

1.1.4 Performance Latitude

It becomes clear that the life of an industrial control system is full of surprises, ranging from environmental assault to specification changes. There's more.

Most of the systems that can be placed into the category under discussion share one basic characteristic, whatever their function. They derive information from their environment, and based on that information as well as some kind of stored algorithms, modify the environment in an appropriate fashion. (This is closed-loop control—we consider it in more detail in Chapter 2.)

Now, this is all fairly straightforward, but our insight into the realities of an industrial environment suggests that there must be some surprises lurking in there somewhere.

What if the algorithm receives inputs it was not designed to expect? Suppose a voltage that cannot possibly be outside the range −10 to +10 volts suddenly becomes 12.3 volts? Suppose some sync pulse gets stuck high? Or a shaft-position encoder indicates an angular displacement of 410 degrees?

These things happen. In a system which controls an IC-socket making machine, an early design oversight led to misinterpretation of unanticipated sync conditions and the subsequent application of a high voltage to a pair of solenoids for a period of many minutes—a voltage that should only have been applied for 7.5 milliseconds (ms) for the sole purpose of actuating them quickly (after which they were maintained by about 5 volts). The solenoids burned up, of course, and it was necessary to redesign the logic so that the "impossible" condition could be accommodated.

That condition, incidentally, resulted from a loose setscrew: an optical shaft position sensor rotated and produced "totally impossible" information about the machine's state. The computer did exactly what it was told, but the design had assumed only what was possible. The software's revision number thus incremented again.

There are thousands of situations in which such misinterpretation of the world can have far-reaching (expensive) results. A computer-controlled lathe with feedback afoul can eat a lot of expensive parts before someone notices that they don't meet specs. A concrete batching system that misinterprets the position of a discharge chute can let fly a few tons of glop onto a plaintiff.

Closely related to the question of handling bizarre inputs is that of dealing with those that are simply wrong, but well within range. If a furnace controller has the high bit stuck low on the analog-to-digital converter connected to the temperature transducer, it might well crank the temperature up to a value that is twice what it should be. This could do some damage. The system should therefore have some redundancy on its sense inputs and should constantly check its own outputs for reasonableness.

All this calls for constant awareness on the part of the system designer that he is not dealing with an ideal world. Everything is suspect, even foolproof switch closures.

1.1.5 Management Information

In any compendium of industrial design requirements, we must somehow include a more abstract need that is not directly related to controlling the process. Perhaps because of the universal availability of data-processing systems, management has an unquenchable thirst for data.

Manufacturing has always been a closed-loop process. Sometimes the loop is not closed quite tightly enough and junk is produced, but in theory there is always some control over the process as a whole. This high-level control derives its inputs from observations made at various points ranging from incoming checks of raw materials to analysis of randomly selected finished samples.

One of the unintended side effects of computerization is an escalation of this closed-loop optimization of a manufacturing process into, in many cases, an all-out love affair with information. Whatever the philosophical and psychological ramifications of this, it has a powerful effect upon the specifications that are handed to an industrial system designer whenever the opportunity arises to add a new control or monitoring system. It is probably fair to observe that any statistical information that can be derived from the system's operation should be made available.

One machine that we have already discussed somewhat (in Section 1.1.1) is a perfect example of this.

The widget maker's original design requirement was fairly simple: keep an eye on the insertion and withdrawal forces of widgets streaming out of the machine, and reject those that are out of spec. "But hey," they said, "since you have a computer in there, why not keep track of the number of bad ones in a given batch?" Sure, no problem. A few days later, a phone call enhanced this further: "Can you maintain running averages and print them out at the end of each run?" Sure, no problem.

"Um," came the call a few days later still, "since you're already keeping running averages on the forces, would it be any trouble to calculate the mean, variance, and standard deviation for each batch of parts and make them available on a communication line to our main computer?"

Well, gee, I think we'd better renegotiate the price.

In Chapters 13 and 14, we will look at one of the more hard-core manifestations of this information-intensive approach to design. A large foundry will be seen to have all of its individual controllers and monitoring systems integrated into a hierarchical network with a DEC PDP-11/40 at the helm, dispensing fairly high-level instructions and receiving reports from the component processes in a somewhat predigested form.

In both of these cases, the "management information" requirement has had a profound effect on the design, changing the systems from insular task-dedicated controllers to parts of much bigger processes—ones which are themselves closed-loop systems on a factory-wide scale.

1.1.6 Serviceability

This one is a lulu. The industrial microcomputer system designer, who according to the foregoing must already be psychic, dazzlingly brilliant, and well insured, must also be able to put himself into the place of the beleaguered fellow with a tool kit who has to scurry about keeping the processors processing. It helps to put on the hard hat and do it yourself occasionally: that elegant scheme you came up with for mounting the interface card might have you cursing when you try to debug the interface drivers mounted directly behind it.

Yup—designing for service is not, like so many other things, intuitively obvious to the most casual observer.

If we accept the unfortunate truth that time, the environment, and the universal perversity of matter are forever conspiring to bring a system to its knees, then we can turn our attention to the task of minimizing the severity of the inevitable.

First, because of the downtime problem lamented in Section 1.1.1, it is desirable to keep the MTTR (mean time to repair) as short as possible. There are various conflicting philosophies about the "level" of field service—unit, board, or chip-level replacement—but whatever the case, the designer must strive to make it easy for the technician. Such niceties as accessible ground points for a scope, test points for meaningful signals, and an extender card if appropriate are a good start. On top of such physical considerations, some method of checking the function of individual system modules independent of the others is a must, even if it costs in hardware. This can take the form of a self-test program which isolates the tested devices one by one, but this has a serious weakness: if the processor is dead, so is the self-test program. Actually, that condition is fairly obvious, but if the processor is just mildly ill—a marginal bus buffer or some other maddening intermittent—then the self-test can be more misleading than helpful.

Design is the first step in service. The field engineer crawling about in the grime of a factory chasing an inexplicable crash syndrome has enough trouble on his hands without having to fight a design that assumes it will never break.

1.2 MICROCOMPUTERS IN INDUSTRY: AN INTRODUCTION

In the process of depressing ourselves with the all-encompassing rigors of the industrial environment, we have already touched upon many of the reasons for the desirability of microcomputers. We also managed to refine the focus of this book—modulating from our opening discussion of what we are not discussing into a somewhat circumspect overview of what we *are*.

But we still haven't talked about microcomputers. This is the late twentieth century, so of course you know what they are ... but let's standardize our mental images somewhat.

Back in the days when everything was done with dedicated hardware, creating a control system of any sort involved statement of the task in some formal representation, followed by a mapping of that defined logic onto whatever hardware resources were available. A closed-loop temperature controller, for example, might have been implemented with op amps: configured as comparators which monitored the voltages from thermocouples, they would switch current drivers on and off as appropriate to keep the temperature within the range specified by a control potentiometer.

It never would have seriously occurred to anybody to take the temperature control hardware, carry it across the factory, and use it to run a time-and-attendance station. The hardware was wholly dedicated to the specific temperature control task for which it was designed.

In those days, such a direct relationship between a task and its hardware was fully accepted as the way things were—how else could you do it? You might, of course, generalize the temperature control problem and manufacture a box which could do the same basic job in many different situations, but it would still be vastly different from, say, an aluminum foil winding machine.

This was not an optimal situation, for such dedicated hardware was also difficult to adapt to variations in the task itself. During the 1960s, people started using minicomputers in those occasional situations where the complexity of the problem could justify an extremely expensive solution, but the bulk of all control logic remained in the form of specialized hardware.

When the microprocessor made its debut in the form of the Intel 4004 during the early 1970s, the "meta-revolution" began. Suddenly, designers had powerful new tools available: a small repertoire of inexpensive standard circuit boards could be applied to a limitless variety of real-world problems. The key was software. For the first time, hardware became general-purpose, serving more as an environment for a program than as an implementation of any particular function.

Now, instead of converting the formal specification of a problem into a configuration of chips, a designer expresses it as a sequence of instructions. These are then performed by the microprocessor, which carries out their bidding via its built-in capacity for logical operations and its interface with the outside world. (As this book progresses, we will be considering not only the structures of these programs, but also the nature of the processor's "built-in capacities" and the means of accomplishing the interconnection with the equipment being controlled.)

This design approach saves money, relegating the bulk of the logic to a program which can be modified both before and after installation to optimize the system's performance. This should not be construed as a statement that software is cheap—it's not—but much of the even greater cost of custom hardware is eliminated, leaving only the interface to be considered by hardware designers.

That, of course, is an unforgivable oversimplification: sometimes the "interface" is more sophisticated than the program. But we have to start somewhere.

Anyway, we can't go much further with this line of reasoning without stepping all over the material of the next chapter, so without further ado . . . .

2

The Intelligent Black Box

2.0 INTRODUCTION

Having established the "rationale," we are in a position to turn our attention to the subject of this book. However, if we are to indeed progress in the manner suggested by Section 0.1, we should wait a bit before leaping wildly into logic diagrams and aggregates of code: there is an important contextual layer yet to be revealed.

We need a real application—an associative "peg" upon which to hang new information in something other than that bemoaned random fashion. How about a Blanchard?

What's a Blanchard?

2.0.1 The Scenario

In various industries which require, for one reason or another, the application of very high temperatures to materials being processed, there is one universal requirement: ovens. There are a lot of problems with ovens, ranging from temperature control to door safety interlocks, but the former has been beaten to death as a control system example and the latter is too easy.

So let's look deeper. Because of the presence of such high temperatures, ovens have to be lined with something that is unwilling to burn or melt—and ceramic blocks made of "refractory" material are the perfect choice. These are molded in a variety of forms, depending on their intended position in the oven, and some have openings to accommodate material flow, fuel injection, exhaust, or measurement.

The problem is that these blocks suffer some pretty heavy abuse. In time, they wear or crack and must be replaced. This creates a reasonably healthy market for new blocks.

Now so far, this all sounds pretty prosaic. What could microcomputers possibly have to do with oven blocks? Here's the catch: no adhesive or bonding material can withstand the heat to which these refractories are subjected. Nothing can be used to fill the cracks or hold them together—except gravity and a damn good fit. Sets of perhaps 30 or more blocks must be produced as perfectly matched and fitted groups, assembled and checked for accuracy, then shipped en masse to the oven-owner's site. It's not a simple process.

We will not be talking about the bulk of that process, consisting of building custom molds and creating the rough blocks from any of a variety of ceramic compositions, of firing them and letting them cool for days at a controlled rate, and of sawing them as necessary to produce odd shapes. We are interested only in the problem of grinding them with great precision so they can be made to fit each other perfectly.

The machines which do this (to a variety of other materials as well) are called Blanchards, in honor of their manufacturer, the Cone-Blanchard Machine Company of Windsor, Vermont.

2.0.2 Basic Blanchard Control

If you go to your local Blanchard store and purchase a Model 42-84, as pictured in Figure 2-1, you will find yourself with the capability of accurately removing "ware" from a workpiece as soon as you provide it with a liberal supply of power and cooling water. You simply place a few blocks on the turntable, level them with wedges, set the machine in motion, and initiate the downfeed of a large rotating grinding head fitted with diamond tooling. When enough material has been removed, you tell the machine to back off, and *voilà*!

Well ... there are a few problems with this (although it has worked quite well for years). First, a careless operator can "overgrind," letting the process run too long. Expensive refractory blocks must then be carted off to the scrap heap (they are all custom made). Second, physical damage to the machine can result from excessive downfeed rates with certain ware combinations on the turntable. These two problems are relatively catastrophic, and are fortunately quite rare in a well-run organization.

But there's more. A third problem is very difficult for a human being to control: rapid wear on the diamond tooling resulting from excessive grinding rates. No problem: just keep it slow, right?

As usual, there is a trade-off. A nice slow downfeed rate indeed prevents excessive wear, but it can also tie up the machine on one job all afternoon. Anybody who owns a rotary surface grinder is probably a big business with a somewhat substantial overhead—not to mention impatient customers—and some poor plant manager is probably saddled with a dilemma: the budgeting department complains

Figure 2-1 The Blanchard grinder allows rapid machining of large flat surfaces by slowly lowering a rotating head onto a counter-rotating workpiece. The control system described in the text adjusts the downfeed rate as a function of the instantaneous heaviness of the grinding effort. (Photo courtesy of the Cone-Blanchard Machine Company.)

about his tooling costs and scheduling complains about the backlog waiting on the machine. Poor guy. He also has to live in fear of the rare but catastrophic problems mentioned above.

Then there are the human factors. A common trick among some operators is to complete the setup, start the machine on the standard fixed downfeed rate, and take it easy. Why? Simple—the process won't start for nearly an hour: the machine is cutting air as it slowly closes the gap between the grinding head and the blocks.

We have just found a good example of open-loop control. The Blanchard grinder can automatically lower the grinding head at a fixed rate; it can also stop the downfeed when a predetermined setpoint is reached. But it can't continuously optimize its feed rate as a function of the amount of work actually being done by the diamond tooling, nor can it recognize the "cutting air" condition and save time by bringing the head down at maximum speed. And an operator, if inept or disgruntled, can quite easily damage both the ware and the machine by simply directing it to plow its way rapidly into a heavy load.

We can do better than that. We have the technology.

2.1 CLOSING THE LOOP

Pictured in Figure 2-2 is a general closed-loop process. Instead of just telling a machine what to do, crossing our fingers, and hoping for the best, we use *feedback* to continuously trim the control signals that we give it. Once we have conceived an effective control scheme and competently implemented it in a processor, we can turn our backs on the process with at least some assurance that it will not go awry.

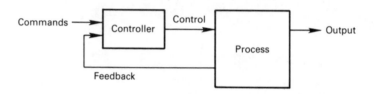

Figure 2-2 A general closed-loop process involves the ongoing modification of control signals on the basis of both input commands and feedback.

Closed-loop control is nothing new—it has been around a lot longer than microcomputers. It is, in fact, what your nervous system is all about. The simple act of turning a page of this book would be a horrendously difficult task if you were running open-loop: the book would probably be torn to shreds. Well below your conscious awareness, a distributed motor-control system acts upon your high-level decision to turn the page by initiating a series of muscle movements. Rather than leave it at that, it monitors the process via a variety of sensors (which provide visual, somesthetic, kinesthetic, and vestibular feedback) and in steps so small and well integrated that they seem continuous, trims those muscle movements and coordinates them into a single smooth action. It does all this on the basis of a "goal state" that is generated by a higher level of your brain.

Germane to this control process is the question of *damping*: the little corrections that are applied to the musculature must be of appropriate values—and they must be made at equally appropriate times (Figure 2-3).

If your hand is being positioned such that your fingers will be able to grasp the edge of the page, some sensing apparatus must be brought into play which makes a comparison between the hand's desired position and its actual one. Ideally, the system will begin issuing "decelerate" commands well before the hand actually reaches the desired position; otherwise, the effects of inertia and control system delays will allow it to overshoot its mark. The reaction of the system to this would be a command to move in the opposite direction in another at-

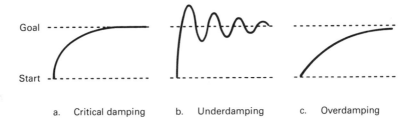

a. Critical damping   b. Underdamping   c. Overdamping

Figure 2-3 Comparison of system responses for three different levels of damping: In (a), the system approaches its goal state as fast as it possibly can without overshooting. In (b), the goal is approached too rapidly and a series of corrections is required. In (c), the system takes much longer than necessary to draw arbitrarily close to its goal.

tempt to reach the goal state. If the condition that allowed the overshoot, called underdamping, were chronic and not just a momentary glitch, the hand would oscillate around its intended target, driven to and fro by ill-timed control signals. In the opposite case, an overdamped system would move the hand in an agonizingly slow asymptotic approach. Clearly, there is more to it than simple "forward–stop–reverse" commands triggered by positional error.

In addition to such exquisite biological systems, the world is full of noncomputerized closed-loop control. Home heating systems regulated by thermostats, automotive fuel-injection units, AVC circuits in communications receivers, and most op amp designs fall into this category. Damping is equally critical in these cases; in the home heating system, for example, underdamping would have the furnace rapidly switching on and off, wasting fuel, while the thermostat attempts to keep the temperature exactly right. But we're here to talk about Blanchards.

How can we control the downfeed rate automatically?

What measurable machine conditions exist that would allow a controller to optimize the downfeed rate such that the material throughput is well balanced against the expensive wear on the diamond tooling?

How about the power required by the motor that rotates the grinding head? It would reflect, in some repeatable if not necessarily linear fashion, the instantaneous "heaviness" of the grinding effort, and could be used as an input to a closed-loop control system that regulates the rate at which the grinding head is moved down onto the workpiece. Using such a scheme, one might conceive the loop depicted in highly schematicized form in Figure 2-4.

In this scenario, a stepper motor of sufficient torque is grafted via a gear reducer onto the existing downfeed drive assembly of the Blanchard grinder, and a Hall effect watt transducer is coupled to the power circuitry feeding the main drive motor. The output of the transducer is proportional to the instantaneous power level of the circuit.

The system is now provided with the two necessary points of interface with the process: a source of feedback information and a means of control. All that

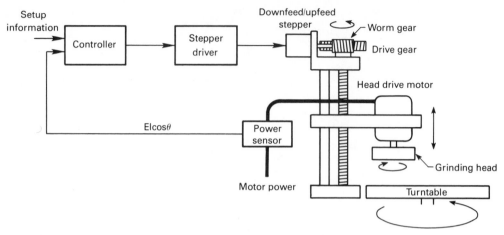

Figure 2-4 In the basic Blanchard control scheme, a controller varies the pulse rate delivered to the downfeed stepper according to the load encountered by the grinding head motor.

remains is the design of an appropriate "algorithm" and its implementation in a microcomputer.

2.1.1 The Blanchard Control Algorithm

We arrive at a key part of the design process. The *algorithm*—that sequence of steps which will perform the desired function—must be carefully contrived such that all possible combinations of operating conditions are accommodated without difficulty. But it's not too hard to break the grinding process into four discrete states: no load (cutting air), load greater than zero and less than a preset lower limit, load between lower and upper limits, and excessive load. Assuming that somebody knows enough about the process to arrive empirically at meaningful upper and lower load limits, the control scheme can be defined as follows:

1. If the peak load attributed to the grinding head motor (after calculation to remove the basic machine loading due to friction, etc.) is zero, meaning that the head is "cutting air," then the grinding head is directed to move downward at the fastest possible rate (defined in the system operating parameters). This will have the effect of minimizing the time required to begin the grinding task after the completion of machine setup.

2. If the load during the previous table revolution was less than the lower load limit, the downfeed rate during the next revolution is increased proportionately. (The grinding head motor's load is observed over each complete table revolution, and the peak value that occurs during that time is used for the calculations. This allows the process to respond to irregularities and high spots in the material that would be ignored otherwise—defeating the whole purpose of controlling diamond wear.)

3. If the peak load was between the upper and lower limits, the rate during the next revolution is maintained at the same value.

4. If the load was greater than the upper load limit, the downfeed rate during the next revolution is decreased proportionately.

This approach pretty well covers the conceivable conditions that will be encountered during a grinding task. A couple of tricks have been built in which smooth the operation—first, in cases 2 and 4, we have made reference to proportional increase or decrease of the downfeed rate. This is determined on the basis of the amount of deviation from the preset power limits, and prevents gross variations in the rate for a minor excursion out of bounds. Also, since the downfeed rate is actually controlled by the stepper motor, the true output of this controller is just a stream of pulses. But instead of thinking in terms of the pulse rate, the system calculates the number of them which should be delivered over the next complete table revolution, and then distributes them evenly, preventing unnecessarily abrupt transitions.

Incidentally, for an appreciation of scale, it is worth noting that in the implementation we are describing, each stepper pulse results in a downfeed of 7.1 millionths of an inch. Approximately.

The net effect of all this is the "containment" of the downfeed rate within the prescribed limits (Figure 2-5), which presumably have been determined by

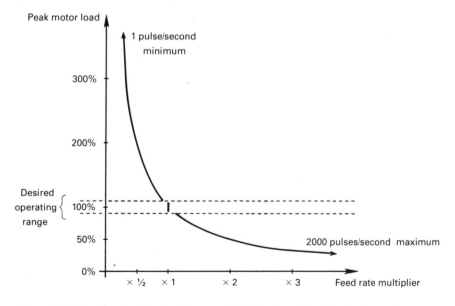

Figure 2-5   The transfer function illustrates the behavior of the Blanchard control system as it attempts to maintain the machine load within the desired operating range. As the motor load increases beyond the target value, the multiplier of the previous feed rate falls below 1; as the load decreases below that value, the multiplier increases accordingly. A deadband exists around the target value to prevent constant trivial corrections.

the plant engineering people on the basis of historical observation and a measure of intuition. In addition, as said people gain experience with the system, they can make use of generated statistical information to refine the operating parameters further. Yes, a machine like this one is a prime candidate for some clever management information programming—but we'll get to that sort of thing later.

2.1.2 Blanchard Control Software

It's a bit early in the book yet for a hard-core plunge into the software that can make the above a reality. Yet we can easily observe its structure without having to deal with any real instructions, and such observation will serve us well as we step gracefully into programming a few dozen kilowords hence.

In fact, if we ignore such details as establishing the limit values, communication with the operator, and other necessary frills, we can express the logic in the simplified form shown in Figure 2-6. This flowchart embodies the essence of the program that is performed by the controller.

As you can see, there is not really much to it. After something called "initialization" which takes care of preliminary setup details loosely analogous to the operator's job of arraying ceramic blocks on the turntable and starting the flow of coolant, it just spins eternally in a loop. On each pass, it waits for a complete table revolution to occur—all the while sampling furiously to determine the peak load that can be attributed to the grinding head motor during that time. Once the revolution is complete, and it has a value in hand, it checks that value to determine which of four parallel paths through the logic will be followed.

These were characterized in Section 2.1.1. Each one results in the application of a different calculation to the number of stepper pulses that were sent during the revolution that was just sampled (X). The result of the selected calculation (a new value of X) is the number of pulses that will go out next time—ranging from one per second in the case of an extremely heavy load to the maximum of 2000 if the head is cutting air.

Once this new value has been calculated, there is no longer a need for four separate paths, so they converge on the next calculation. Here, the pulse rate in hertz is calculated on the basis of X and S, the speed of the table's rotation. At this point, the frequency of the pulse train that is transmitted to the stepper motor drive circuitry becomes equal to "RATE." The system then checks, in some unspecified fashion, for task completion, and either goes around again or quits.

One of the most elegant things about this approach is its simplicity. In the Old Days, a typical analog solution to the problem might have entailed a box of complex, heat-sensitive circuitry, astrewn with potentiometers which would have been used with maddening regularity to calibrate the skittish servo loop. But here, the whole process can be understood as a series of straightforward discrete steps, which can even be isolated for testing without automatically rendering the results meaningless. And, since everything from the power sensor out is digital, problems with drift are nonexistent.

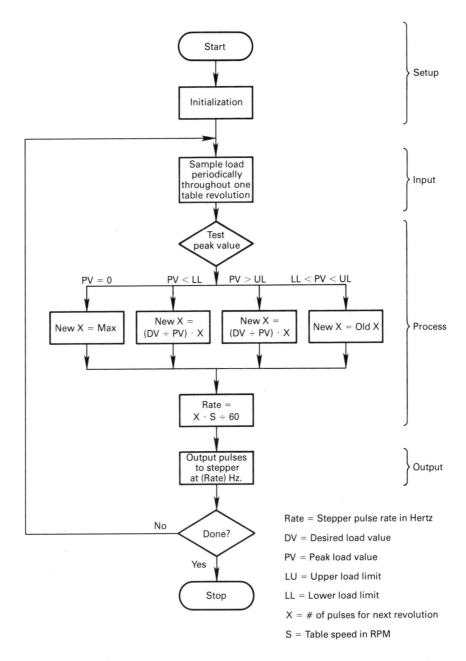

Figure 2-6 Flowchart of Blanchard control system operation, showing the four possible responses to the peak motor load measured during a given table revolution. The newly calculated stepper pulse rate is maintained until the next pass.

Somewhere at the heart of all this, there is a computer. It dutifully spends its life following simple instructions in sequence, skipping among groups of said instructions to act upon logical decisions. The computer is merely a servant to its program. As a system designer working with such a tool, you have a striking advantage: all you really have to worry about is the expression of your design as a sequence of instructions and the interconnection of the black box with the machine being controlled.

2.1.3 Implementation

Figure 2-7 is a sketch of a typical factory installation of our enhanced grinder. Bolted to the machine in the foreground is a padlocked steel box containing the microcomputer and its various interfaces with the machine, the plant computer, and the human operator. On the front panel of the enclosure, there is a single-line text display which continually indicates such data as downfeed rate, amount left to grind in thousandths, and system status. There is also an array of pushbuttons which allow the operator to enter the parameters for a given task (although not the upper and lower limits described previously—these are maintained in the plant computer), and a small badge reader that restricts operation of the system to certain employees.

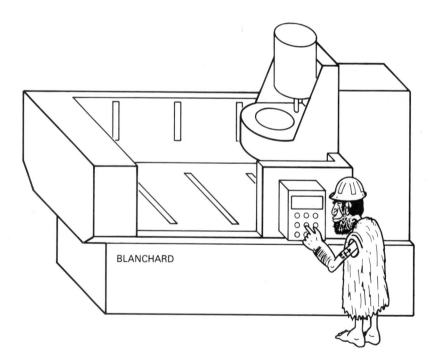

Figure 2-7   Typical Blanchard installation.

The presence of the communication line to a big system somewhere else says something about the role of this controller. As we shall see much later in this book, very reliable and efficient operation can be achieved by distributing tasks among a variety of specialized, local systems. Problems with long-haul communication of sense and control information can be reduced, and the whole system can be designed such that the satellite processes are not seriously disturbed by a glitch or even the outright failure of the main computer. Many of the systems currently in use out there are quite capable of bringing production to a (groan) grinding halt if they should happen to die temporarily.

We already spoke of the input side of the system—the Hall effect watt transducer—in enough detail to communicate its function: it is simply coupled to the motor drive circuit through a transformer, and its output is filtered, amplified, and applied in analog form to the controller.

But the unassisted microprocessor can make no sense of this continuously variable voltage, and an input interface is called for. Called an *analog-to-digital converter* (ADC or A-D), it produces binary values of appropriate precision corresponding to the voltage present on its input.

Consider Figure 2-8, a block diagram of the Blanchard control system. We have deliberately left the microcomputer at the black-box level to avoid cluttering this discussion with internal details.

The interface with the Blanchard, as we have said, consists of power sense and stepper drive. The latter is handled in either of two fashions, depending upon how much else the computer is being asked to do. Since, for example, the process of determining the peak load during a table revolution is performed by sampling the A-D's output as frequently as possible and subjecting the value to a series of calculations, it might follow that the processor is too busy during that time to deal with many other realities. It already must make periodic checks for the arrival of data over the communication line from the host computer and for an operator input; perhaps the actual generation of stepper control pulses is too much of an additional burden. This is one of those tough design decisions.

The solution is the addition of more hardware, in the form of something we could call a "rate generator." Its task is to accept a number between 1 and 2000 from the processor and generate that many stepper drive pulses per second until directed otherwise. In this fashion, the computer can arrive at the desired rate as depicted in the flowchart of Figure 2-6, then simply hand that number off to its assistant and forget about everything but its analysis of the load and the ever-present possibility of communication from man or machine.

This approach extends the idea of distributed processing, already implemented in the implied heirarchical structure of the factory-wide network of which this is a part. It has the additional advantage of keeping the design requirements of the processor and its program at a somewhat more manageable level, perhaps saving as much in development time as is spent on the additional hardware of the rate generator. That's where the tough design decision comes in, and we will have more such questions to ponder in greater depth as we go along.

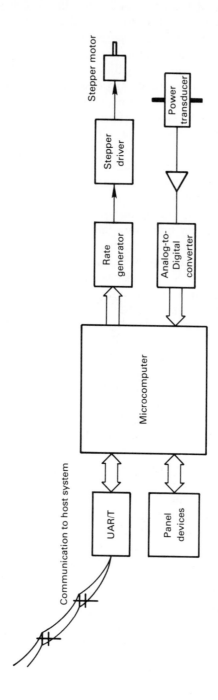

Figure 2-8 The hardware of the Blanchard controller is comprised of a microcomputer and its associated interfaces with the equipment being controlled, the human operator, and (optionally) a host computer system.

The rest of the hardware design is reasonably straightforward. The communications line is handled with a universal asynchronous receiver/transmitter (UAR/T), a device which, much like the rate generator, takes care of all the overhead associated with shuttling serial data between its host machine and another one, somewhere. Because of the noisy factory environment, the communication line itself is very likely a shielded current loop—reasonably immune to all but the most persistent electromagnetic interference (near an immersion-arc furnace, even current loops have to be encased in low-resistance conduit). The panel devices are interfaced via a small logic board with a few decoders and lamp drivers.

And that's about all. It boils down to a box, weighing perhaps 50 pounds (most of which is steel), bolted to a convenient surface and interconnected with the Blanchard via a cable. Presumably, the unit has been designed with a high tolerance for the environmental abuses we mentioned in Chapter 1.

2.2 RHAPSODY

OK. We have gotten a major stumbling block out of the way: we have managed to establish a realistic framework for much of the material to follow. This should be interpreted not as an obsession with Blanchards, but as a convenient reference point. Besides, as both a stand-alone system and part of a factory-wide "closed network" (an extrapolation of closed loops), it embodies a healthy reportoire of design features and concepts which are applicable to all sorts of problems waiting to be solved out there.

It's also a real system.

3

Inside the Box

3.0 INTRODUCTION

Now that we've examined a typical industrial control system in enough detail to reveal the role of the microcomputer around which it is built, we have enough contextual information to permit a look inside without suddenly becoming obscure.

It is worth noting at this point that we will have little occasion to consider specific CPU chips in any detail. There are at least three good reasons for this.

First, the technology will probably not accommodate us by pausing long enough for such product-specific information still to be current by the time this book has completed its production and distribution cycles. (I have a selfish as well as an altruistic interest here—the greater the book's longevity, the happier *everybody* is. Writing details about the 8086, Z-80, or whatever is tantamount to printing a large notice on the front cover: "WARNING—OBSOLETE AFTER 19xx!")

Second, the literature fairly abounds with data sheets, application notes, and other such detailed information, and an attempt herein to duplicate that with enough depth to be useful would crowd our major objectives into one or two chapters. For product-specific data, you should consult the manufacturer's literature or a regularly updated compendium.

Third, everybody has his own favorite processor chips, and if we devoted our efforts to one that you don't like—well, you might never have bought this book in the first place. That would be intolerable.

These points lead to a suggestion that you may find worthwhile. As you read this, keep some documentation handy for a processor family that particularly interests you, whether academically or practically. You will find that such information will plug right into this book without excessive gaps or overlaps, tailoring the material herein to your own applications in the process.

If you have been exposed to micros in any depth prior to your perusal of this text, then much of the information in this chapter will already be familiar to you. Try to fight the compelling urge to skip ahead—the internal model of microcomputers that we assemble herein will be a component of future chapters.

3.1 GENERALIZED SYSTEM OPERATION

We might as well begin by considering the fundamental requirements which underlie the architectures of microcomputer systems. We can safely lump them all together at this level, because as long as we think of CPU chips as still smaller black boxes, the various systems are all more or less the same in principle. A few timing variations, a few different signal names, assorted methods of addressing input and output devices—yes, there are differences. But they all have the same general objectives.

Primary among these is the execution of instructions. Microprocessors spend their entire lives (as long as power is on . . .) either frozen in the HALT state or performing instructions as fast as they possibly can. The instructions, of course, are stored in some form of memory, so the first requirement of a micro is the ability to access its stored program. And together with reading information from memory, it needs to be able to write to it, allowing data to be stored and retrieved at will.

Communication between a microprocessor and its memory is indeed the essential requirement of a system design, but it is difficult to make use of a machine that does only that. No matter how clever the program and efficient the processor, it is totally isolated from the rest of the world without a facility for input and output (I/O). Given some arbitrary number of "ports," a system can interact with its environment, whether it involves a Blanchard grinder or an operator at a CRT terminal.

With those very general requirements in mind, look at Figure 3-1. At the left, running the show, is the CPU itself. For this discussion, we don't care if it's 4-, 8-, or 16-bit, if it's CMOS, NMOS, or bipolar, or who made it. It's just a central processing unit, connected via buses to memory and I/O.

3.1.1 The Tri-state Bus—an Aside

But before we can deal in any depth with the architecture of this typical system and the types of things that go on within it, we must establish a solid image of the key communication scheme central to virtually every modern computer.

Much of the activity of a computer system involves the "simple" transfer of data from one place to another—from a memory location to a register, from a register to an arithmetic logic unit, and so on. Most potential data destinations must

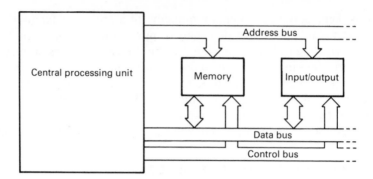

Figure 3-1 The microprocessor CPU interacts with its memory and input/output logic via three buses: Address, Data, and Control.

be capable of receiving data from any one of a number of sources. This creates a problem: how can the proper data source be selected without interference from the others? Clearly, if you just tie the outputs of a number of standard logic devices together, you have mayhem.

In the Old Days, this was most often taken care of with a *multiplexer* (mux), as shown in Figure 3-2. Here, under control of two bits called "Select," the mux switches one of its four inputs to the "data sink." It is perfectly analogous to a rotary switch, and, in fact, many TTL multiplexer chips are specified using traditional switch nomenclature (2P4T for a 2-bit-wide mux with four positions—a 74153). The function in Figure 3-2 could be performed with four 74153s if the data paths were 8 bits wide.

This is a workable solution to the communication problem, but it suffers from some drawbacks. It requires additional hardware, it introduces propagation delays that slow the system down, and it makes future additions of other "data sources" very awkward without a hatchet job. There's a better way.

It is normal to think of logic devices as strictly two-state circuits—they always output 1 or 0 as indicated by a high or low voltage. Many TTL devices generate these two states with the output structure depicted in simplified form in Figure 3-3. Here, it can be seen that the output pin is connected to a pair of transistors such that when the top transistor is on and the bottom one off, a high level is produced, and in the opposite case, a low. The internal logic of the IC assures that they are never both on at the same time (which would have the immediate effect of melting them, thus implementing a self-destruct function).

If we modify the internal logic of the chip such that both transistors can be off at the same time, however, we produce a useful effect. As far as anything connected to the output pin is concerned, the device has just disappeared. This third state, in which the output line shows only a high impedance, makes possible the types of bus structures around which all microcomputers are built.

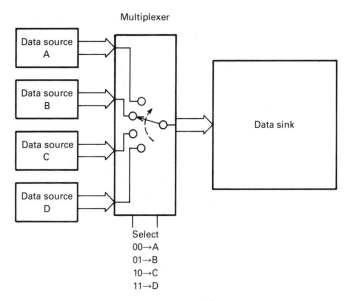

Figure 3-2   One method of data selection makes use of a multiplexer, a device that is logically equivalent to a rotary switch. Here, any of the four "Data sources" may be connected to the "Data sink" by the application of an appropriate 2-bit select code.

The beauty of *tri-state* logic is that we can implement "party-line" data paths over which signals from a variety of sources can be sent without the relatively cumbersome multiplexing circuitry described previously. New data sources can be added with a minimum of fuss long after the system is built—a feature that helps

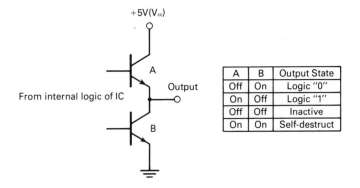

Figure 3-3   Classical digital ICs, to produce a 1 or 0, always have one of the two transistors of the "totem pole" output stage on. Tri-state chips allow party-line use of a data bus by allowing both transistors to be off at the same time, making the device electrically disappear.

lower the cost of computers. Early systems had to accommodate in the original design all of the anticipated add-ons; unanticipated ones called for substantial rewiring jobs.

By way of comparison with the multiplexed approach, let's briefly consider Figure 3-4. Here, we have accomplished exactly the same function that we did in Figure 3-2: on the basis of a pair of *select* bits, one of four "data sources" is connected to a "data sink." It is easy to see that the design is by no means limited to four—decoders of arbitrary size can be constructed to select one of any number of data sources as the "talker" on the tri-state party line that we have created. In the multiplex scheme, things would quickly get out of hand in a vast tree structure which is not only architecturally rigid, but rough on system performance since it introduces additive logic propagation delays.

Tri-state logic can be found everywhere—in TTL devices such as the 74367 and 74368 buffers, in RAM and ROM chips, UAR/Ts, and CPUs themselves. It has phenomenally simplified computer design, but complicated computer service. Consider: with any one of perhaps a hundred different chips dragging down a line to which they are all connected, how can you tell which one it is just by wandering around in the circuit with a scope or logic probe? You can't. It is more difficult than the classic series Christmas tree lights problem—at least in that situation, you can unscrew bulbs to successively divide the string in half, homing in on the culprit within a small number of iterations. Tri-state logic calls for a service approach as revolutionary as the design approach that engendered it. Not the least of the problems is the fact that a scope trace of a typical bus line is next to useless without a lot of contextual signals on the same screen. It is hard to make much sense out of electrically "floating" signals which can look high, low, or somewhere in between.

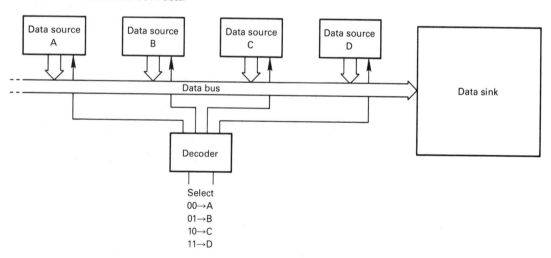

Figure 3-4 The application of tri-state logic to the data selection problem allows all data sources to be directly wired to the data sink.

3.1.2 Bus Communication

Returning to Figure 3-1, which is now inconveniently on another page, we can see how the use of tri-state buses has reduced a block diagram of a microcomputer system to its basic essentials. There is no clumsy overhead circuitry to tie it all together—just three buses onto which any addition to the machine is grafted.

What are these three buses? Let's consider them in no particular order.

First, we have the *address bus*. This carries the binary codes that specify any of the available memory locations in the system, or any of the I/O ports. In a typical 8-bit system such as the 8085, 6800, or Z-80, this bus is 16 bits wide, giving it 65,536 possible unique states. This corresponds precisely to the "address range" of the system, and defines how much memory can be directly addressed by the CPU. Actually, there's a bit of trickery that can extend that considerably, such as the use of an output port to switch in different banks, but let's not get carried away.

(Incidentally, 65,536 is usually called "64K." In the computer world, a "K," or kilobyte, is 1024 bytes.)

Normally, you can think of the address bus as being unidirectional, with data flow always away from the CPU and toward the memory and the I/O. After all, the CPU never has any occasion to receive information on the address lines—it uses them to specify what memory location or I/O port will be involved in the operation about to be performed. But there are system configurations in which the CPU is not the only source of address information. A technique generally referred to as "DMA," for *direct memory access*, is frequently used in such subsystems as disk controllers which depend for their effectiveness upon extremely high data rates. This approach allows transfers into and out of memory without the involvement of the CPU. In these cases, the CPU agrees to disappear for an appropriate length of time while some other controlling device seizes the bus for its own purposes.

But the address bus is unidirectional from the processor's viewpoint. Think of it as such.

Second, we have the *data bus*—the most famous of the three. Its width (number of lines) is the key system specification ("it's an 8-bit machine . . ."). It is the main data path of a computer, and unlike the address bus, is bidirectional. All instructions, data to or from memory, and I/O information make their way through the data bus as they meet their appointed rounds. We will look at some examples in a moment.

Third, there is the *control bus*. This one is not much of a bus in the traditional sense, but we call it that because of its general behavior and the fact that it is usually tri-stated to allow other CPUs or controlling devices to be added to a system. It is a collection of control lines (five in the case of the 8080-style processors) which inform the various hardware elements of the system what type of event is about to occur: memory read, memory write, I/O read, I/O write, or interrupt. This scheme varies widely from system to system, but they all have some method of tying the behavior of the other buses to the function being performed by the CPU.

Let's consider, totally out of operational context, a memory read operation. For purposes of this discussion, we don't care if it is an instruction fetch or a random data read operation—we are just going to copy a byte from location 82E9 of memory to some register in the CPU.

The processor places the binary address "82E9" (a hexadecimal value corresponding to 1000001011101001) on the address bus, and more or less concurrently activates the line in the control bus that specifies a memory read operation. The address, simultaneously applied to all the addressable hardware in the system, is decoded successfully in only one area, and that "block" of memory finds itself selected. The lower bits of the address are further decoded within the selected chip or chips, and the data stored at location 82E9 are placed onto the data bus. After a discreet delay, which allows for the access time of the memory as well as the electrical "settling" of the bus, the value is read by the CPU, and the read operation is complete. Net elapsed time: substantially less than a microsecond. That's all there is to it.

The use of the bus structure indeed simplifies things. Not only is an operation like the one just described a thoroughly straightforward affair, but all conceivable operations within the system are very closely related to each other. (The simplicity of individual data transfers would be pointless if there had to be dedicated hardware for each of 20 or 30 equally simple actions: the net effect would be a mess.)

Before proceeding, it would be wise to spend a moment or two chatting about some of the more esoteric implications of a bus structure that carry its meaning well beyond the simple party line described above. In particular, we should note that the bus is far and away the most important part of a system. Even more important than the CPU—CPUs can be replaced.

If we think of all the various elements connected to the bus as *nodes*, then we can state that the design of the bus defines the interface structure that exists between those nodes. If it is a flexible design, the precise identity of the nodes can be quite variable, and can in fact change freely as the technology matures and provides more desirable functions. This naturally requires a healthy measure of industry standardization, and it is provided by some of the major bus structures on the market: Multibus, S-100, STD, and GPIB. Those all address completely different kinds of application requirements, but have in common the objective of formalizing and simplifying the task of adding new nodes to the system. Among other advantages, this keeps you, the designer, from being tied to one specific system manufacturer, for each of the named buses is supported by dozens of vendors.

There are a lot of implications in this, many of which will appear later in this book. Keep in mind while reading more about system designs that a system is only as good as its bus.

3.1.3 Basic Microcomputer Behavior

Now that we have had an introduction to the bus, we can observe, briefly, the basic operations of a micro. (Briefly, because we have not yet looked within those three boxes in Figure 3-1 in any more detail than can be expressed by a casual sentence or two.)

Assuming that the CPU is, indeed, the driving force behind all this (ignoring the philosophical idea that it is, in reality, the program stored in memory that has that distinction), we can generalize system operation as follows.

First, the CPU activates a signal in the control bus that notifies other elements of the system as to the type of event that is about to occur. In the example above, this would have been the "memory read" line. (For some device families, like the 8080 and its descendants, it is explicitly decoded as such; in others, it may simply be a signal indicating any kind of read operation.)

Second, the CPU places a binary code on the address bus which specifies the memory location or the I/O device that will be involved in the activity (such as "82E9"). Sometimes the I/O devices are nothing more than memory locations (as in the 6800 family) and are treated no differently.

Third, a data transfer takes place between the CPU and the "occupant" of the address, possibly followed by an internal logical operation in the CPU (such as an ADD).

Fourth, the operation is completed, leaving some kind of presumably useful aftermath, and the sequence begins anew with the next instruction or data transfer.

It is worth noting that many actual instructions involve more than one iteration through this basic four-step loop—especially if you count the *instruction fetch* (the transfer of the stored command itself into the CPU for decoding and execution).

There are a lot of microprocessors on the market, and most of them are backed by manufacturers who claim that their particular baby is better than the rest. This is all quite reasonable, and perhaps they are all correct. The point is that there is an impressive variety of techniques for accomplishing the basic objectives we are discussing here, and although we are trying to avoid obvious bias toward our particular favorites, those intimately familiar with others will doubtless find an occasion or two for an accusatory "aha!". Even if the four-step sequence above looks strange in the light of your preferred system, the intent is the same—and that's what we are trying to communicate. There are not too many places in this book where device type matters, especially after we manage to satisfy ourselves concerning the contents of the "black box" and refasten the lid with a sigh of relief and a dollop of epoxy. In most of our design examples, general processor "horsepower" and class will provide enough design information without the need for a potentially misleading parts list.

So. With that whirlwind explanation of system operation and its associated disclaimer, let's consider the innards of the three elements that we have identified as the components of a microcomputer system.

3.1.4 Memories

Memory technology has progressed dramatically since the early days of computers, and continues to do so today. Before the first microprocessors, in fact, solid-state memories had already appreared in the form of the venerable Intel 1103 to challenge core—which for years had remained the traditional solution to the problem of providing a computer with someplace to keep its programs and data. Core had the distinct advantage of nonvolatility with respect to power failures, but was expensive to build and clumsy to interface. It didn't take long for semiconductor memory to push core into the fringe area of specialized applications.

As time went on, cost per bit dropped further and further as more and more storage cells were crammed onto single chips. It became possible by 1980 to house the entire 64K address space of 8-bit machines on eight chips, each of them one bit wide. And that was hardly the ultimate in density—even as manufacturers battled over standards for the 64K chips and struggled with production yields, the promise of 256K parts was enchanting system designers mercilessly.

We need take only a brief look at the breeds of memory devices. They have become so easy to use since the grim days of multiphase clocks, oddball supply voltages, and sense amps, that close analysis is hardly necessary for our purposes.

First, we can handily split them into two major categories: RAM and ROM. Those of the first group, *random access memory*, are characterized by the ability to read and write with equal ease. Those of the second group, *read-only memory*, do just what their name suggests. Obviously, however, it must be possible to write to them at least once, or there would be nothing to read. This process, depending upon the type of ROM, happens either at the time of manufacture (by a masking process) or anytime the user wishes, if the ROM happens to be an EPROM (for erasable, programmable ROM). We'd be lost without EPROMs, and we'll talk more about them shortly.

RAMs can be further split into two categories, *dynamic* and *static*. The former use capacitors and very high impedance gates to hold the data, although only for a few milliseconds, necessitating periodic data refresh. This adds to the operating overhead, but they have the advantage of being substantially cheaper and denser than statics, which essentially use a dedicated flip-flop for every bit. The latter are very easy to design with, but do have a higher cost per bit.

RAMs of both types are available in a variety of shapes and sizes, both physically and architecturally. They can also be had in a number of different device technologies, allowing the designer to address problems of extremely low power dissipation with CMOS and extremely high speed with bipolar or ECL—although the "mainstream" technologies such as NMOS, HMOS, and others have pushed access time and power consumption back dramatically. Sizes range from 256 bits to 256K bits (at least), and said bits can be organized in numerous ways. In some applications, for example, only a small *scratch pad* may be required, and the use of 8 or 16 chips (depending on bus width) to provide it might be inefficient. For situations like this, RAMs already organized into bytes may be used, allowing the problem to be solved with only one or two chips.

Also, as the ever-maturing technology has allowed die sizes to shrink, it has become possible to build RAM (and ROM) right into the microprocessor itself. We touch on these devices in Section 3.2.1.

ROMs have taken on considerable importance as microcomputers have found applications in countless out-of-the-way places. In "traditional" computers, the use of read-only memory is pretty well limited to "bootstrap" programs (that need to be available when power is first applied) and utilities executed so frequently that it becomes objectionable to pull them in from disk each time. But when a computer has no disk, and is instead built into a piece of equipment (like a Blanchard, or an automobile), then there must be some assurance that it will not suddenly forget its program. In these cases, which comprise the vast bulk of microcomputer applications, ROMs are the answer. A typical system will have a field of ROM for program storage and a field of RAM for variable data.

When a piece of micro-based equipment is being manufactured in quantity, with the identical program stored in each one, it is cheapest to submit a binary image of the program to a chip vendor and have her create ROMs custom-masked to contain it. This is fine in quantity, but the setup charges are horrendous and the designer had better be sure the program is correct—there is no turning back. Any changes require discarding the old chips and replacing them with new ones. For the myriad applications that could not possibly support such stringent limitations, EPROMs are perfect.

And wondrous devices they are! They work so beautifully, it's easy to take them for granted, but consider: You take a blank EPROM (2708, 2716, 2732, 2764, or any of their variations), and plug it into a programming socket. Appropriate hardware hits it with elevated voltages and programming pulses as you simply copy an image of your program or data over to it. You then unplug it, and there, in your hand, is a totally nonvolatile and quite rugged block of memory with your code "permanently" locked inside. You store it on the shelf, ship it to a factory in a Venezuelan jungle, subject it to a few thousand power cycles, then get it back when it is time for a new version of the code. But you don't discard it—you just zap it with a brief dose of ultraviolet light and it's as good as new. A fresh program, and it's out the door again. And an even newer class of parts—EEPROMs (electrically erasable PROMs) replace the UV erasure with on-board modification under system control.

Elegant devices. CPUs get too much of the credit for this microprocessor revolution: the available alternatives to EPROMs are not very exciting.

Before leaving our quick overview of memory devices, we should make at least passing mention of a couple of other types which either find application in specialized areas or hold great promise for the future.

Both of the types we have discussed—RAM and ROM—are random-access.

The CPU, or whatever happens to be addressing them, can pick any specific location with ease and do with it whatever the device type allows. There is no significant difference in access time between any arbitrary pair of locations.

Not all memories are like this, however, and one of the more recent technologies is a case in point.

By applying a magnetic bias field to a thin film of epitaxial garnet (upon whose surfaces more or less conventional lithographic techniques have been used to create a permalloy pattern), it is possible to generate and then manipulate little isolated domains of magnetization that are opposite in polarity to their surrounding regions. Shrewd handling of these ''bubbles'' via the design of the patterns and external controlling logic allows very high density serial memories to be created. They have the added fillip of nonvolatility—the data doesn't evaporate upon power failure.

The implementation of these *bubble memories* in a system is not quite as trivial as that of RAM and ROM—partly because of the sophisticated driving logic required to make it work in the first place. This is all available in the form of LSI chips, however, so the real problem is operational: making efficient use of a serial-access memory. As a replacement for floppy disks and other file storage media, it is fine, but it quickly becomes awkward as a replacement for random-access addressable memory. For this reason, bubbles are seeing their greatest success in virtual memory systems and file storage devices, as well as applications in which the access mode can conveniently be sequential (like data loggers and communication buffers).

Another serial memory device that has found numerous applications is the CCD—the *charge-coupled device.* Unlike bubbles, CCDs are volatile—but they are faster and can be built with standard IC fabrication techniques (bubbles typically require field coils, bias magnets, and Mu-metal shields—though one newer design eliminates the coils by using a ''current-access'' technique).

But CCDs have had trouble penetrating the main-memory markets, instead finding extensive use in signal- and image-processing applications. They possess an interesting characteristic: the ''charge packets'' that propagate within them are not limited to the two states that express binary information. As a result, they have been put to work as acoustical ''echo chambers,'' allowing once-clumsy audio signal processing tasks to be accomplished in silicon. They also can be made light-sensitive, allowing their use as optical scanners and replacements for traditional TV cameras.

But we stray.

We will not examine any of the other classes of memory devices at this time. Those that we have discussed will adequately fill our needs herein, and though there are others of academic and possible practical interest, they have not arrived in any force at the doorstep of the microcomputer. So armed with RAM and ROM, and at least aware of bubbles, we move on.

3.1.5 Input and Output

The use of RAM and ROM is so straightforward in most cases that any examination of the circuitry involved would amount to little more than a restatement of our opening comments about the system bus: you just take the memory chips and hang 'em on. Actually, there's a little more to it, but not much: it is necessary to decode the high-order address bits to create memory select signals, and at some point, you must connect the "write" line coming out of the CPU to the RAM in the system. When that line becomes active, it tells the memory that it should store the data currently on the data bus at the addressed location, rather than copy said location to the bus. But that's about it, at least in principle (yeah, yeah, with dynamic RAMs something must handle refresh, and with fancy architectures something might need to take care of DMA, index segmentation, paging, or virtual memory, but let's keep this reasonable).

So how does I/O differ? Well, first, we can't embody the whole subject of I/O in a few chips—since the very nature of the term "input/output" implies an outside world, not the inside of an integrated circuit. This raises a number of interesting problems, which fall into the general category of interfacing. That is why there will be a whole chapter on the subject—soon.

But let's get the basic concept out of the way. Consider Figure 3-5.

Here, we have a simple I/O configuration attached to the bus structure of a microcomputer. We don't care what else is on the bus; we can just assume that the "big picture" is something akin to Figure 3-1. Shown in the diagram are one input port and one output port.

Making use of the context that we established in the last chapter, let's assume that the input port is reading a binary value corresponding to a Blanchard grinder's motor load and that the output port is passing a new value to a stepper rate generator. Since both of the numbers involved in these I/O operations require more than 8 bits of precision to provide the performance we require, we will assume that the the data bus is 16 bits wide. (An alternative approach is either the use of a pair of ports on an 8-bit bus or two successive operations with the same port.)

We will look at the input operation first. Coming in from the top, presumably from an analog-to-digital converter, is a 16-bit data word. In all likelihood, only about 12 of these are being used—16-bit A-Ds are not only expensive, but probably overkill in a situation like this. The remaining 4 bits, then, are either tied high or are being used for something else. But whatever the organization of the data, there exists a 16-bit value that we wish to place on the data bus at an appropriate time.

When the CPU decides that it is interested in that value (a "decision" that is simply the result of encountering an INPUT instruction), it issues the port number (35) on the address bus and an "I/O READ" signal on the control bus.

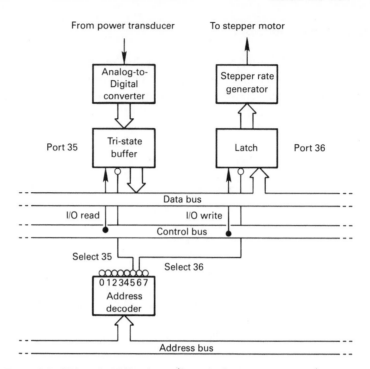

Figure 3-5 This typical I/O scheme (from the Blanchard controller) shows one input and one output port, each selected by the simultaneous presence of the proper port address and the appropriate control command.

The combination of these events has the effect of enabling the tri-state buffer, thus placing the input data on the data bus, where, presumably, it finds its way into the CPU.

It is worth pointing out that there is nothing sacred about this particular way of doing things. By deciding to avoid processor-specific information, we have accepted the burden of generalization, and in some cases, that can be confusing. There are a number of alternatives to this approach: the port can be viewed as a single memory location rather than something special called "I/O," the read line from the control bus might be used as a "chip enable" to the address decoder (which then singlehandedly enables the buffer), the A-D converter might (in fact, probably does) already have its own tri-state output, eliminating the need for the separate buffer shown here, and there may be no address decoder at all, with that function performed instead by an LSI interface chip that takes care of everything. There are dozens of ways to go about this, but the principles are all the same, and it's easier to grasp if all the boxes in the diagram map handily onto standard TTL functions.

So. After a slight propagation delay, we have succeeded in getting the A-D's value into the computer. Upon subsequent processing as described in Chapter 2, the system has a new RATE value for the stepper motor that takes care of grinder head downfeed. This calls for an output port.

In Figure 3-5, it is provided in the form of a 16-bit latch, something like four 74LS175s (quad D flip-flops) strobed as one. The output to the rate generator remains unchanged until a strobe occurs as the result of coincidence between the correct port address (36) and an I/O WRITE command, at which time the output suddenly changes to reflect the data that were on the data bus at that precise instant. Once the strobe goes away (in a few hundred nanoseconds), the contents of the latch are again frozen until a new strobe comes along.

All that was said a moment ago about different ways to perform the input function goes for this as well.

And that's the essence of microcomputer I/O. But every system has different requirements, and we will consider the broader subject of interfacing with the outside world in Chapter 5.

3.1.6 The Central Processing Unit

In Figure 3-1, the generalized architecture of a microcomputer system, we saw three boxes, interconnected via tri-state buses. We have briefly discussed two of them, memory and I/O.

But the key element is the CPU. Therein lie whatever "smarts" the system might be said to contain; therein may be found the controlling logic that orchestrates the behavior of everything else.

The innards of the various CPU chips that populate the microcomputer marketplace are well documented, as we have stated. So let's consider instead the general requirements that underlie the majority of chip designs.

The primary business of a microprocessor (we try to maintain some sort of semantic distinction here: the microprocessor is the CPU and the microcomputer is the whole system ...) is the execution of instructions. Somewhere in memory, in any working system, is a sequence of primitive operations which together comprise a program—hopefully designed well enough to properly dictate CPU activity in all situations. The primitive operations of which said program is assembled all belong to something called the *instruction set* of the microprocessor.

The instruction set, which in most cases is built into the CPU chip at the hardware level and is not modifiable, is a repertoire of functions selected by the chip designers to provide the programmer with enough flexibility in data manipulation to render the device competitive in the marketplace. As the technology continues to mature, the amount of flexibility that people consider essential grows, making some of the modern instruction sets (and the processor architectures in which they reside) masterpieces of design. When the 8080 supplanted the 8008 in the mid-1970s, designers accustomed to a lack of external stack, a maximum of 16 I/O ports, and cumbersome interrupt handling rejoiced at the wondrous features that the new part had to offer. People began to chuckle at the 8008. But it wasn't long before programmers began grumbling about some of the

8080's shortcomings, and soon, even better devices appeared. None of them were the ultimate, and none, probably, will ever be. But in the process of making a stepped asymptotic approach to that ever more elusive "ultimate," some surpassingly clever CPU designs have come about.

But enough oohing, aahing, and historical perspective. How does a CPU chip go about executing instructions? What, in fact, is implied by that term?

Accepting for a moment the general premise that we want a microcomputer to do something involving data, we immediately begin to suspect that there are some central data-handling "tools" that would be of considerable use. We certainly need the ability to move information to and fro within the confines of the system—from any place in memory to any other, into or out of the CPU, and to or from some interface with the outside world. Examining the problem more closely, we note that no matter how communication between the CPU and memory is accomplished, it involves some overhead (we have discussed it, in fact: address and control bits out, data onto the bus, etc.). This suggests that some temporary storage locations should be provided (within the CPU itself) which can be accessed quickly and conveniently without the need for external memory addressing. We'll call them *registers*, and add a class of data movement instructions which allow the unrestricted flow of data among them.

So far we have imagined a device with some unknown number of registers, with instructions that can move any piece of data from any place in the system to any other. We need more.

At some point in most practical microcomputer applications, it is desirable to do something with data besides move it about. Perhaps we simply want to add pulses from a sensor into a total; maybe we want to monitor a number of individual status bits and take various actions depending upon their conditions; we might even want to gather a sample of time-domain information from a signal source and perform a Fourier transform upon it to extract its frequency components or implement a digital filter. All of these tasks require data manipulation—not just movement—and we should add to our hypothetical processor a collection of basic functions that can be combined to do just about anything. We might find in the process of doing this that it is desirable from a design standpoint to take one of the registers and make it somewhat special, defining it as the location of most operations' results. Let's call it the *accumulator*, honoring tradition. Given all this, perhaps we append to our basic data movement instructions ones which add, subtract, increment, decrement, rotate and shift left or right, complement, and perform such Boolean operations as AND, OR, and Exclusive OR.

Great. If we write a program with what we have so far, however, we very quickly encounter an embarrassing problem: once the processor starts executing a program, it just keeps on going, one instruction after another, with no way to alter its path or even quit when it reaches the end. Let's fix that.

The address that the CPU puts on the address bus can come from one of at least two places, depending upon the type of operation that inspired the event. If the processor is fetching an instruction for execution, the address represents the current position in the program. It is maintained in a special-purpose register, usually called the *program counter*. The other possibility, of course, is that the address is involved in a data transfer.

The program counter is a rather critical part of a processor, for obvious reasons. It is also the focus of our next group of instructions, which we might call the "jump, call, and return" group.

In the most trivial case, we might wish to abruptly change the region in memory from which the CPU is sequentially taking instructions. If we provide a JUMP instruction which has as its operand an address which uniquely specifies the location where we want the processor to "go," then we have solved the basic problem of keeping execution within bounds—the CPU just places that operand into the program counter and proceeds as if nothing has changed. At the end of the program, we can just put a JUMP back to the beginning. Or to the middle. Or to some other program, out there.

But it still can't make decisions. Think about it: for all the data manipulation capability that we have designed in, there is still no way for the processor to modify its behavior as a function of the data. Useless, unless you need a sequencer. How about this: add to the hardware a group of *flags*, which are set and reset in predictable ways as data are manipulated. Whenever the accumulator becomes equal to zero, for example, a ZERO flag could be set. Another could indicate a CARRY (or overflow) condition—another the arithmetic sign of a number—yet another its parity. A few others could be tossed in which reflect special conditions of various sorts.

Armed with these flags, we can increase the processor's power by orders of magnitude if we simply use them as conditions for jump instructions, and the other transfers of execution we will discuss momentarily. This makes it possible to perform a certain sequence of instructions some arbitrary number of times by simply placing the number in a register, decrementing it upon each pass through the loop, testing its new value, and jumping if it is zero to another part of the program. Suddenly our processor has the ability to make—and act upon—logical decisions by altering its paths through its program as a function of various operations. At this point, it is a practical computer.

Now let's carry it a little further, and escalate it to a realistic one.

A universal programming technique involves the use of *subroutines*—well-defined groups of commands that are isolated from the main program. The value of this becomes obvious when you consider that a typical program such as the Blanchard controller may require, for whatever reasons, a standard operation such as a multiply at perhaps a dozen different places. The insertion of a sequence of instructions that performs the multiply at each of those places is an absurd waste of memory space.

Instead, the multiply operation is defined as a subroutine, and is "called" whenever it is needed from anywhere in the main program. The CALL is somewhat analogous to the JUMP—it places the target address into the program

counter, immediately resulting in the transfer of execution to the code in the subroutine. At the conclusion of the subroutine, however, execution resumes at the
instruction following the CALL. The general structure of a typical program with
three subroutines is suggested by Figure 3-6. Note that A, B, and C can be called
from any location in the program, and, in fact, from anyplace else: C is shown being called by B. The latter phenomenon is called *nesting*, and in most computers
can occur to any depth (limited only by memory size).

    Yes, we have left out a key. How on earth does the processor know where to
return to after it completes a subroutine? How, in fact, does it know when it has
completed one?

    OK: In addition to JUMP and CALL, we need one more instruction type
that causes a transfer of execution. We shall call it RETURN. Unlike the other
two, however, RETURN does not require (or even allow) the specification of an
address, for the next instruction executed by the CPU is entirely dependent upon
the location of the CALL that caused the subroutine's execution in the first place!
The trick is this: we need to create another register, one which, like the program
counter, is capable of addressing all of memory (data registers such as the accumulator need not have this many bits). We will call this new one the *stack pointer*.

    Now, when a CALL instruction takes place, the CPU must perform one
small chore before wandering off to execute the code in the subroutine. It must
increment the program counter (by the appropriate value, depending on the

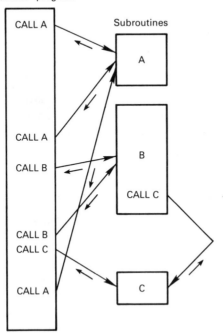

Figure 3-6 A subroutine may be called
from any point in a program—or from
another subroutine (even itself, in some
"recursive" applications).

number of locations occupied by the CALL instruction ...) to the next instruction in the program—just like it would during straightforward sequential execution. But then, it must store the new address in the program counter at the location "pointed to" by the stack pointer, then decrement THAT. Once this has been accomplished, it can perform the subroutine with impunity, for upon encountering a RETURN instruction, it will increment the stack pointer, take the address stored there and place it in the program counter, then simply continue processing. The location at which it does so is the one immediately following the CALL.

The clever thing about the use of the stack pointer for all this is that a list of return addresses (of arbitrary length) is stored in memory. The main program can call a subroutine, which in turn calls a subroutine, which in turn calls a subroutine ... and as long as the proper number of RETURN instructions is executed, it will eventually end up back in the main program, right after the original CALL. In typical operation, one of the first instructions in the program sets the stack pointer to the top of memory, and the design sets aside an appropriate amount of space for the stack of addresses that will alternately grow and dwindle as the program executes.

We should note, just in case it isn't obvious, that not only JUMPs, but also CALLs and RETURNs, can be made conditional on the basis of flags. All types of execution transfers are thus available as a means of acting upon logical conditions.

If we implement all of the foregoing in a CPU chip, we have a typical microprocessor (of the Intel style). There are a few other things that round out its capabilities, however, so let's note them briefly before moving on.

First, we should provide a variety of "addressing modes." It is highly desirable to have more than one method of specifying a given piece of data in the system: we would like to be able to include the datum itself in the instruction, reference a particular register that contains it, explicitly identify a memory location that contains it, and point to a memory location that in turn points to a memory location that contains it. There are a lot of variations upon this theme, and a healthy stock of them available without too much overhead can make certain types of programming tasks almost pleasurable.

Next, since we have that elegant structure called a "stack" for saving our subroutine return addresses, we can add a few instructions that let us put other things there as well—in particular, registers. This comes in handy when we allow asynchronous external events to interrupt the execution of a program, forcing a special form of subroutine call to a location in memory wherein the interrupt is serviced. Since we have no way of knowing when this will occur, we have no guarantee that the processing in progress can be safely disrupted by the register and flag modifications that will result from handling the interrupt. So—we throw in some PUSH and POP instructions that let us save and recall, respectively, the registers (and flags) on the stack. (One must be careful not to RETURN to a piece of data: a common programming error involves careless intermingling of PUSH,

POP, CALL, and RETURN instructions. The result usually is called a *stack crash*.)

While discussing the PUSHs and POPs, we might toss in an instruction or two that will let us maintain more than one active stack. This is begging for trouble, but sometimes it is handy.

And then, there are the endless refinements on the existing classes of instructions—we can add ones that allow direct set, reset, or test of individual bits, ones which move entire blocks of data without the programmer having to write loops which do that, ones which exchange registers with each other, etcetera, etcetera, etcetera. We need one called HALT, which does just what its name implies (an interrupt or hardware reset will make it go again), and one called NOP, which does absolutely nothing.

We have just spent a lot of pages talking about the innards of a microprocessor, and of course, it quickly became a software discussion. If the whole business of a micro is indeed the execution of instructions, it is somewhat pointless to talk about its internal hardware design without establishing its objectives first. But we have just done that, so keeping all the foregoing in mind as a general set of microprocessor design criteria, let's look into four major classes of devices to see how they integrate those essentials with their applications categories.

3.2 DEVICE FAMILIES: AN INDUSTRY OVERVIEW

3.2.1 Single-Chip Microprocessors

In the beginning of Chapter 1, we rather offhandedly categorized "consumer" systems as relatively disposable items whose production economics exclude them from our serious consideration herein. So it would seem that single-chip microprocessors, a healthy segment of the overall micro market that was spawned by the needs of mass production, would fall well outside the range of our focus.

Well ... maybe a few years ago, we could get away with that. Single-chip systems with RAM, ROM, and I/O built in were originally developed for such applications as electronic games, calculators, and appliances. Their horsepower was sorely limited by the boundaries of the technology, and any "serious" application required multichip designs.

Single-chip processors have historically focused on large-volume units, of course, with a late 1980 market survey showing them outselling their big brothers by a 5-to-1 margin. But the advances in VLSI technology that have taken place are dramatically changing the roles that are being played by these devices.

In the preceding section, where we hypothesized a microprocessor design, we created a set of registers to allow speedy manipulation of data whenever the programmer could arrange things to avoid extensive memory addressing. Crossing the chip's outer boundaries is always more time-consuming than internal activity, since the bus must be time-shared, buffer and latch delays must be accommodated, and control signals must be coarse enough that designing with the part is not too much of a pain. Our intent with the registers, then, was to reduce these external activities at least a little, allowing as much of the "bit crunching" as possible to take place internally.

But suppose that the technology allowed us to carry this one big step further, and place the entire memory inside the chip? Suddenly there would be very little need for registers (except possibly as a programming convenience). Now all memory accesses would be fast, and interaction with the world outside the 15 or 20 nanoacres of silicon comprising the chip real estate would be limited to I/O operations with the peripherals. This could lead to some very fast processing, and a range of applications far removed from hand-held games and washing machine timers. The discrepancy between internal and external speeds will continue to increase as the size of the basic FET building blocks continues to decrease along with the line widths and spacings that interconnect them.

So—we already see an architectural change from register access to memory-oriented designs. What is it good for?

One of the classic problems with computers in general is that they are single processors, restricted by the information bandwidth implied by sequential operation through a bus structure. We go into this in much more depth in our chapter on artificial intelligence (AI), but this is a good place to examine a significant trend in the war on technological limits: *multiprocessing.*

When confronted by a design problem requiring extremely high "throughput," such as image recognition, real-time structural modeling and the like, the "von Neumann bottleneck" alluded to above imposes immutable limits on the efficacy of the solution. All too often, it is necessary to resort to tedious sequential scans which move the task well out of real time. Even music synthesis, which seems at first glance to require only the 20-kHz bandwidth of the human aural system, is nigh impossible with uniprocessor real-time techniques when the complexity of the musical passage moves beyond one or two voices.

The answer is multiprocessing: the simultaneous activation of widely distributed processing resources toward the solution of a single problem. That, in fact, is how our brains do it. We do not yet have the technology to produce a true distributed network, with no elements called "CPUs" identifiable, but we can effectively model that sort of architecture with a large number of serial processors. It is convenient to do it with single-chip devices, for it otherwise gets expensive and complex very quickly.

The evolution of single-chip processors, then, began with I/O intensive controllers (the Texas Instrument TMS 1000 is a classic big seller here) and moves now toward optimization for roles as nodes in multiprocessor systems. And because of the speed of internal access, numerous interesting design features can be implemented, such as in the Intel 8051 that is essentially a Boolean processor capable of converting random-logic expressions directly into code without all the standard clumsy programming problems.

In this book, most of our detailed discussion will assume the use of the "standard" processors described in the next section. But it is worthwhile to keep an eye on these single-chip devices, for they will take on more and more substantial roles as our understanding of multiprocessing techniques matures to match the problems before us.

3.2.2 8- and 16-Bit Multichip Systems

Here is the "bread and butter" of the industry. Robust general-purpose CPUs, marketed alongside intelligent support chips that handle the details of parallel and serial I/O, disk access, memory control, floating-point processing, and a lot more, are the most visible denizens of the microcomputer world. It is not only the breadth of market support that keeps the 8-bit multichip systems healthy, but also the universe of users that have, over the years, become locked in to some of the standards as their primary design tools. A huge population of personal computers, business systems, and industrial controllers are based upon 8-bit multichip systems (8080, 8085, Z-80, 6502, 6800, ...), and hardware and software support has been extensive.

The advantage that systems of this category have over the single chippers we just discussed is generality. For a variety of reasons, it is worthwhile to focus single chip processors on certain types of tasks; no such biases exist in the world of multichip micros. But there are a few significant subdivisions.

Eight-bit systems brought the technology into the 1980s, and have bootstrapped 16-bit systems into prominence. The wider bus allows not only twice the precision and bandwidth for a given execution rate, but lends itself very nicely to the implementation of high-level languages, some of which are a bit awkward on 8-bit machines. These new tools, far from being dedicated to the (yawn) world of business DP, are extremely powerful allies in the development of industrial application programs. PASCAL, ADA, and C, to unfairly name only three of particular prominence, have effectively obscured the old standbys such as FORTRAN and (choke) BASIC. And *assembly language*, used to explicitly define a program one machine instruction at a time, is rapidly becoming less attractive as the old saw concerning its superior coding efficiency rusts in the wave of new timesaving tools offered by high-level languages. It will be a long time going, for there are legions of assembler addicts, but it is happening.

We won't dwell further on the general-purpose multichip systems here, for we'll be dealing with them from a variety of angles as time goes on.

3.2.3 The High-End Systems

This business of categorizing systems is dangerous. History has shown that there are no fixed boundaries: once upon a time, minicomputer makers sat smugly in their East and West Coast offices, certain that microcomputers would remain forever limited to hobbyist and certain small, dedicated applications. Hm.

So here we have the gall to identify something called a "high-end system"?

Well, let's just use this section to follow very briefly the continuum upward from the systems we have just discussed.

If you take a general-purpose processor and append some sophisticated support chips, people will be hard-pressed to tell you that it is not a high-end microcomputer—especially if the support chips are things like coprocessors and I/O subsystems. The machine will run circles around a lot of minis. Also, there is such a wide range of trade-offs that it becomes difficult to decide in which domain the title "high-end" should be applied: raw speed, array processing ability, generality, extended addressing, or ....

Whatever. It should be noted that a few "micros" are showing up which very definitely fall not just within the performance range of minis, but within that of mainframe systems as well. Intel's iAPX 432, for example, is architecturally optimized for the high-level ADA language, and can be elevated to higher performance levels with the simple addition of more CPUs. The system automatically redistributes the computing task to take advantage of the added processing power.

We will be seeing a lot more of this as time goes on, as Section 3.2.1 hinted in its mention of multiprocessing. Very likely, the high-end systems that continue to emerge from the labs will be constructs of this sort, with any present advantage in cramming a lot of power into a single CPU overshadowed by the architectural desirability of distributed systems.

But that's not the subject of this book. Let's press on.

3.2.4 Analog and Special-Purpose Microprocessors

Certain applications that call for the use of microprocessors are characterized by such specific requirements that they spawn dedicated CPU designs. If it is just an unusual instruction set that is required, accommodation is rarely a problem— many modern CPUs implement their instruction sets as *microcode* in ROM instead of random logic, and are thus modifiable (if the price is right).

But digital signal processing (DSP) is one of those areas in which an enhanced instruction set is not quite enough. DSP is finding more and more volume applications, now that some of the diehard analog designers out there are finding that functions like filtering really can be performed digitally with dramatic performance increases. One of the first of such volume applications involves telecommunications voice filtering and the related processing tasks associated with the phone lines. Intel, again, was the pioneer in this area, introducing the first analog microprocessor—a chip with built-in analog channels and an architecture oriented to DSP, in particular the fast computation of sum-of-products terms characteristic of the field.

We will have occasion to look more deeply into digital signal processing much later in this book, but the vast majority of industrial design requirements within our accepted task domain can be addressed by the midrange, multichip systems.

It is on these that we will focus for at least the next 0.1 million words.

4

Hardware Tools

4.0 INTRODUCTION

Chapter 3 attempted to convey something of the essence of microprocessor operation through discussion of bus communication, memory, I/O, and the CPU itself. This is the heart of the technology, and must be well understood by a designer charged with the task of implementing "smart" controllers.

But if you really had to design every system from the ground up, you would end up spending at least as much time getting CPU chips to work as you would actually solving industrial control problems. It would be analogous—one level removed—to constructing logic gates from transistors and building op amps from scratch. Every now and then, for an unusual application, it may be necessary, but the range of tools available to the system designer encompasses far more than chips.

For example, need a microcomputer with about 256 bytes of RAM, 4K of ROM, four I/O ports, a couple of interrupt lines, and a real-time clock? You can build one fairly handily, given a few weeks of intensive development time, either wirewrapping it or going all the way and designing a custom PC layout. You can probably keep total engineering cost under $10,000—which is perfectly acceptable if you are going to manufacture a lot of boxes with that processor card nestled snugly inside.

But you only need one? Or a dozen? Suddenly the much-touted cost effectiveness of the micro looks like something of a fraud, eh? (Especially if you are just trying to replace a handful of random logic with a computer in order to build in a little more flexibility.)

Well, the answer to that problem can be found in any one of hundreds of different *single-board computers* (SBCs), offered by every microprocessor manufacturer and many others besides. In most small-volume applications, SBCs offer a far more efficient solution than the "homebrew" design, with the added advantages of factory support: if the board is dead on arrival or fails in service, there's someone out there who can ship you a fresh one.

The hypothetical computer we proposed a moment ago exists in a variety of forms, both off-the-shelf and cheap. One of them is called the Intel SBC 80/04 and is the bottom of their SBC product line. Yet it is a full-fledged microcomputer system, needing only power and a program to reach its rendezvous with destiny. A couple of the systems we discuss later in this book will make profitable use of this low-cost card.

Another pleasant aspect of single-board computers (and their various companion support cards) involves some relaxation in the amount of microprocessor design wizardry required of the user. As the technology has continued to mature, a variety of interesting features have become available, including very high speed operation. Creating a CPU board that can take advantage of this speed capability without noise problems is not exactly trivial, and the use of SBCs can eliminate any need to make the attempt. Further, it's easier to have a variety of different processors in your design repertoire, offering various software and architectural features, if you don't have to design each one from scratch. Many SBCs hosting different CPU chips appear identical at their edge connectors.

In this book, reflecting the realities of industrial design, the SBC will be our basic hardware tool. This chapter is concerned with them, their external buses, and the integration of the SBCs with support boards to form complete systems.

4.1 DESIGN, DEBUGGING, AND OTHER MATTERS

When you set about to create a system for some application, you are suddenly confronted with something more than the old "crank-turning" engineering problem. Typically, you have in hand something called "specs," very likely a pile of stained coffee-shop placemats (or their formal equivalent) astrewn with flow diagrams and random ideas, as well as a mandate to keep the cost under some fixed number of dollars. Hopefully, this latter figure (really the key specification) was arrived at by something more relevant than the classic rectumological projections that have landed so many small engineering firms in bankruptcy court. Hopefully.

Anyway, the customer (or your boss, the sales department, or whoever) wants a universal, flexible, reliable, programmable, easily modified, fully tested, serviceable controller that not only meets the placemat specs but also solves his business problem. You, the designer, see the task as the implementation of a transfer function or control algorithm; the customer (unless he is very enlightened) sees it as the reinforcement of a weakness in his production environment.

That's your first problem. Right away, you're not communicating. A box that meets the specifications invented during that long-ago ill-fated lunch may not, after all, reduce plant costs or decrease the percentage of scrap that is generated by the manufacturing process.

But you press on. You scratch your head, lean back in the chair, fill the wastebasket with stupid ideas, get burned out on coffee, take a long break, and eventually find yourself with a solution so elegant and simple that you are a bit embarrassed that it didn't occur to you immediately.

It doesn't take long to arrive at a plausible hardware configuration—a stock processor card, some wirewrapped interface logic, a box, a power supply—the usual. The software, hmmm, maybe a couple of 2716 EPROMs, power, gee, probably about 6 amperes or so. It more or less congeals along those casual lines, unless you are one of those formal types who somehow manages to avoid empirical processes, preferring to generate a finished package of drawings and specifications before centerpunching that first LED hole or typing that first tentative TITLE statement into your software development system.

But it gets off the ground either way. Vendors fail to sense the urgency of the project and ship a few weeks late of course, but you planned for that. The software problem contains a few surprises and you vow to multiply your estimate by four instead of three next time—if there is a next time. Whenever you polish off a piece of code with a flourish, you suddenly remember some major routine or control structure that hasn't been written yet. It's getting old.

And where the hell is the edge connector that had to be ordered to fit that "hamfest special" board that your tech went ahead and wirewrapped all the interface logic on? Winter sets in.

Deadline looms.

But in the space of one glorious afternoon, the part comes in and you emphatically add the long-lusted-after END statement to the bottom of your source file. Your harried technician wires in the connector, applies power, and something smokes. You link all your code together and find that it will not fit into the ROM space provided. You go home early to fix the dishwasher, which dumped water all over the kitchen.

Sooner or later, of course, those problems are solved, and it is time to put everything together and try it. You have spent the first 90% of the available time designing and building the system.

You spend the next 90% debugging it.

"Never again," you mumble, as you watch it go out the door four months later.

"Why me?" you ask under your breath on the first five service calls (on which you take EPROMs bearing steadily-incrementing revisions of the program).

"What is reality?" you whimper one night at 2:15 A.M., staring bleary-eyed into an incomprehensible jumble of blue No. 30 Kynar wirewrap wire, a scope probe dangling listlessly from your limp and sweaty right hand, the floor about you littered with chips, tools, smudged schematics, stained vending machine coffee cups, and the accumulated crud of an 80-year-old factory which somehow makes everything the same color—sooner or later. You painfully shift from squatting to kneeling, completing that cycle for the twenty-eighth time, and contemplate farming as a viable career option.

Suddenly you see it: a stray bit of wire buried in the tangle of the interface card. Your heart stands still—you probe gently. Could this be it? Yes? Yes! A twisted little piece of No. 30 that broke off and fell when an erroneous connection was unwrapped—how long ago? Incredible. Could this be that stubborn intermittent glitch that you had blamed on the ac power, telling the customer to bring in a special conditioned circuit for your controller? Ever so carefully, you extract it with the hemostats, then after joyfully venting your emotions upon it you fling it into a murky corner and close up the box. It seems solid.

You don't hear from him all week. You're dying to call and actually get him to tell you it's OK, but you don't dare. Time passes.

Six months. You've gotten yourself into another project, but still you jump each time the phone rings, for you dread the inevitable call . . . .

"Hey, how ya' doin'? Look, we've modified our production setup—when can we get together to talk about a few changes in your controller design?"

4.1.1 "Analysis"

Hm. What a mess. First, it is worth noting that the designer above (whom we cruelly cast in the second person—hope it wasn't too traumatic) used a single-board computer, but still had problems. So we can observe that it is not a panacea. It probably sped up the project, of course, since he didn't have to design his own processor, but it didn't keep him from lying awake at night.

The reason we have launched into this colorful but depressing scenario in what is theoretically a chapter on single-board computers is that we need to establish a useful paradigm for project development. The example above is fairly typical, and much of this book will be involved with techniques and ideas which can take some of the unpleasantness out of the process. Some of the tricks are specifically involved with software—with structure and organization of programs—and the rest have to do with everything from component choices to overall attitudes.

One of the first points, reflecting back to our opening comments about SBCs, is that designing with off-the-shelf modules is nice when possible. If you are doing all this as a hobby, then the added time and uncertainty of custom circuit designs might be acceptable—possibly even fun. But there is some kind of unquantifiable geometric relationship between the number of unknowns in a system and the time needed to debug it. If most of your industrial design can be broken down into purchased modules, then you're ahead of the game.

The same logic applies to software, too, but we will wait a few chapters before considering that.

In a typical case, like the one described above, the system will end up being a collection of modules—with one exception: an "interface" card that mates the whole mess to the particular corner of the universe which will become its home. It follows that most of the hardware problems will be in that card, unless it is designed and tested as a module itself.

Most of the problems in our woeful tale can be traced to design philosophy. It is tempting when starting a new project to just sit down and start designing whatever part seems most obvious up front, but such an approach can create a lot of difficulty. For one thing, a fair amount of patching may be necessary to compensate for incongruities between independently designed modules. For another, difficult parts of the project may end up being put off until later, when they will be subject to an added set of constraints: the characteristics of the system components already designed. This isn't an area in which surprises are welcome—you might suddenly be confronted by an impossibly awkward set of requirements. According to the foregoing, one should design—or at least consider—all parts of the system at once. Although brains are indeed parallel processors, they have a hard time managing parallel job streams. There must be a trick.

There is. One popular buzzword for it is *top-down design.*

4.1.2 Top-Down Design

The idea of approaching a design project from the highest level of abstraction and working downward to the lowest level has validity in many fields of endeavor— even writing and education, as noted in our introduction.

A top-down approach may be contrasted with a bottom-up approach, in which the component parts are all assembled, then integrated together to form the whole.

If all this appears cumbersome, you can always take the position shared by the second-person character in our system design paradigm and use an amorphous, middle-out philosophy. But we all know where that leads.

Anyway, you can often symbolize your initial top-down design with the use of a structural sketch, shown in Figure 4-1. With this graphic tool, you can start by isolating the general design concept of the system (the control algorithm or whatever) and placing the various support functions underneath it, detailed as finely as you like. It must be emphasized that the resulting diagram will have very little to do with the software structure that will be developed later—this is just conceptual. (In actuality, the highest level of the software will probably be some sort of event- or clock-driven testing loop which determines via flags the need to perform any of a variety of subtasks.) But at this level, before you really know how it's all going to fit together, a drawing like this is a big help, for it lets you establish an overview of the design and begin to identify modular functions.

Then you can simply start at the top and specify the operation of the system. Wherever connections emerge from the bottom of a box, define the characteristics of the link and forget everything below. For example, in the figure, the con-

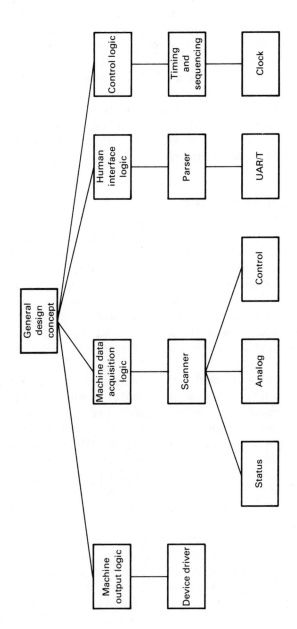

Figure 4-1 This typical "structural sketch" allows the designer to isolate functional modules from each other and visualize their interrelationships. Such an approach reduces the confusion that results from attempting to view a complex system as one large body of hardware and software.

nection between the GDC and the "machine data acquisition logic" might be expressed as follows: "On command, the data acquisition logic returns a 'snapshot' of the external process, including machine status, measured values, and operator control modes." The value of this approach is that you can then ignore all the boxes involved with that process when you design the GDC, for their collective behavior is completely defined at the interface. You have converted that branch of the logical "tree" to something called, appropriately, a *stub*.

At this level, it must be noted, the process is still pretty tentative. The same technique is later applied to software design with a degree of formality inversely limited by your imagination. But it is useful here, for it allows you to begin conceptualizing the overall architecture of the machine: Should the data acquisition logic be done on a hardware card, or can it be software controlled by a few ports?

4.2 TYPICAL SYSTEM HARDWARE

The system described in Section 4.1, whether it congealed or crystallized, somehow ended up with a hardware configuration more or less typical of micro-based industrial control systems. Let's save further elaboration on the design process for later chapters and start looking at the box. For the time being, the software can safely remain a mystery—as long as we accept its role in making the machine do something other than sit there and dissipate power.

We never decided what the lamented system was for, so we're free to pull a configuration out of the air. How about Figure 4-2? That's a fairly typical collection of parts.

In the middle is the card cage, containing four circuit boards. We'll assume that they are all off-the-shelf standard products which, together with the backplane, we purchased from a system vendor. The boards are the processor, some additional RAM, a disk controller, and an I/O interface. (Yes, much of that is available on a single card, but let's assume that the system is a fairly robust sort, and needs the additional resources of the extra circuit boards to do its job.) The backplane, or motherboard, is the card into which all the others are plugged, and represents one of the most important single characteristics of the system: the bus.

Wait a minute. Didn't we already talk about buses? You know, address, data, and control?

Yes, we did. But terminology in this business is at least as ambiguous as it is in any other. A bus, in addition to the above, can be considered as a standardized collection of pin definitions, including power, expansion signals, interrupt lines, and all sorts of bells and whistles besides the three processor buses we have discussed. It, in its finest form, is something that is agreed upon across the industry and used as the basis for a variety of board designs, resulting in a wide selection of available modules that can be purchased and plugged into your system without extensive patching. Not only does this protect you against becoming orphaned by the untimely demise of a manufacturer, but it also brings market competition into

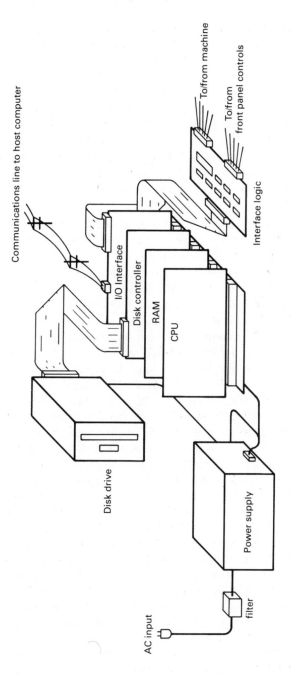

Figure 4-2 A possible hardware configuration for an industrial control system. There are four boards plugged into the system bus and one outboard interface card (which in a custom system may well be the only "scratch-built" component).

play and lowers your costs. Three of the biggies are called Multibus, S-100, and STD.

It will not matter in this discussion, but let's pretend that the system in Figure 4-2 is based on the S-100 bus. Originally introduced by MITS in the mid-1970s, this bus began as the first "standard" in the personal computer market, finally became a true standard near the end of the decade, and then found itself being pushed more and more into business and industrial systems when "appliance computers" appeared to make such relatively expensive configurations less attractive to the home market. The little comment about becoming a true standard was somewhat understated: until the IEEE and others finally agreed upon a precise definition of the 100 signals in the bus, various manufacturers felt free to use some of the undefined lines for their own purposes. You can imagine where that led. People felt safe in assuming that all S-100 boards would work in all S-100 systems. Tsk, tsk.

Not only that, but the S-100 only works well with 8080-style CPUs, it's rife with cross-coupling, logic polarity is inconsistent, and there is no provision for master-slave operation. But it works.

Anyway, in Figure 4-2 we have an S-100 system. We don't care what the processor is, how much memory it has, or any of that. We just have four 5×10 inch boards plugged into a backplane, presumably mounted in a cage with card guides to keep them from flopping about.

Power for the system is provided by a dc supply which receives filtered 115 volts from the line. S-100 bus specifications call for three supply voltages: $+8$, $+18$, and -18. The first is the big one—it is "spot-regulated" on all the cards to yield the 5-volt V_{cc}. (Spot regulation is the alternative to central distribution, and they each have their advantages. In a nutshell, on-card regulation provides cleaner power with less *IR* drop—but the regulators dissipate heat among the boards rather than elsewhere, creating cooling problems.)

The disk controller is connected to a disk drive, presumably a floppy. Floppy disks have gained a well-deserved bad reputation for poor reliability in dirty industrial environments (and it is hard to convince nontechnical users that it really is worthwhile to handle the media carefully), but let's assume that the system is in a relatively clean place and that there is only infrequent need to swap disks.

All of the interaction with the outside world takes place through the I/O card. One port is a UAR/T (universal asynchronous receiver/transmitter) and handles communication with a "host system" located at the other end of the stylized data link. Others connect with the random-logic interface card, which we can presume is the wirewrapped agglomeration of chips which gave the designer such fits in Section 4.1. It handles all the input and output with both the process being controlled and the front panel devices, probably taking care of some signal conditioning, data conversion, and "convenience" logic as well.

It is worth noting that the use of a standard I/O card to talk with the interface logic is somewhat arbitrary. An equally feasible option is the use of a blank S-100 card to host a custom interface circuit that plugs right into the bus. That would have the advantages of reducing chip count (probably) and eliminating the need to contrive a mounting scheme for the one oddball logic board that is not in the card cage. It would also reduce cabling requirements. But it would be harder to test.

Now let's think about modularity. Looking at the card cage again, observe that there is nothing preventing us from ripping it out and replacing it with a single-board computer—or a PDP-11. (That's another good reason for the outboard logic!) It is just a black box called a computer, and as long as it can be made to host appropriate software, it's OK. This flexibility makes development a lot easier: when you are designing the software, the computer in the picture can be a development system that provides appropriate tools (assemblers, compilers, linker, debugger, etc.), but when the software design is complete, the code is burned into ROM and plugged into an SBC, which then replaces the development system as the heart of the machine.

There is something pleasantly elegant about such flexibility. In the Old Days, the hardware comprising a system was generally an absolute—an immutable mass of components whose configuration established the machine's present and future capabilities. But with today's tools, we have freedom in two orthogonal domains: the software can be manipulated to define the operation of the hardware, and the hardware can be changed to define the scope of the software.

This realization immediately raises the observation that one of the key design steps is the establishment of some sort of hardware/software balance. This subject will pop into our awareness repeatedly as this book progresses. One example has already come to our attention: the rate generator of the Blanchard control system. Somehow, we must decide whether it makes more sense to crank out the stepper drive pulses using a "subtask" of the program or to let the program pass a value to a chunk of TTL and let it do the work. Such a decision certainly affects the software design, and might even impact the choice of processor type.

Let's take a quick look at some prototypical SBCs and their support cards to gain some appreciation of the range of hardware possibilities.

4.2.1 Bare-bones Boards

Beautifully exemplifying much of the economic rhapsodizing we have committed at various points herein, the low-cost, bottom-end SBCs on the market have enabled the much-touted power of the microprocessor to be put to work in some pretty unglamorous places. Generally providing little more than a CPU, a touch of RAM and ROM, and perhaps three or four ports, machines of this class are perfect for applications in which significant expansion and flexibility are unnecessary. They're cheap.

We mentioned, at the beginning of this chapter, one of the products in this category, the Intel SBC 80/04. It has been around for a few years, and sports no razzle-dazzle technology, but it is an excellent example of the value to be found in even a minimal micro system. Take a look at Figure 4-3.

On a board of 6.75 × 7.85 inches, we have an 8-bit processor, 256 bytes of RAM, 4K of ROM, 22 programmable I/O lines, a programmable 14-bit binary timer, a serial interface, four levels of vectored interrupt, and space for a +5-volt regulator and a little bit of kluge logic. Fully packed, the board has 20 chips. It works on a single 5-volt supply (as long as you don't use the RS-232 interface for serial I/O, or old three-voltage 2708 EPROMs).

The SBC 80/04 is intended for applications in which the designer has no interest in adding more memory or I/O cards, for there is no provision for external access to the bus. The only edge connectors are for the 22 I/O lines, power, reset, serial interface, and interrupts. Since it is not intended to be mounted in a card cage, there are holes in the corners which allow it to be screwed directly to a piece of equipment.

Since we will be using this card in later examples, let's look at it more closely than the relatively general tone of this chapter would otherwise suggest.

The CPU chip is the Intel 8085, a single-supply part that executes the same instruction set as the 8080 but which does not require nearly as much support circuitry. There are six general-purpose registers, which can be considered singly or as 16-bit pairs, and an accumulator. It has a 16-bit program counter (addressing 65,536 memory locations) and provides for an external stack which can be used both for subroutine nesting and last-in/first-out storage of registers and flags.

This device evidently was conceived as an answer to the Zilog Z-80, which stole much of the 8080's thunder sometime in 1976 or 1977, but Intel marketing wisely realized that the 8085's original design, falling squarely between the two top contenders (within the Intel-style philosophy, that is), would probably end up orphaned. As a result, they marketed the device as if it had precisely the same instruction set as the 8080, with the addition of only two commands which handled serial I/O. The interesting part of all this is that the 8085 really performs 10 other instructions as well, ones that help overcome some of the most frustrating shortcomings of the venerable 8080. There are also a couple of new flag bits.

This is all terrific, of course, and so tempting (once some clever folks in West Germany discovered and published it) that the undocumented instructions have been put to use in more than a few system designs. The only catch is that Intel makes no guarantee that they will still be there in later versions of the part; nor, of course, do second-source manufacturers. So it is dangerously pathological to yield to this temptation.

So much for interesting historical notes on the 8085. The SBC 80/04 provides a small scratch-pad RAM of 256 bytes. It's nothing to write home about, but it is surprising how much can be done with such a small amount of space. Program storage is provided in the form of two 24-pin sockets, which can be jumper configured to accept either 2708 or 2716 EPROMs. In Chapter 12, we present a very sneaky technique that makes use of one of these sockets to save about $20,000 in capital equipment costs.

The board's parallel I/O is clever; instead of hardwiring some input and output ports, the designers left the behavior of the 22 lines up to the programmer.

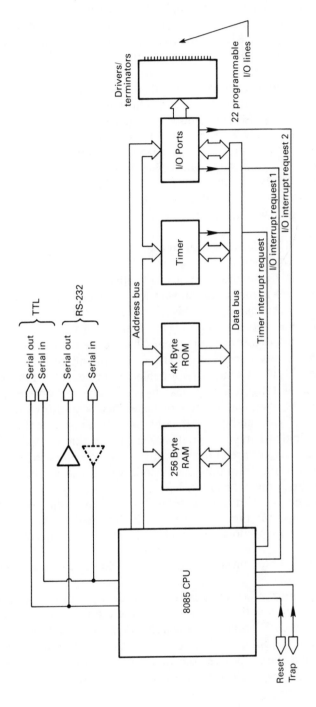

Figure 4-3 As the block diagram shows, the Intel SBC 80/04 board is a self-contained computer system, needing only power and a program to interact with its I/O devices in a useful fashion.

There are four ports, and when the application program is initialized it must issue an output instruction which defines their functions. The hardware allows such choices because it uses an 8155 as the I/O interface and provides only empty sockets for the drivers or terminators, allowing the user to configure them as necessary.

There's also a timer, which is basically a down-counter that can be written to and controlled via two other output ports. This has a variety of uses, including interval measurement, pulse generation, rate generation, and the reference for a real-time clock. Although the time taken by software loops can be measured (or predicted) with accuracy, they are a dangerous basis for real-time operation since interrupts, actually unpredictable subroutine calls, can befoul the synchrony originally intended. A hardware timer like this solves the problem—in the data collection terminals, for example, it will be seen to be used for a variety of interval and delay functions which would have been clumsy or impossible without some external reference.

The timer, incidentally, resides on a single chip along with the RAM and the 22 programmable I/O lines.

The SBC 80/04 provides a serial I/O facility, eliminating the need to use a hardware UAR/T for communication with a terminal or a host system. But there is quite a trade-off associated with this: the baud rate of the "software UAR/T" that is thereby generated is limited by the amount of free time available to the program, and is under control of the timer. If the code running in the SBC is of a busy sort, then there is a possibility that the system will become confused if you try to communicate at any kind of reasonable baud rate. Also, the software to accomplish it is something of a pain, leading some people to prefer the hardware UAR/T despite its overhead. Matter of taste. It is a useful option, however—given the patience, you can put together a system that resides entirely on the SBC with no support logic whatever. And it is clear that any kind of volume application would generate a strong preference for the software approach, since the added nuisance of writing the communications code would quickly become amortized over multiple units, unlike the UAR/T.

The board comes with hole patterns designed to accept 1488 and 1489 RS-232 interface chips, for those applications in which the TTL-level serial lines would be eaten alive by noise. In industrial environments, you can practically count on that.

The only remaining physical feature of the board that is of interest here involves the interrupts.

Each of the four interrupt lines provided corresponds to a specific address in system ROM, an address that is generated by the interrupt arbitration logic of the 8085 when an "event" occurs. This facility is typically used by placing jump instructions at these four addresses, directing the processor to blocks of code which appropriately save status, process the interrupt, restore status, and return. The software can mask any or all of them except one called TRAP, an external input that is useful for such emergencies as power-failure detection. The other three are generated by I/O strobes and the timer.

That, in a nutshell, is the SBC 80/04. It is only one of a whole marketplace full of low-end single-board computers.

If you take the circuitry on one of these boards and put it on a slightly larger one along with an interface to something like the Multibus, then you have an SBC 80/05. You also have all the capabilities alluded to above in addition to the freedom to add many kilobytes of memory, more sophisticated I/O, disk drives, and even additional processors in either managerial or subordinate roles to strengthen the system and increase its throughput. Such additions can allow the machine being designed to be its own development system, given appropriate software resources. (With the SBC 80/04 and others of its ilk, the code has to be developed on another system and then burned into ROM and plugged in. There are ways to cheat a little, but it is still awkward.)

4.2.2 Robust Boards

All it takes is a bit of enthusiastic extrapolation to conceptually progress from the SBCs of our first category to those that are heralded with expensive full-page ads in every issue of our industry's glossy trade journals.

One such unit is the MSC 8009 from Monolithic Systems Corporation, created with the general intention of cramming as much on a single board as possible (Figure 4-4). There's more to this idea than a marketing edge, more even than the obvious reduction of costs associated with simpler packaging requirements and fewer components in the system.

The intent here is the reduction of bus traffic. Designed to be fully compatible with Intel's Multibus, the MSC 8009 can coexist on a system backplane with up to 15 other bus "masters" to provide very high information throughput and task concurrency. The fact that a whole system is on the board reduces the need for Multibus accesses to a minimum—memory, I/O, interrupt logic, and even arithmetic processing are independent of the other residents of the bus. CPU boards that do not have their own resources must access them through the backplane, thus reducing the system bandwidth.

Let's take a look at what this unit has to offer.

There's a built-in floppy disk controller based on the Western Digital 1793, capable of handling up to four drives (5- or 8-inch, single- or double-sided, single- or double-density).

There's an AMD 9511 floating-point processor, providing up to 32-bit precision (in integer mode) over a full repertoire of arithmetic and transcendental functions. It uses an internal stack architecture, and can introduce a further measure of concurrency by performing mathematical manipulations while the CPU attends to other matters.

There's a Z-80 CPU.

There's provision for 32K of EPROM and up to 64K of RAM. A memory-mapping ROM allows software-controlled reassignment of memory address space so that booting and initialization can occur under ROM control, with execution then handed over to disk-based software in RAM.

There are eight levels of prioritized interrupts allowing asynchronous interaction with driving events in the real world, and a "nonmaskable" interrupt which is useful for the invocation of a power-failure routine.

There's a pair of RS-232 serial interface ports with associated baud-rate generators, and there is a timer like the one on the SBC 80/04.

The board even accommodates direct memory access to its devices by external bus masters.

Clearly, this thing is no slouch. It is entertaining to think of such capability on a 6.75 × 12 inch circuit board and reminisce about IBM 704s and Bendix G-15s. But in the industrial system design milieu, such logical horsepower represents vast untapped potential for intelligent control and information processing. Often the choice of processor hinges on the information bandwidth—the throughput—and techniques such as those represented here which localize the considerable overhead associated with CPU operation free the bus and vastly extend the applicability of "micros."

There is hardly any point in a truly complete description of specific products in this category for we are, after all, only trying to convince ourselves that "black boxes" exist in a wide enough range of capabilities to serve our diverse needs in

Figure 4-4   The MSC 8009 Multibus-compatible single-board computer. (Photo courtesy of Monolithic Systems Corporation.)

the industrial design world. The board market is as active as the chip market—they are inextricably linked—so all of our rationalizations for avoiding chip details apply here as well: the tools of this topsy-turvy business are far more volatile than the techniques.

4.2.3 Multicard Systems

There exists another whole family of microcomputer configurations that we might as well mention in passing—those systems whose CPU boards are incapable of standing alone. It's sort of a hazy distinction, since net system performance is loosely comparable, but it is occasionally worthwhile to think in terms of segregated board functions.

The system that was used for all the text processing in the preparation of this book will serve as a useful example. It is a Cromemco Z-2D, an 8-bit Z-80 system that is based on the S-100 bus (Figure 4-5).

Mounted in a rack-mount enclosure, there is a hefty power supply [believe it: 30 amperes of +8, and 15 amperes each of plus and minus 18 (remember, the

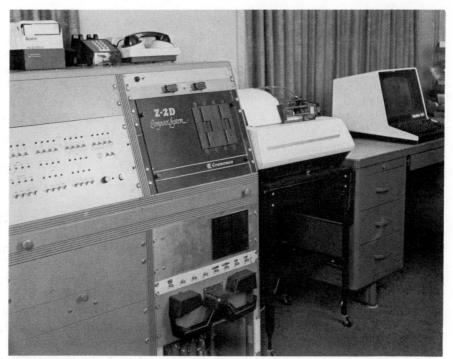

Figure 4-5 The author's computer system, a rackmount Z-80-based processor hosting 128K of RAM and four floppy-disk drives, serves as a software development facility, a word processing system, and a general-purpose tool. (Photo by Frankie Foster, Lenslife Photography.)

S-100 uses on-card regulation)], a card cage with 21 slots, and a pair of minifloppy disk drives with about 190K each of storage space.

The processor ... is a processor. It works fine; lasts a long time. So instead of talking about its specs, let's get a more useful picture of the system's character by chatting about the various things that have been grafted on to it.

One of the boards on the bus is called a 4FDC, and it takes care of the two disk drives mounted in the cabinet as well as two more mounted in a separate enclosure. (How we change! A pair of disk drives would have been considered the ultimate luxury not too long ago, but during the creation of this book the others were added because, well, two weren't enough. Cramps the style, you know. Now, of course, we lust after a hard disk.) The 4FDC also provides a bootstrap and monitor ROM, as well as a serial communication line which currently runs at 19,200 baud. On the other end of the serial cable is the much-pounded-upon CRT terminal (a Hazeltine 1500) that is the system console. The CRT contains its own processor—an 8080—which is more or less invisible to the user.

Another board in the Z-2D is an old IMSAI four-port parallel I/O board, an unspectacular little thing that just provides four channels in each direction, with no programmable options. Its major use is interface with the Diablo daisy-wheel printer.

The Diablo, a wondrous contraption, produces fully formed characters on paper at the rate of about 45 per second. It prints bidirectionally to save time, and can handle "microspacing" in both dimensions, allowing proportional spacing as well as a bit of graphics fun when no deadlines loom. It is sinfully slow compared to a matrix printer, of course, but it provides good excuses for coffee breaks as it methodically churns out formatted text. The interface with the processor is complex—this particular member of the Diablo product line is flexible but dumb, and the system has to explicitly tell it what to do. It requires three output ports and one input port, together with about one-half K of code to control it.

There are two RAM boards in the machine at the moment—both are 64K Measurement Systems & Controls DMB 6400s. The 128K replaced a once-adequate 48K, which replaced a once-adequate 16K, which seemed positively awesome in the light of the 2K that graced my first home-built, wirewrapped 8008 system.

Let's see ... there's an analog I/O board which provides eight channels each of 8-bit resolution A-D and D-A. It isn't being used for anything right now, but did handle joysticks for awhile until economic realities forced games permanently into the background in favor of writing and software development.

There are three boards, all tightly cabled together, which are collectively known as the Cromemco SDI with 48K two-port RAM. This, driving a Conrac RGB color monitor, provides some very pleasing graphics with a resolution of 756 × 484. It can display any 16 of 4096 different colors at one time (via a color map), and can handle intermingling of graphic and textual information. It lives in this system as support for a software development project which can be loosely characterized as a real-time frequency-domain oscilloscope.

Then there are the homemade wirewrapped cards. Occasionally, something is needed that either doesn't exist on the market or can't be afforded at the time. In the latter category, there is a programmable serial I/O board which is dedicated to a pair of modems, allowing software and text communication over the telephone line at either 300 or 1200 baud.

In the former category, there are the RAMPORT and the kluge interface. We will save the RAMPORT for Chapter 12. The kluge interface just isolates the S-100 bus and makes it available on the end of a shielded flat cable. This has proved useful for all sorts of bizarre projects, usually involving late-night obsessions with obscure real-world interface tasks and other things involving blinking lights and positively reinforcing emotional feedback. What good is a computer if you can't play with it now and then? There's nothing like a few bells 'n whistles to make it all worthwhile.

OK. We will similarly overview the software later. The key point of this "chautauqua" about trusty BEHEMOTH II is that it is just another system, not unlike the ones that you may find yourself integrating into various sorts of contexts. Yet it has been used here for writing, code development, music composition and synthesis, home security, signal processing, games, communications, testing, and a host of frivolities too numerous to mention. That seems vaguely suggestive of flexibility—something for which we must thank the universality of a bus architecture. With no significant changes, this same box (although not "industrial grade") could be converted into a Blanchard controller: just write some code, provide appropriate interface circuitry, and protect it from the environment in some fashion that will extend its survival.

A humorous note, with apologies to the manufacturers involved: BEHEMOTH is an acronym, originally coined in honor of this system's predecessor—the home-brew 8008-based machine. It stands for "Badly Engineered Heap of Electrical, Mechanical, Optical, and Thermal Hardware." It has fortunately become a misnomer.

4.2.4 Sexy Custom Jobs, and Closing Notes

With microcomputer hardware configurations covering the range suggested by the foregoing, with processor types spanning the spectrum between simple 4-bit controllers and 32-bit super-high-speed multiprocessor systems, and with a wealth of bus-compatible support circuitry, there are not too many jobs that cannot be handled by some combination of available products. We certainly won't encounter any in this book, although we will have occasion to drag out the wirewrap gun now and again when particularly bizarre manifestations of industrial reality confront us and call for a custom interface.

If you get sufficiently esoteric, however, you might encounter all sorts of situations in which the requirements of a design can't be met with special interface logic. High-speed digital signal processing, for example, could call for some advanced math capability on the processor card. Or perhaps the requirements of a real-time graphics project require a hardware vector generator. Possibly, some fast dedicated process could be approached more neatly if there happened to be a coprocessor that responds to the same instruction stream that feeds the CPU.

Such oddball problems are so widely diverse in character that generalization is foolish—aside from the offhand comment that such things are often best handled with a distribution of resources. (Why, with processors so cheap, should a single one have to do all the work?)

So. We have another level of black box to use in our various efforts: first the CPU chip, now the SBC. We are ready to shift our focus from the hardware tools to the techniques of their use, first with a discussion of interfacing and later with an investigation of software. Then, at last, we can consider some real problems.

5

Interfacing

5.0 INTRODUCTION

We have intimated at various points so far that much of the art associated with creating a microcomputer-based system is in the *interface*, that body of hardware involved with the relationship between the micro and the rest of the universe. In addition, the interface circuitry is subject to a much heavier set of environmental and performance constraints due to its greater exposure to the environment. Compounding this still further, said circuitry normally must be assembled from relatively small-scale components which lack the ruggedness and high level of integration enjoyed by their host CPU.

That set of realities suggests some of the reasons behind the dedication of a chapter to the subject of interfacing. It would be easy, in fact, to do a whole book on the subject, as some have done. But let's keep it in the perspective of our intentions: we are interested in accumulating a broad body of heuristics into which the entire industry can be fit. It wouldn't do to get too carried away, would it?

The design of interface circuitry cannot be handily summarized as the simple selection of a few black boxes, unlike our earlier discussions concerning processors. No, we have off-the-shelf computer hardware to provide an "intelligent" environment, but it is entirely up to us to create two necessary things—a program to make it go and an interface with the outside world to make it useful.

If there can be said to be a general philosophy of microcomputer interfacing, then perhaps it would go somewhat as follows:

The objective of a control system is the continuous adjustment of external conditions on the basis of sensed inputs and a stored algorithm. The inputs with which it is provided are chosen by the designer from an infinite range of possibilities, and they are selected on the basis of their collective ability to accurately represent the conditions that are meaningful in the context of the control algorithm. The effect, then, is a limited internal model of the world that tracks reality via these sense inputs. The software, which presumably has built into it some ideal plan for this "world," thus spends its time attempting to modify external conditions to make the real and the ideal as nearly alike as possible.

In the Blanchard control system, the processor's internal model of the world consists of the loading on the grinding head motor, a train of pulses from the drive gear, and a set of conditions and commands presented by the operator and the host computer. It is not really very interested in the temperature of the room, the weight of the operator, the flow rate of the coolant, or the local magnetic variation from true north. These factors, although equally viable as indicators of the world's current status, fall outside the domain of meaningful data in the Blanchard control task. Part of the design process involves the identification of the appropriate subset of real-world data which will be presented to the system, since every information channel costs time and money.

This viewpoint helps us approach the subject of microcomputer interfacing, and, in fact, just a little more thought reveals four major categories of interconnection between the processor and the rest of the world.

They are: operator and intersystem communications, system overhead, sensing, and control. The techniques overlap mightily, but let's try to take them one at a time.

5.1 OPERATOR AND INTERSYSTEM COMMUNICATIONS

In this category we include communication between the system and various human beings who might in some way be involved with it, as well as chitchat between different systems. It would be pleasant to split those into two categories, since they are philosophically distinct, but the techniques overlap so heavily that it would be pointless.

Let's start with one of the most ubiquitous techniques of all.

5.1.1 Serial Data Communication

There are a lot of good reasons why it is undesirable to ship data in parallel over long distances. Wire is expensive. TTL-level signals tend to get lost in noise and reflections. And in some cases, such as communication over phone lines, it would be exceedingly awkward—or impossible.

All this led, long ago, to the development of some standard techniques for serial communication. The underlying principle can be understood with the aid of Figure 5-1.

Here, in simplified form, are a pair of shift registers. Assume that the sending system, on the left, wishes to transmit a byte (hex 96) to the receiving system. It first loads it into a shift register, an operation identical to the output data latching operation we saw in Figure 3-5. In the process, it activates control logic which, based on the frequency of an applied clock signal, begins shifting the data out, least significant bit first. Meanwhile, the corresponding device at the other end begins shifting the arriving data into its register, bit by bit, until the entire byte has been received. It can then notify the receiving system that the data are ready to be read, again a straightforward bus interface operation.

That is the basic trick, but there are a few things missing. Most notably, there is no way for the receiver in the diagram to stay synchronized with the transmitter, or for it to recognize boundaries between successive bytes. That can be handled a variety of ways, but the one that has achieved prominence is the asynchronous technique, usually embodied in the form of UAR/T chips (a development that took most of the pain out of designing serial I/O logic—it otherwise involves a fair amount of overhead).

The UAR/T, in most incarnations, is a 24- to 40-pin chip that contains everything needed to both transmit and receive (independently) bit-serial data over a communication line. (Its serial inputs and outputs are TTL levels, in most applications requiring additional line driving and receiving circuitry, but the logic is complete. We will consider the other part in a moment.) It is designed to mate directly with a tri-state bus, accepting and delivering parallel data and delivering as well a "status word," which allows the host CPU to keep tabs on its operation and detect any error conditions. The device is capable of operating over a wide range of communication rates, up to at least 19.2 kilobaud (19,200 bits per second).

UAR/Ts solve the problems noted above by prefacing each transmitted character with something called a *start bit*, which just looks like a 0 bit. Since the resting state of the serial line is high (mark), the receiving device can react to this start bit as an indication that transmission is about to occur. First, however, it

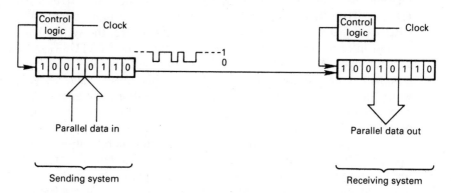

Figure 5-1 The basic principle of serial data communication involves the conversion of parallel bytes into bit streams (and back again) via shift registers. A UAR/T is merely an embodiment of this scheme that includes all the control logic and interface circuitry in one chip.

keeps an eye on the line to make sure that the start bit is of reasonable length, and not just a noise spike. Following this, the transmitter begins sending the data, bit by bit, at a rate equal to $\frac{1}{16}$ of its input clock frequency. As long as the receiver's clock is reasonably close, it will be accumulating the character without any need for actual synchronization.

Once the 5, 6, 7, or 8 bits (jumper selectable) have been sent, the transmitter sends one or two *stop bits*, which are really just high bits indistinguishable from the quiescent "mark" condition. But this period is used by the receiver as an assurance that no slippage of the data took place, an event that would be flagged as a "framing error." According to most UAR/T manufacturers, this scheme will guarantee error-free reception of data with up to 42% distortion.

There are a few other features thrown in, such as the ability to generate and detect even or odd *parity*, a scheme whereby an extra bit is added at the end of each transmitted character such that indication is given concerning the odd or even number of "1" bits in the rest of it. The receiver can check this bit against its own calculations, giving it another method for detecting random errors from noise or other pernicious influences.

First, building a UAR/T into a bus structure is simple. Take a look at Figure 5-2, a typical pinout. At the top, we see eight data lines entering the device. These, of course, carry the data that are to be transmitted and, depending upon the type of system design, are connected either directly to the system data bus or to an output port. In order to tell the UAR/T when the data are ripe for acquisition, the CPU issues a brief pulse on the XMIT DATA STROBE line.

When that takes place, the sequence described previously occurs, and data emerge serially from the SERIAL OUT line at the right.

Proceding in a clockwise direction, we next come to the SERIAL IN line, whose function is obvious. Once the character is assimilated, it becomes available to the processor—but it does not yet appear on the output lines, since the system may not be ready to accept it. Instead, one of the status signals becomes high (DR, for Data Ready),but it doesn't appear on the output either. Why would the CPU want that if it wouldn't want the data itself? No, the system has to explicitly request information from the UAR/T, by taking the STATUS ENABLE line low. This removes the high-impedance state from the five status lines (which we show externally connected to the data lines, but this doesn't have to be the case), at which time the processor can do its input operation, test appropriately, and discover that data are ready. Its logical next step, of course, is the enabling of the data via the RXCV DATA ENABLE line.

There are variations on this theme, which depend entirely upon the configuration of the hardware with which the UAR/T is interconnected. There is nothing to stop you from using it with random logic, permanently enabling the outputs, and ignoring all the buslike features. But we're in a micro frame of mind.

Continuing further clockwise, we pass a RESET line, which just initializes everything (applying power to one of these devices and attempting to use it without a power-on reset signal often results in problems). There is an output line labeled TEOC (for Transmitter End of Character); it is high during the interval between one character's stop bit and the next character's start bit (in case your logic needs to know when the serial line is inactive).

Then, on the left, we have six lines that are used to define the device's operating characteristics. It is via these that such details as number of data bits, number of stop bits, presence or absence of parity and, if the former, whether it is odd or even, are established. In most applications, they are hardwired according to the communication specs and forgotten until debugging time, when it seems that even the most innocuous static conditions are hopelessly screwed up.

Then, there are the clock inputs for the transmit and receive sections. They are usually tied together, but you never know. That leaves only the power inputs,

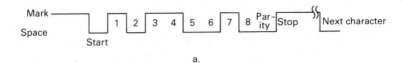

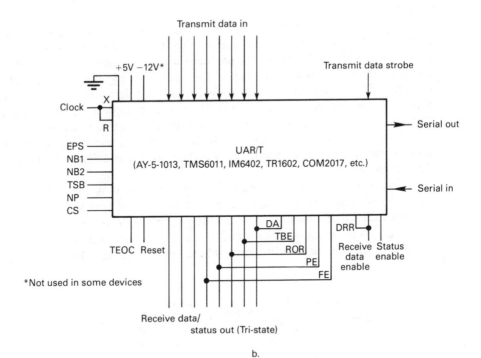

Figure 5-2 (a) The serial data produced by the asynchronous communication scheme include a start and stop bit, as well as an optional parity bit. (b) Pinout of the standard UAR/T devices. These are "dumb" UARTs; others, such as the Intel 8251, require a software setup but offer additional flexibility.

which need no elaboration except for the comment that some parts, such as the CMOS IM6402 from Intersil, don't need the −12. It is very pleasant to eliminate a power supply from a design.

We were going to show a picture of a UAR/T actually connected to a bus, but it looks exactly like Figure 5-2 with some of the wires going off the page. That's really all there is to it. Besides, we'll see it a time or two in real systems later in this book.

Now. That brings us to the question of getting those serial lines from one UAR/T to another without undue hardship. If you are attempting to communicate over a distance greater than, say, 10 feet, the use of straight TTL levels is contraindicated. Oh, you can play a few tricks with Schmitt triggers and the like but, especially in industrial environments, it is a reliable source of unreliability.

There are two universal alternatives that come close to solving the noise problem and allow communication over fair distances. One, known as *RS-232*, is implemented in its simplest form with the circuit shown in Figure 5-3. Using the 1488 and 1489 interface devices, the TTL levels are converted to roughly +12 volts for a "0" and −12 volts for a "1," with anything between −3 and +3 undefined and ignored by the receiver. This approach vastly improves the noise immunity.

Actually, the scheme shown in the figure is not really RS-232—it is just a circuit that makes use of the signal levels that are specified by the RS-232 standard. A full implementation of the standard involves numerous handshaking and status lines, and is most often seen in the context of big systems and high-speed modems. The most common use of RS-232 in the microcomputer world is probably CRT and printer interfacing, since it allows very high speed. But though freer from noise than TTL, it is not nearly as quiet as the other alternative.

Look at Figure 5-4. Here, we have depicted a communications scheme called a *current loop*. This is what we will find ourselves using in nearly every industrial application.

The first noteworthy characteristic of this circuit is the optical isolation: There is no dc path connecting system A, the loop, and system B. This is of critical importance in an installation with widely separated systems (or terminals, sensors, etc.), since the ground-loop resistance of the power distribution system can be enough to obscure even the most robust of bit streams. This way, the ground

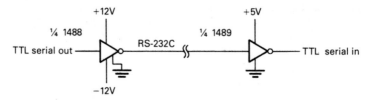

Figure 5-3 A basic RS-232 data link involves the conversion of TTL levels to those with greater noise immunity for the trip through the cable. The term "RS-232" implies considerable handshaking protocol, but the standard signal levels are often used in this simple form.

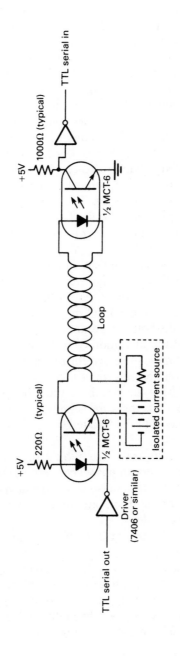

Figure 5-4 Serial data can be transmitted over long distances with little degradation using an optically isolated current loop such as the one shown here. Two of these can be implemented back to back if bidirectional communication is desired, or the scheme shown in Figure 5-5 can be used.

references of the two systems can be even 1000 or more volts apart without negatively affecting the data communication.

It works like this: Whenever the polarity of the data input is such that the driver's output is low, the LED in the optocoupler is turned on. The resultant flood of photons on the base of the phototransistor optically wedded to the LED causes saturation, and current flows in the loop—and since the current source is striving to maintain a constant flow (typically 20 mA), the loop can be very long, even a mile or more. This current, of course, forward-biases the LED at the other end, a similar phenomenon takes place there, and the transistor's output goes low. After inversion, we end up with a duplicate of the original input data.

This scheme is trivial to implement in a system, with the exception of the isolated current source. To be truly isolated, it must be either battery operated or a separate line-operated dc supply. You can usually cheat a little here, by replacing it with a connection to a local dc supply with a series resistor trimmed to accommodate the overall loop resistance. There is still net isolation between the systems, thanks to the opto device at the far end, but the loop no longer floats. We leave that to your discretion—it seems to be equally reliable and it is certainly cheaper.

When bidirectional communication is desired (most of the time), there are a couple of ways to go. You can either implement two of these loops back to back, running a four-wire cable between the systems (this is how we'll do it in the data collection system, since we need full duplex communication), or you can induce the single loop to handle both send and receive functions as shown in Figure 5-5.

Here, we accept the limitation of half-duplex operation (we can't send and receive at the same time), in return for operation over a single twisted-pair line. The control logic at either end must take its DATA OUT line high when it is ready to receive, allowing the system at the other end to control the loop. Other than those differences, it is the same scheme as above, and you can get away with a dc power supply instead of the isolated current source if you're not feeling like a purist at the moment.

Well, we can't quite leave the subject of serial communication without at least mentioning a few other things. There are numerous situations in which it is awkward to run wires. There are also some in which the wires are no problem, but certain people might object if you attempted to use them as a current loop—the telephone lines, for example.

That poses no real problem. All you need do is take the serial data and convert them to audio: whenever you say "0," one frequency of sine wave is generated, and when you say "1," another is. The result of this is sort of a two-tone warble which, when presented to a complementary gadget at the other end, produces the original bit stream again. This communication technique is obviously much slower than the ones we have been discussing, because the tones have to be maintained long enough for the receiving equipment to make some sense out of them.

The devices that accomplish this function are called *modems* (for MOdulator/DEModulators), and come in a variety of flavors. You can get sexy,

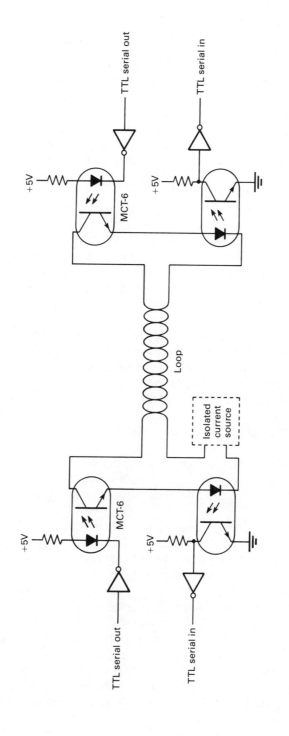

Figure 5-5 Half-duplex current-loop design allows a single twisted pair to carry two-way communication with straightforward logical control. Each end of the link is isolated by a dual optocoupler and converted directly to TTL-compatible levels.

high-speed models that operate at 1200 baud (given a decent phone line), and incorporate complete handshaking logic to render their use more or less automatic once the system is appropriately designed. For less money, you can obtain modems of the widely used "103" style, which operate at 300 baud over voice-grade telephone lines, with variants that connect directly to the line or couple with it acoustically. These can even be had as LSI chips, if you don't mind having to do a little external analog circuit design.

The low-speed modems operate in such a relaxed fashion that there is a fair amount of flexibility and noise tolerance. Typically, the receiving circuitry is built with phase-locked loops which, at 30 characters per second, have little trouble locking on to the received tones. But the faster ones, such as the Bell 202 and 212 styles, have to be high-performance circuits, especially when you consider the frequencies involved. In a recent project involving a 202 modem, for example, it was noted that there were something like 30 op amps on the board. Huh? It turned out that they were high-Q active filters that were exceedingly fast; in fact, the lower of the two frequencies used was 1200 hertz, and at 1200 bits per second, the device had to respond reliably to a single cycle of that sine wave. Observation with a scope revealed that the unit was unambiguously detecting the bit after only about one-third of a cycle.

Try that with a PLL.

Anyway, most modem applications can make use of off-the-shelf units sporting DB-25 connectors on their back panels, allowing direct connection to a standard RS-232 cable identical to the ones expected by CRTs and their brethren.

Incidentally, once you have this audio incarnation of your serial data, there is nothing stopping you from tape recording it, transmitting it over a two-way radio, or piping it through an intercom system. We will have occasion later to look at a radio link; it will let us mount a microcomputer-based terminal on a fork-lift truck.

But you want blinding speed and there's no way to run wires? You are attempting to link a control system in a plant with an office across the street, and the local utility people get little dollar signs in their eyes when you inquire about use of the poles?

All is not lost. Serial data can be crammed through all sorts of information channels. You might find, for example, that a small solid-state laser diode, appropriately pulse-width modulated and blinking infrared across the street to a phototransistor, will solve your problem. Or a polarized helium–neon laser shining through a Pockels cell that is driven by your data. Or a microwave link. Or an FM transmitter tucked away in some obscure corner of the electromagnetic spectrum.

A hot, up-and-coming technology that belongs in this category is *fiber optics*. After years of intense research, low-cost digital links have become available that can be used as modules with little or no engineering effort.

Fibers used in this context offer significant benefits: communications link susceptibility to noise from EMI and RFI is nearly eliminated, data transmitted optically are secure from "compromise," and a fiber link is surely the ultimate opto-coupler, providing nearly-absolute electrical isolation between the two systems. Data can be shuttled through the fiber at many megabits per second, with no degradation from nearby high voltages, explosive environments, petrochemicals, and so on.

Numerous plug-in links are available which can be treated as TTL-compatible components at both ends. In addition, there are many special-purpose units, such as the Canoga Data Systems CBE-100, which can transfer 16-bit words at the rate of 200,000 per second, making it useful for fast data acquisition systems or optical bus extension.

The technology is rife with possibilities, but let's move on.

5.1.2 Parallel Data Communication

After all we have just said about serial techniques, it would seem that there is not much room left for parallel, at least in the domain of operator and intersystem communication.

Well ... it is all over the place—so much so that it defies categorization. The mechanics of the parallel interface itself have already been more or less stated, back in Chapter 3. What you choose to do with it, however, is entirely open, and includes things as diverse as printers, floating-point processors, lights, you name it. So we will not even bother with "parallel interfaces" per se, instead proceeding directly to a variety of specific techniques that make use of them in one way or another.

5.1.3 Keyboards and Switches

There, that's better—something a bit less nebulous. Aside from CRT terminals connected via RS-232 links, how might one provide keyboards, pushbuttons, and switches for a user? It is not at all uncommon for an industrial system to have a control panel somewhere, designed (hopefully) with enough ruggedness to withstand vengeful punching by large, gloved, hamlike paws with all the motive force of biceps conditioned by decades of lifting locomotives. We're not talking about delicate little switches used in keyboards designed for gentle writers, petite secretaries, myopic programmers, and timid clerks; we're talking about hard-core Switches. There's a difference.

Some interesting demands are placed upon contacts by environments that might contain gaseous corrosives and airborne conductive particulates, that present temperatures ranging from winter wonderland to the land of the pharaohs, and that include brutal vibration in addition to human beings such as those characterized above. In essence, they have to sacrifice some of the low-bounce and feather-light actuation forces that those in the relatively isolated world of electronics have come to expect.

Perhaps the most obvious problem to be confronted when interfacing these things to a micro is their *bounce*—that mechanical reality that prevents a switch closure from looking like a zero-rise-time square wave. The processor—or something—must keep asking, "Are you sure? Are you really sure?" while keeping track of time (so it will know when to conclude that it is sure) and watching for illegal conditions such as simultaneous depressions. In addition, the circuitry associated with the switches is not necessarily nestled in the comparative safety of the card cage, if there is one, but is Out There. The logic, or at least the cable coming from the switch panel, is exposed to various pernicious influences, not the least of which is electrical noise.

It's not always that evil, of course, but you might as well plan for it. So what do you do about it?

Consider Figure 5-6a. To get an idea of the severity of the switch bounce problem, think very carefully about the physical behavior of the armature when it is suddenly toggled from position A to position B. It has inertia, it is under spring tension, and, well, it just bounces. The problem is that during all those little intervals when it is not actually touching the contact, it is electrically floating.

Figure 5-6b shows what happens. The first "waveform" is an attempt to show the armature's physical bounce. The second waveform shows an unambiguous low level only at those points of contact in the armature's exponentially damped succession of little paraboloids—those points when the "output" of the switch is reliably low.

If you were just turning a light on and off with the switch, this would obviously be of no concern. But if the wiper is connected to a logic gate (which quite happily responds to transitions of a few nanoseconds) that in turn is connected to the clock input of a counter, then a depressing phenomenon would occur. Since the input of the gate tends to float high, the output will look something like the bottom waveform, stabilizing only after the bouncing has ceased.

The counter would probably increment at least a dozen times.

This is obviously unsatisfactory. One possible solution might be the circuit in Figure 5-6c, bizarre though it may be. Here, we just short the input to the output of a CMOS noninverting buffer, and hang it on the bouncy line. Presto. Think about it.

Somewhat more realistic, and certainly easier on the components involved, is the scheme in Figure 5-6d. It takes a lot of hardware—half a 7400 and two resistors per switch—but very solidly eliminates bounce in the manner demonstrated by the accompanying waveforms (Figure 5-6e). But we presumably have a microprocessor to play with, so why do all that wiring?

Shown in Figure 5-6f is the original configuration of the switch, with a pullup resistor added for good measure. But instead of direct connection to random logic, it appears as one bit of a microcomputer's input port.

In order to determine the switch's position reliably, the program running in the system can periodically check it—say every 5 ms. Whenever it finds a change of state, it sets a tentative switch position in some memory location and continues to "poll" it as before, but now incrementing the value of a counter every time it finds that the switch bit is the same as it was on the previous test, clearing it otherwise. The value of the counter is tested each time to see if it has reached 40 (which corresponds to 200 ms).

Eventually, the bounce dies out, the counter makes it all the way to 40, and subsequent logic in the program can safely assume that the closure is a real one.

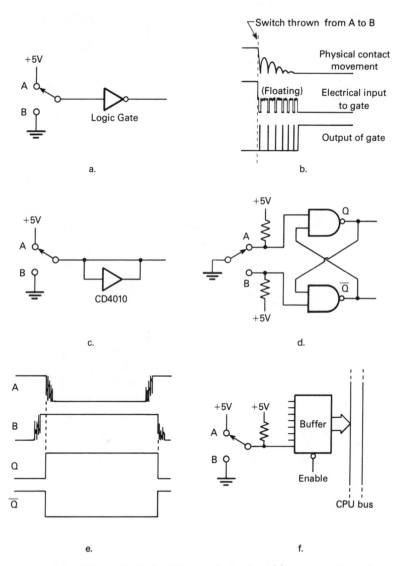

Figure 5-6   Switch debouncing. The simple circuit of (a) generates the garbage shown in (b) due to the mechanical characteristics of the switch armature. Adding a CMOS buffer (c) can help, or a full debouncing circuit can be implemented with an R-S latch (d), yielding the waveforms shown in (e). Yet another alternative is software debouncing (f), simplifying the hardware but adding processing overhead.

There is a possibility that the bouncing effect would be synchronous with the 200-hertz scan rate for a while, but certainly not for $7/10$ second!

Incidentally, the switch could just as easily be an SPST normally open push-button, since the pullup resistor creates a reliable "1" state if the contact closure is not actively causing a "0."

When the design moves beyond a small number of switch closures, how-ever, this approach starts to become somewhat cumbersome.

Consider Figure 5-7. Here, we have an array of keyswitches or equivalent normally open contact closures, and we are interested in knowing when one has been depressed and, of course, which one it was. The circuit that accomplishes this is called a *keyboard encoder*.

Essentially, it consists of a counter with enough bits to address all the switches (4 bits for 16 switches, 5 bits for 32, etc.), and a multiplexer that allows logical "selection" of whichever one corresponds to the counter's current value. The line emanating from the mux represents the position of the addressed switch.

Normally, the clock signal is allowed to drive the counter, but when the clock control logic detects a switch closure, two things occur. First, the clock sig-nal is removed from the counter, freezing it at its current value. Second, after watching the mux output long enough to eliminate the effect of bounce, the logic issues a DATA AVAILABLE signal to the processor and then waits for an ACK-NOWLEDGE signal indicating that the counter's value has been read. Once that "handshake" is completed and the key has been released, the logic restores the clock input to the counter and operation continues as before.

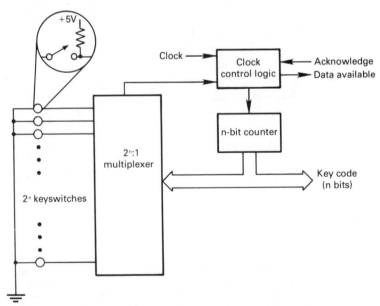

Figure 5-7 A basic keyboard encoder involves a multiplexer and binary counter, producing an output code corresponding to any key closure.

This basic concept underlies a variety of different keyboard encoder designs, although when the number of switches becomes large, it loses efficiency. A scheme for dealing with an ASCII keyboard will be shown in a moment, but first let's look at one of the more interesting variations on the design we just discussed.

Shown in Figure 5-8 is the schematic of a board developed in 1975 by the author, initially as an interface between a home computer system and a standard organ keyboard. Later, it proved to have a variety of industrial applications.

The original problem here was that music keyboards attached to synthesizers were, at that time, almost always monophonic—incapable of responding to two or more simultaneous key depressions. This limitation was admirably overcome by early electronic music artists, but they typically had 16-track tape decks which allowed painstaking construction of chords, harmonies, and even symphonic passages from these sequences of single notes.

As digital techniques became available to replace the old voltage-controlled oscillators, filters, ring modulators, and the like, the problem of polyphony became approachable. But how do you interface the keyboard? A circuit such as the one shown above would respond to the simultaneous depression of three keys by delivering the code of the first one encountered by the scanner, waiting until it was released, then delivering the code for the next, waiting for it to be released, and so on. It would be useless in a music application, as well as an industrial control task wherein the processor needs to monitor a large number of independent and asynchronous binary conditions.

The classic approach to the problem involves a "bit map" interface, wherein all of the input lines are connected directly to microcomputer input ports. This uses a lot of wire, and also requires the CPU to spend a lot of its time in the keyboard scanning task—a very demanding operation if anything approaching real time is desired. The music application, in particular, quite effectively highlights any system response delay over a few milliseconds. Of course, the CPU is perfectly capable of performing the scanning operation with acceptable speed, but if it does, it just might not have much time for anything else. This suggests either a dedicated processor, which seems ridiculous even in these days of cheap computers, or a different design approach. Perhaps we could offload the bulk of the scanning operation on a smart interface?

That is just what has been done in Figure 5-8. At the top you can see the multiplexers, which perform a task identical to those in the previous circuit. Note that they have been arranged in a two-level "tree" to accommodate the number of bits required: eight 8 : 1 multiplexers present their outputs to another one, effectively resulting in a 64 : 1 mux. This rather neatly handles the 61 keys of a five octave (plus high-C) organ manual, when driven by a 6-bit counter.

So far, we have the same setup as the design in Figure 5-7. But there's a bit of trickery.

The chip labeled "74206" is a 256 × 1 bit RAM, which is used here to maintain an image of the keyboard's present status. As the scan progresses, each key is compared with its last known condition, and as long as there is no difference, the

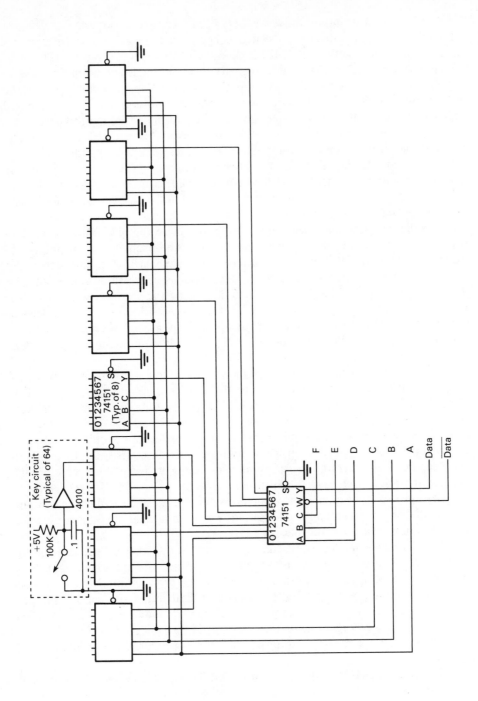

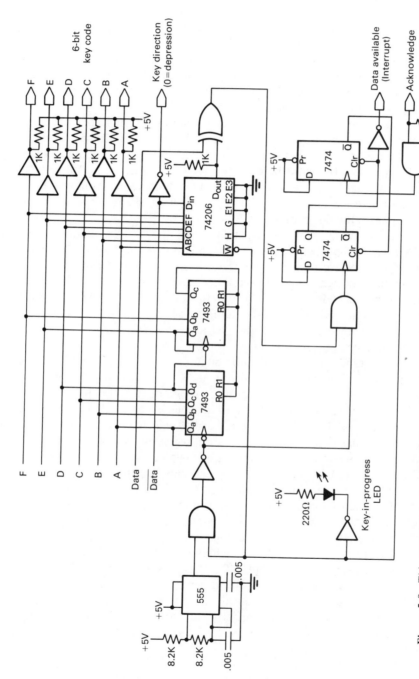

Figure 5-8 This scanning keyboard interface, originally designed for a polyphonic synthesizer, maintains a map of all key positions in the 74206 bipolar RAM. Any change of state is communicated to the processor as a parallel code accompanying an interrupt, allowing an arbitrary number of simultaneous depressions and releases without tying up the CPU with software scanning activities.

circuit moves on to the next one. If, however, the key's state has changed, as detected by the exclusive-OR gate IC17, then it is time for some action. Specifically, the processor is interrupted with the DATA AVAILABLE signal, and the key's binary code, together with a "direction bit" indicating whether the key was depressed or released, are made available for a CPU input operation. When this occurs, the ACK signal is generated, informing the interface that the CPU has the data, and scan continues. Somewhere in the midst of all that, the bit in RAM is changed to reflect the new status of the key.

The result of this design approach is a series of change-of-state notices to the host computer corresponding to keyboard activity. If the performer suddenly strikes an eight-note chord, then the processor receives a quick succession of eight key values (all with the direction bit = 0), which it presumably stores in a table somewhere for use in the control of note-generation hardware. If the performer then releases one of those keys, the processor receives a ninth interrupt, this time with the direction bit = 1.

In each case, scan is suspended until the CPU responds to the DATA AVAILABLE signal with an ACK (which is simply the read strobe from the input port). Net interface performance, then, is a function of CPU response time, but an intelligent design would place the keyboard interrupt at a high level of priority and, surely, the code is written in something other than BASIC. (The author tried that, just for grins, and noted that a forearm laid upon the keyboard resulted in a sweep up the musical scale lasting over a second. The real code is written in Z-80 assembler language, and the delay is unnoticeable.)

A few random notes about the circuit design: The LED is lit during the interval between DATA AVAILABLE and ACK, just to provide a visual indication of CPU response time—or failure (if the program dies, the light will come on and stay on when a single key is struck). The clock is provided by a 555 timer IC free-running at about 12.5 kHz (a value that was selected partially on the basis of the physical resonance of the "J-wires" comprising the keyboard switches—a sneaky way to evade some of the bounce problems). The key inputs are conditioned by *RC* networks and buffered by CMOS, an approach that effectively smooths the rough edges. The data lines to the processor are isolated by open-collector buffers to prevent the evil effects of reflection into the outputs of the flip-flop and the counters. Among other things, this is associated with a phenomenon called "collector commutation," perfectly capable of resetting a flip-flop that was just set.

The entire interface circuit, including the CMOS buffers, requires 30 chips and runs on a single 5-volt supply. The design is readily expandable to 256 inputs.

We take the time and space to describe this circuit in detail for a couple of reasons. First, we feel that a practical design is necessary now and again to keep us firmly rooted in, or at least appreciative of, reality. Second, this technique proves to be quite useful in a variety of industrial sensing applications and, well, we immodestly think it's kind of cute.

Moving right along ....

Interfacing an ASCII keyboard is a different sort of problem—an easier one but for the number of keys involved. It would be a bit of a pain to use the straight multiplexing approach, especially since we don't need the "polyphonic" capability of the last design.

Fortunately, this is a common enough design requirement that there are a number of chips on the market that take care of it. It is unlikely, in fact, that you will ever have to deal with it, since any time you need an ASCII keyboard, you can just buy one. But just for perspective's sake, let's take a very quick look at Figure 5-9.

The chip is called a *keyboard encoder*, and does everything associated with ASCII keyboards, including appropriate handling of the CONTROL and SHIFT keys. The switches are simply arranged in a matrix which is then scanned by the encoder, resulting in the same old thing—data to the CPU and a strobe line indicating data available.

The nice thing about devices like this is that you, the system designer, don't have to think about them very much.

More commonly, however, you might find yourself with problems that do not map handily onto the resources of a mass-produced integrated circuit. Let's look at one example before moving on to the next subset of operator interfaces.

Later in this book, we will have occasion to talk in some depth about something called the IDAC/15 Industrial Data Collection System, a contraption that

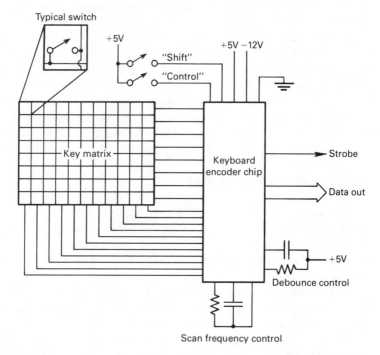

Figure 5-9 Interfacing an ASCII keyboard is easily accomplished with off-the-shelf encoder chips.

we have already alluded to once or twice. One of the design requirements of the terminals involved the interconnection between an array of 20 Microswitch PW-series lighted pushbuttons and the Intel SBC 80/04 processor at the heart of the box. The multiplexing scheme would have been wasteful (and specifications required the system to take no action in the event of multiple pushbutton depressions, a condition that is somewhat awkward to detect without implementing something like the polyphonic design).

The solution, which turned out to be trivial, is shown in Figure 5-10 (and, in more detail, in Figure 14-7). It does require the processor to do the work, but it is not a terribly demanding task and the system was spending too much time loafing anyway. It goes like this: Whenever the CPU becomes curious about the status of the pushbuttons, it simply places low levels, one by one, on the three horizontal lines of the matrix (LONUMS, HINUMS, and ALPHAS). During the time each one is low, it then does an input operation on the eight vertical lines, yielding three successive byte-wide maps of switch activity. All debouncing, multiple depression rejection, and encoding is then done by the program.

This highlights one of the standard trade-offs in the industry, one that we have already mentioned: hardware versus software. There is occasionally a temptation to do everything with the program (after all, that's what micros are for) even to the point of complicating the design dramatically. Unless production economics have to be considered, it makes more sense to take the design approach that minimizes the number of engineering hours required. Only the peculiarities of the specific project can tell you whether that would suggest, for example, software or hardware keyboard encoding, but in the case of the IDAC/15, the relatively low CPU load allowed use of the software approach without concern for serious timing problems. The music system represents exactly the opposite situation—encoding the keyboard with the program would add such a processing load that other activities would suffer, and the money saved in the cheaper interface would have to be spent in additional software development time or even additional hardware someplace else.

5.1.4 Displays of Various Sorts

There's nothing like a panel full of blinking lights to add enjoyment to a design project, simplify debugging, and convince the customer that the box really is doing something that justifies the price. And a liberal supply of status indicators can take much of the pain out of field service, sometimes pinpointing the source of a problem at a glance.

LEDs can be trivially interfaced to TTL circuitry in the manner suggested by Figure 5-11a. The driver here is a 7404 or similar, and the resistor value of 220 ohms will allow about 16 mils to flow through the LED.

If you need an incandescent bulb or other higher-current display, the 7404 can be replaced with an open-collector, high-current driver like the 7406 (40 mA), or a transistor driver stage can be added as in Figure 5-11b. Also, there is a

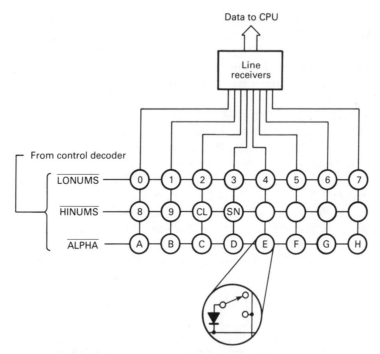

Figure 5-10 The keyboard interface scheme used by the IDAC terminals depends upon software scan, trivializing the hardware design. In this system, the CPU is not loaded so heavily that routine tasks such as this pose a serious overhead problem.

variety of "interface driver" devices on the market that eliminate the need to handle things other than DIPs—one of them, the 75451, is shown in part c of the figure. It includes a NAND gate which allows the addition of a lamp test function with no extra logic, as we shall see momentarily.

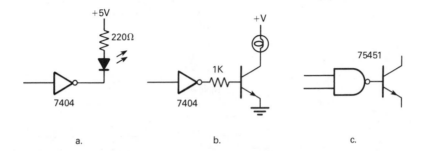

Figure 5-11 Methods of driving indicators. In (a), a TTL gate sinks sufficient current for a single LED. In (b), an external driver transistor has been added to handle the heavier load of an incandescent bulb. In (c), half of a standard driver package is shown schematically.

There are any number of seven-segment displays on the market which look like seven LEDs with either their anodes or their cathodes tied together. Talking to them is simplicity itself, and is accomplished with the aid of a BCD- or HEX-to-seven-segment decoder/driver. Lest we flounder about considering a rash of philosophically equivalent TTL devices, we will go no further than the sketch of Figure 5-12.

Here, we see a four-digit binary-coded decimal display suitable for use, say, as a parts count indicator. It is assumed that the processor has converted its binary count into a form suitable for human consumption. It takes the form of "packed BCD," four 4-bit values in the range 0 to 9.

In operation, the CPU places the first byte, containing the low two digits, on the data bus, and strobes the latch (a pair of 74175s?) with the simultaneous presence of the proper address and an I/O write command. Once the data have replaced the previous contents of the latch, the decoder/drivers (7447s?) perform their code-conversion function, sourcing or sinking current to the appropriate LEDs. An identical operation is then undertaken with the high-order byte, and

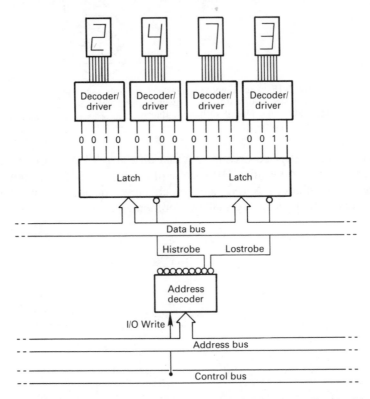

Figure 5-12 Interfacing seven-segment LED displays is normally accomplished with the use of latches and decoder/drivers, reducing the problem to a simple bus-interface task. Many multidigit display packages include the logic in the same device as the LEDs, totally eliminating the need for outboard logic.

the display indicates the count.

It's worth noting that our tired old trade-off applies here as well. Most of the logic in this circuit can be eliminated if you don't mind doing it all in software, but it's a pain.

Now let's consider some more IDAC/15 circuits. First, each of the pushbuttons that we so handily interfaced in the last section has a small incandescent bulb under its cap, allowing such features as operator prompting and subsequent rejection of any selection that was not enabled and illuminated. The circuit in Figure 5-13 (also in Figure 14-8) shows the solution to the problem.

Here, the output data bus from the CPU (this system uses a split bus, which we discuss later) is connected to the inputs of three latches—two of 8 bits and one of 4 bits, since there are 20 lights. When the processor feels the urge to modify the configuration of illuminated bulbs, it places its updated "lamp maps," one at a time, on this bus, and strobes the appropriate latch with one of the three control signals, LONUML, HINUML, and ALPHAL. This, of course, has the effect of storing the current status of the bus in the selected latch, whereupon a (possibly) new set of logic levels is applied to the lamp drivers.

The drivers are the 75451s (TI) that we mentioned before. Since the transistor needs a high level at its base to turn on, the NAND gate is performing an active-low OR function, and this allows us to add a lamp test that does not enlist the aid of the processor. (The 74175 latches that are used have inverted as well as noninverted outputs, so we don't have to think in terms of negative logic.) This design incorporates one extra discrete component per lamp, by the way, the 150-ohm "keep-alive" resistor. This allows a few mils to trickle through the filament when the bulb is dark, keeping it warm and preventing the thermal surges that are

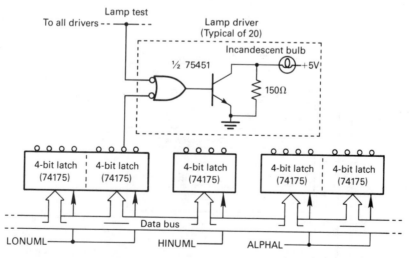

Figure 5-13 The lamps in the pushbuttons of the IDAC terminals are driven by 75451s on the outputs of latches connected to the data bus.

the primary cause of burnout. In an environment wherein service is expensive, this technique is a cheap way to forestall some of the nuisance failures.

So much for lamps and LEDs. Let's look at one other IDAC circuit before leaving the display subject.

The simple diagram of Figure 5-14 represents the interface between the processor and a pair of Burroughs Self-Scan Displays (SSDs). The displays, although very complex devices, have been thoughtfully provided with all of the timing, refreshing, and character generation logic, making their use "refreshingly" trivial. Just hang 'em on the bus and strobe 'em.

Flat-panel displays provide a useful set of features in an industrial setting. Unlike CRTs, they are not overly fragile and bulky. They can be had in a variety of configurations from various manufacturers, from 16- or 32-character single line displays to full-page "flat CRTs" that have graphics capability. In the IDAC/15 system, they are used to prompt operator action, display the results of system diagnostics, and verify entries. The fact that the SSDs display the uppercase ASCII subset allows substantially greater flexibility than a corresponding number of seven-segment or 5 × 7 readouts, especially considering the control logic involved. And of greatest interest to the designer, they are easy to use. They do require a +250-volt dc supply (at 30 mils per 32-character display) for the plasma discharge, but that is available as an off-the-shelf supply or as a dc-to-dc converter.

We wouldn't be so cocky as to imply that we have covered the entire field of operator interface techniques with this whirlwind tour of the subject. But we're getting impatient to get to the meaty stuff. Let's cleanse the palate with some of the more entertaining I/O devices, spend a few obligatory pages on system overhead, then look into sensing and control interfaces.

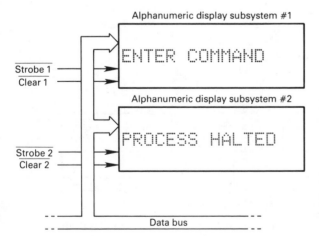

Figure 5-14 Burroughs Self-Scan Displays and others of similar design can be interfaced to processor buses with little difficulty. The scheme shown here is used in the IDAC terminals.

5.1.5 Esoterica

We wouldn't want to be limited to switches and displays, no matter how nice some of them are. Sometimes you don't have hands free, can't look at a screen, need to express spatial information, or wish to be amused.

This latter point, although it sounds frivolous, really has an application in an industrial setting. At a certain foundry in southern Mississippi, management was having a serious problem with vandalism on the data collection terminals. The employees, not overly thrilled with their lot, found it desirable to ram them with lift trucks, douse them with liquids, cut their cables, and otherwise do things to them which underscore much of our dramatic commentary about industrial environments. No amount of ruggedizing could completely solve management's problem.

A consultant was consulted. "Make 'em funny," he said, and sent them a bill.

The result was a new FORTRAN task, called JOKE, which was invoked at random intervals during each shift. Without warning, a randomly selected terminal would suddenly beep and display the message, "DO YOU WANT TO HEAR A JOKE?" on its SSD. Two buttons would light up, one for YES and the other for NO. Invariably, of course, someone would rush over to it, mash the YES button, and be rewarded with the latest in an endless series of carefully chosen gags stored on disk by the company's full-time joke manager. Vandalism dropped abruptly to near zero and stayed there.

All of which has very little to do with esoteric I/O devices, but it's interesting. The closest we can come is probably the complex sound generator chips.

Wondrous devices, they are. One, the General Instrument AY-3-8910, appears to the microcomputer bus as 16 successive ports. By writing control bytes to it appropriately, the programmer can generate a variety of sounds ranging from crude speech to phasers, from the moos of lovestruck bovine enchantresses to the cacaphony of colliding 747s. In most systems that for one reason or another require human interaction, it is desirable to have (at the very least) some audio feedback for keystrokes and, perhaps, a raspberry for errors. But with the complex sound generators, you can extend this considerably (making it only as frivolous as the application environment and politics permit). The bandwidth of the audio information channel can be expanded to the point where it is far more useful than simple "yes/no" feedback.

The next level in this progression is speech synthesis, finally coming into its own as an affordable and desirable technique for operator communication. Once possible only with arcane analog techniques (*LC* formant filters cascaded after a wide-bandwidth buzz generator, together with an avalanche diode for fricatives and sibilants, a capacitive discharge gate for plosives, and various circuits for nasals, glottal stops, and inflection—it was horrible back then), speech synthesis techniques have evolved to the single-chip level. These have become available in a variety of forms, with the primary trade-off being intelligibility and programming ease versus flexibility and vocabulary size.

Many of the potential applications for speech synthesis are large-volume propositions, resulting in a vigorous and competitive drive on the part of various manufacturers to be the first in the marketplace with a high-quality, single-chip synthesizer. They are going about it in many different ways: National uses waveform digitization, TI likes linear predictive coding (modeling the vocal tract with a multiple-stage lattice filter network), Votrax, one of the pioneers in the field, uses formant coding, and some, like General Instrument, combine linear predictive and formant coding techniques by deriving the coefficients for an LPC algorithm from the frequency domain. Each scheme has its own peculiar system overhead requirements, and some of them allow unlimited vocabulary size. It is a vigorous endeavor, and by the time you read this, there will already be some well-established talkers on the market in the $5 to $10 range. Of course, the availability of speech synthesis systems suggests the need for their opposite, systems that inhale human audio and derive appropriate intelligence from it. This, it turns out, is significantly more complex than the first problem, requiring techniques of artificial intelligence to apply some understanding of context to the stream of input utterances, unless, that is, you are willing to accept a very limited vocabulary.

Intense research is going on in these areas, since the ability to provide voice communications would allow interaction between complex equipment and relatively untrained people. This has obvious implications in industrial control systems, where even the installation of a simple data collection terminal meets with employee resistance due to its perceived complexity. If a person could say, "Hey, computer!" and be recognized by voice; if the system could then ask, "Whaddya wanna do, Joe?" and carry on a casual conversation to establish the various parameters of its next task, the whole problem of operator I/O could be relegated to the history books. It would also revolutionize the JOKE program.

There are some other interesting techniques in the realm of I/O esoterica. One, shown simplified in Figure 5-15, allows the operator to select options displayed on a screen by simply pointing his or her finger at them. The display, ideally a large-format flat-panel unit, is set back in a frame. Arranged around the borders of the frame are infrared LEDs and phototransistors such that there is a grid of beams directly above the glass surface of the display. It is a simple matter for interface logic to detect the X-Y coordinates of a pointing (or moving) finger on the basis of beam interruption.

This device can be hermetically sealed, requires no training, and could even operate underwater or in an explosive environment. Further, the various options displayed on the screen could be expressed graphically, eliminating the need for the operator to be literate or fluent in any particular language. Commercial units provide 16 × 16 or greater resolution, and some newer units eschew the beams and sense the operator's finger with a matrix of transparent thin-film switches.

Infrared emitter

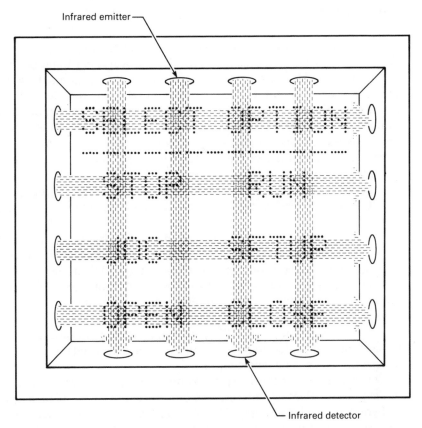

Infrared detector

Figure 5-15 A "touch matrix" technique, such as that used on the PLATO terminals, is based upon the interruption of infrared beams by the operator's finger. Such an approach is extremely rugged and can be hermetically sealed.

We could go on, discoursing upon joysticks, trackballs, optical eyeball-followers, "tank" controls, "heads-up" CRTs, tactile Braille displays, three-dimensional graphics techniques, and Morse-code interfaces, but it is not really productive at this point. Clearly, just about anything can be interconnected, somehow, with a microprocessor, and as time goes on, more and more methods of expanding the information bandwidth of the human–machine link are being developed. That's the real purpose of all this, of course. We think relatively slowly, but with great coding depth; computers think extremely fast—but when we attempt to communicate effectively with them (or with each other, for that matter), we immediately run into problems. We shall talk about the philosophical aspects of this in Chapter 9, but until then, let's consider our three other classes of interface problems.

5.2 SYSTEM OVERHEAD

Funny thing about a subject as broad as the one "encompassed" by this book: it is perforated by wormholes leading into other universes of thought. Over and over again, we stumble onto (into?) them, encountering compelling forces which threaten to draw us off into equally lively but relatively remote subjects: hierarchical CPU architecture design, natural language comprehension, optical computing, cam design, employee relations, rainbows, mating rituals of the speckled grouse, the acoustics of Italian virginals, the ecology of compost ... there's just too much interesting stuff out there! We wouldn't have that problem if we were content to call this a textbook, but nooo ... we had to make it a *meta*-textbook. That places us squarely in the realm of human thought instead of engineering technique. If we didn't firmly believe that the latter could be approached most successfully via the former, we'd be in trouble.

Ever notice how these little asides crop up in the strangest places? We could spend the next 50 pages talking in a most animated fashion about clever ways to distribute processing resources and augment the CPU's power, all without affecting the problems of industrial design in anything more than a tangential fashion. But instead, we had better just indirectly suggest the importance and depth of the field, then hit a few high points.

Given some kind of a CPU performing some kind of a task, let's consider some of the hardware interfaces that fall outside the domains of operator and intersystem I/O, sensing, and control.

First, there are the myriad add-ons that can be loosely classified as *direct memory access* (DMA) devices. Among these are disks, high-speed data links, and graphics display systems. The basic principle of DMA involves the presence of another source of address and data besides the CPU. There are numerous schemes that accomplish this, the conceptually simplest of which is shown in the block diagram of Figure 5-16. Here, tri-state buffers, under the control of some "arbitration logic," have been used to determine which of two logical entities gets to control the RAM. This setup allows for very high speed communication between the CPU and the external device, since it does not have to be funneled through a port. In reality, of course, the net speed of data transfer is limited by the rate at which the CPU can address sequential memory locations, but the speed advantage is significant nevertheless. DMA also allows the luxury of random access.

Fortunately, the use of DMA has been simplified by the availability of chips that do all the work, such as the Intel 8257, and the Zilog Z80-DMA.

We spoke earlier of some of the advantages inherent in distributed processing, wherein a task is shared by two or more processors in order to soften the impact on throughput of the von Neumann bottleneck (the interconnection between a CPU and its memory which allows only one data word at a time to be transferred). This is a good place to point out that similar results can be achieved in certain applications by pairing the processor, not with another CPU, but with a specialized support device. A good example is the floating-point processor.

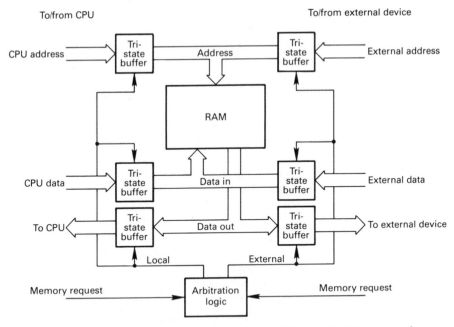

To/from CPU   To/from external device

CPU address — Tri-state buffer — Address — Tri-state buffer — External address

RAM

CPU data — Tri-state buffer — Data in — Tri-state buffer — External data

To CPU — Tri-state buffer — Data out — Tri-state buffer — To external device

Local   External

Memory request — Arbitration logic — Memory request

Figure 5-16 Direct memory access involves the ability to select the source of address and data to a field of memory, allowing extremely fast data transfers between two systems. The technique shown here has further applicability in a development context, and will be discussed later.

It is not uncommon in the design of a system to discover that mathematical operations, frequently taken for granted, have a curious tendency to monopolize system resources if they are invoked with more than casual frequency. In one system (the "widget machine" controller), the need to perform high-speed operations with 32-bit integer precision rendered the SBC 80/04 somewhat inadequate—but in typical fashion, the project had progressed too far and deadline loomed too near for a change in CPU cards, especially since the upgrade would have to have been to a very high performance, expensive processor.

The solution to the problem is depicted in Figure 5-17, and is based on the AMD 9511 floating-point processor chip. This device, dubbed "NUMCRUNCH" in the system schematics, handles a robust repertoire of arithmetic processing operations, including trigonometric and exponential functions. It has a stack architecture reminiscent of HP calculators, but with implied data formats (the binary data occupying the locations in the stack might be 16- or 32-bit, fixed- or floating-point—it is up to the host processor to keep track of them).

There have been many attempts at hardware math processors over the years, many of the early ones being little more than scientific calculator chips grafted clumsily onto a bus via a set of keyboard and display interface lines. But the 9511 and some other devices specialized for digital signal processing applications have elevated this to a level useful for real-time tasks, with a 32-bit sine taking 1.9 to

2.4 ms (depending on the clock frequency and the value of the data) and arithmetic operations requiring only a few microseconds.

Most of the hardware of Figure 5-17 is involved with outside world interfaces and, in fact, we will be referring back to it in the next couple of sections as we discuss sensing and control interfaces. But note the 9511 connected to the buses, located next to the CPU. The schematic is about as complex as the block diagram.

During the execution of the program, the system must, among other things, perform calculations on the data from the load cells to determine their acceptability. It uses this information to make accept/reject decisions concerning the components whose insertion and withdrawal forces are being tested. But together with that, it is engaged in "management information processing," as described in Section 1.1.5, and the combined load is manageable only with the aid of the support device.

Floating-point processors and similar math devices are not the only chips that can be considered in this class of CPU support. There are also smart peripheral controllers, counter/timers, interrupt and priority arbitration controllers, and memory mappers. We will not belabor the point by pursuing them—although they are all similarly important, information about them is generally well provided by their manufacturers. All we needed to accomplish here was the presentation of the general architectural philosophy which embodies the strengthening of a processor with the addition of relatively "internal" support hardware.

5.3 SENSING

With all that behind us, we can turn our attention to the more specifically "industrial" classes of interface techniques. At some point in the integration of a microcomputer system into a process, it becomes necessary to bridge, bidirectionally, the gap between the outside world and the comparative calm of the logic.

Sensing techniques span the entire spectrum of observable phenomena. Certainly it is obvious that we can measure such standard parameters as voltage, temperature, pressure, shaft angle, and things like that, but we can also devise sensing apparatus for magnetic flux, stress, radiation, fluid flow, vibration, spectral distribution, and ionic concentration. It's a big business: it has been predicted that the 1985 market for transducers is on the order of $1 billion, and that doesn't include optical devices, servos, and many other types.

More or less arbitrarily, let's begin with the optics.

5.3.1 Optoelectronic Sensing Techniques

Perhaps the most common sensing problems involve mechanical conditions: shaft or cam angles, interlock status, and so on. A system that we just discussed is, in fact, an excellent example. Look again at Figure 5-17. All the interfaces with the

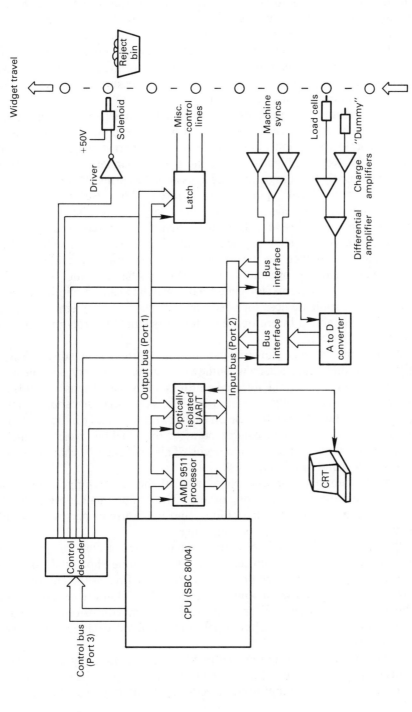

Figure 5-17 The block diagram of a real-time machine control and QC system shows a variety of real-world interfaces. Additionally, it contains a 9511 floating-point processor which handles all the mathematical operations, dramatically simplifying the system software.

production machine are shown at the right side of the diagram. At the top is a solenoid, which kicks defective parts into a reject bin. Next are some miscellaneous control lines, which stop the machine as necessary and take care of some minor functions. Below that are the "machine syncs," inputs from three optical sensing devices which keep the controller advised of machine timing, and at the bottom is a pair of load cells, which are used to measure the forces that are the basis for the system's control algorithm.

The problem of communicating the state of a mechanical system to a computer used to be a major nuisance, since it had to be done with switches. As we know, contact pairs exhibit numerous distressing characteristics, one of which is hysteresis. To avoid ambiguity and arcing at the boundary conditions (with the contacts just barely touching), most designs added a spring to the switch such that the "make" and "break" points were different. The resulting hysteresis loop indeed eliminated flaky operation at the boundary, but resulted in an even more annoying problem: it was very difficult to identify the precise point at which the contacts would change state. Typically, sufficient means for mechanical adjustment would be provided by the designer to allow empirical calibration—which solved the problem until age and wear changed the characteristics of the switch.

Most of the problems associated with switches can be eliminated with the use of optoelectronic devices for position sensing. In the system of Figure 5-17, the syncs are generated by emitter–detector pairs straddling rotating slotted disks. Look at Figure 5-18.

In the simplest case, shown in part a of the figure, a single disk is fastened to a shaft with a collar and a setscrew. Somewhere on its periphery is a hole or notch; when this passes between the infrared emitter and its mated sensor, a pulse is produced which can be handily interfaced to TTL, as we shall see shortly. The pulse thus appears at the same shaft angle on each revolution, and can be used by a control system to establish a repeatable point in the machine cycle at which some other event or measurement should occur. This angle can be changed by loosening the setscrew and rotating the disk.

The emitter–detector assemblies are off-the-shelf components, available from TRW and Spectronics, among others. They come in a variety of shapes and sizes, including reflective configurations, and make use of narrow-band infrared to minimize interference from room lighting.

Occasionally, it may be desirable to delay and/or stretch this pulse, and the circuit in Figure 5-18b will do just that. The first one-shot establishes a delay period from the sensing of the hole; the second one fires at the end of that interval and produces a pulse of known length. This allows a certain amount of machine speed independence (and also avoids logical confusion if it should happen to "park" on the hole), but generating the fixed delay is risky if speed is highly variable.

A better way is depicted in Figure 5-18c. Here, both angular displacement and pulse duration (as a function of rotational velocity) may be adjusted mechanically, guaranteeing the output to be directly representative of a mechanical condi-

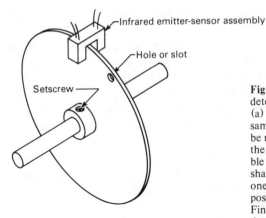

—Infrared emitter-sensor assembly

—Hole or slot

Setscrew

a. Single-pulse disc

Figure 5-18 Methods of optically determining shaft positions. The disk in (a) will produce a single pulse at the same point in each revolution, and can be mated to system requirements with the circuit shown in (b). The more flexible approach shown in (c) eliminates the shaft-speed dependence exhibited by the one-shot approach, accurately reflecting position regardless of angular velocity. Finally, the device shown in (d) will produce a pulse for each 15 degrees of shaft rotation.

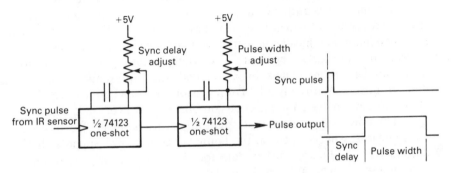

b. Electronic variable pulse generator and waveforms

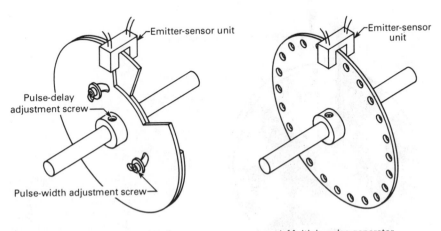

c. Variable-pulse-width disc

d. Multiple-pulse generator

tion. It is this style of sync sensor that is used on the insertion/withdrawal tester of Figure 5-17.

But what if you need, say, 24 pulses per revolution? You could always tack on a 1 : 24 gear train and place a disk on its output, but that is sheer overkill. Instead, the configuration of Figure 5-18d could be used. This is just a disk with 24 holes (or a gear?) — used in the same way that the others were used. Alternatively, you could allow the output from the single-pulse disk of Figure 5-18a to drive a phase-locked loop — if the special requirements of the application can justify the expense.

It is not uncommon for design requirements to call for explicit identification of shaft position. One approach to this might involve the trick shown in 5-18d, with the addition of an extra hole to define the zero-degree position. Logic in the system could then count the pulses and thus keep track of the shaft angle. But if the shaft is free to turn in either direction, this becomes a nuisance, since the logic then must derive a direction signal with a quadrature detector.

But more straightforward is the use of a *shaft position encoder*, a low-resolution version of which is sketched in Figure 5-19. Here, we have a disk that has been photographically prepared with five concentric circles of alternating transparent and opaque segments. The arrangement is such that any arbitrary radius from the center will describe a 5-bit code that represents the angular displacement from zero. With only 5 bits of precision, the unit can resolve 11.25 degrees; 10 bits would provide a unique code combination for every 0.35 degree.

The only real problem that might arise with a device like this is one of the old standards: what happens at the boundary conditions? Suppose that the radius described by the line of emitter–detector pairs falls precisely along a line between two valid sets of bits? What is the output?

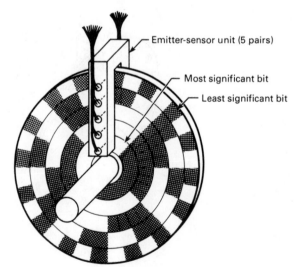

Emitter-sensor unit (5 pairs)

Most significant bit

Least significant bit

Figure 5-19 A shaft position encoder provides an explicit indication of shaft angle, regardless of velocity or changes of direction. This one is binary coded for clarity, but most commercial units use coding schemes in which only one bit changes at each boundary, eliminating false codes.

Random, that's what. Between slight mechanical misalignments in the opto assembly, skews in the coding disk, and uneven sensitivities of the detectors, the ouptut code will be some combination of code 'n', code 'n+1', and the logical OR of the two. This problem is largely due to the nature of binary coding, wherein any number of bits can change at the same time (the change from 15 to 16 involves all five bits: 01111 becomes 10000). Most shaft position encoders, therefore, make use of special codes such as the Gray code, in which only one bit changes at a time. Another option, useful in upgrading cam-timing systems or implementing random logic without a processor, is the custom encoding of the disk with patterns that directly represent implied functions. That sort of thing eliminates flexibility, of course, and could be done much more elegantly with straight encoding and a ROM, even if no microprocessor was involved.

Determination of shaft position can, as we have stated, also be accomplished with the "gear" approach, more accurately called an incremental encoder. Although it requires a little more support logic, it allows greater resolution and introduces fewer boundary ambiguities.

It can also be "unwrapped" from the disk and implemented in the form of an arbitrarily long strip, as suggested by Figure 5-20. This technique, used frequently in matrix printers for timing and in disk drives for position feedback, is based on the same principle: an emitter–detector pair straddles a transparent substrate photographically endowed with a succession of dark vertical bars. When relative motion occurs between them, each cycle of the pattern produces a pulse. Logic can keep track of the absolute position of the moving sensor (or moving strip) by setting a counter to zero at a reference location (detected by another optical sensor, no doubt) and then incrementing or decrementing it as a function of relative motion and an independently derived direction signal.

Let's look at some other kinds of uses for optical sensing devices. Figure 5-21 shows a technique for measuring speed—an optical tachometer. Here, the IR emitter and detector are arranged with their axes at a 45-degree angle, requiring reflection from a nearby surface to complete the optical circuit. It is easy to see that if the reflective surface appears to come and go periodically, a circuit could count the pulses and determine the cycle rate. Attaching a reflective strip to a dark shaft (or vice versa) provides the measurable condition.

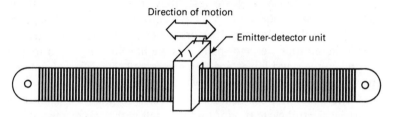

Figure 5-20 A linear position sensor provides an output pulse for each increment of relative motion between the sensor and the coded strip. Direction can be sensed with a quadrature scheme, or derived independently from other parts of the mechanical system.

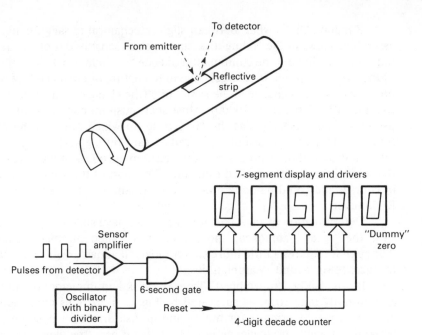

Figure 5-21 An optical tachometer can be easily created by simply totalizing pulses from a sensor into a counter for a known interval. The unit shown here reads directly in RPM.

In the figure we have used random logic for demonstration purposes, but the function can be even more easily accomplished in software. Essentially, a "gate" is established which defines a measurement period during which the pulses from the detector are accumulated into a counter. If, for example, the gate is 6 seconds in duration, then the number remaining in the counter at the end of that time will be one-tenth of the shaft speed in rpm (assuming that it is rotation that is being measured). In a machine interface task, this technique might be useful as rate feedback from a rotating or reciprocating process. It can also be built into a probe and used as a convenient piece of test equipment.

Optoelectronic devices lend themselves well to a variety of general switching applications because they are fast, they interface conveniently with logic, they don't arc or wear out, and they have zero hysteresis. Although hysteresis is frequently useful, as described before, it can seriously complicate some interface tasks. The emitter–detector pairs that we have already used to good advantage in sensing angular and linear position can be put to work as simple switches, in various ways. Figure 5-22 shows a few of the many possibilities.

The door interlock scheme deserves a comment. Often, in the design of an industrial control system, you find yourself with a set of responsibilities not really connected with implementation of a control algorithm. Among other things, these might include safety interlocks which inhibit machine activity if certain access doors are open. The configuration in Figure 5-22b can be put to use throughout a

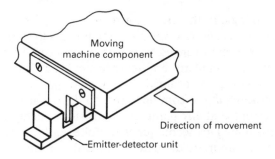

a. Solid-state "limit switch" detects machine position.

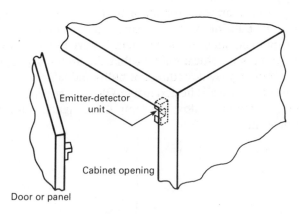

b. Door safety interlock

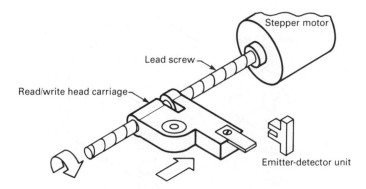

c. Floppy disk "Track Zero" sensor

Figure 5-22   The use of optical emitter–detector units makes available to the system designer a wide range of mechanical sensing possibilities.

machine, with logic in the processor requiring a specific combination of optically detected conditions before it will allow the process to start.

Want to get really carried away with this concept? Try Figure 5-23. Here, the IR LED of our standard emitter–detector pairs has been replaced with a helium–neon laser, its beam woven into a net around a restricted area or dangerous process, at last coming to rest on the surface of a phototransistor. If the beam is interrupted at any point, the transistor switches and the system can take appropriate action. This technique is useful even for very large areas with a total path length of hundreds of meters, since the beam divergence of even cheap lasers is typically well under 1 milliradian.

Optoelectronic devices are so wondrously convenient to use that it is sometimes tempting to put them to work in strange ways just for the pleasure of doing it. Figure 5-24 shows one such application—a monitor for an indicator lamp.

Although it sounds about as useful as mammae on a male wild hog, a photosensor used to keep an eye on bulb operation can prevent potential confusion. After all, an unlit bulb on a control panel means one of two things: either the lamp circuit is not activated or the filament is burned out. How's the operator to know? If the process is critical enough that the ambiguity represents a potential threat, you can either provide redundant bulbs and lamp test circuitry or the little circuit shown in the figure. The Exclusive-OR gate perks up whenever the signal from the detector fails to follow the lamp drive signal, informing the processor that something is amiss.

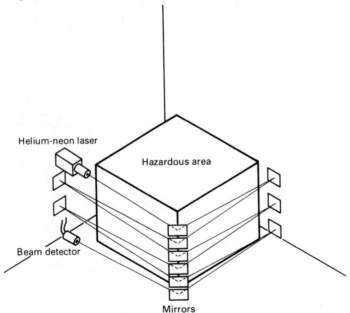

Figure 5-23 If a machine presents life-threatening conditions to nearby human beings, it can be enclosed in a laser "net" which can sense beam interruption at any point and automatically shut down the dangerous process.

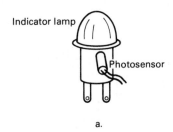

Indicator lamp

Photosensor

a.

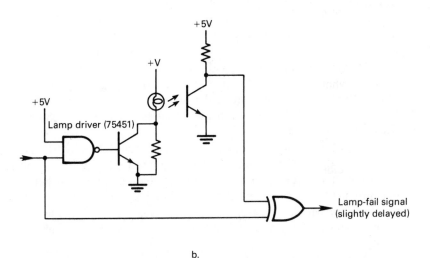

b.

Figure 5-24 In situations wherein a burned-out indicator lamp can have dangerous implications, this circuit can monitor the bulb's activity and return a "lamp fail" signal to the processor if it fails to light upon command.

Another bit of trickery comes in handy when you need a 60-hertz reference signal, whether as the time base for an interrupt-driven real-time clock subroutine, the sync source for a zero-crossing triac control, or some other line-synchronous operation. Shown in Figure 5-25, the scheme is based on the fact that a small neon bulb (NE-2 or similar) will ionize twice during every cycle of the applied ac. How convenient: by hanging a phototransistor on the side and applying its output via a Schmitt trigger to the clock of a flip-flop which divides it by 2, we end up with a 60-hertz square wave. This eliminates the transformer winding technique, which is inelegant, if not of lower performance.

After all this chitchat about using opto devices, perhaps it would be useful to take a quick look at the means by which they can be interconnected with processors and other manifestations of MOS and bipolar technology.

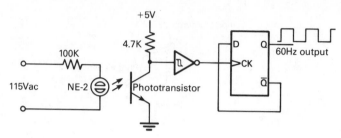

Figure 5-25   Optical method of generating a line-synchronous 60-hertz square wave.

In all the applications we have described, by the way, the LEDs are on continuously. The circuits associated with them, therefore, consist only of a current-limiting resistor and a source of power (see Figure 5-26). If for some reason it is desirable to switch the LED on and off, it can be done in exactly the same fashion as the switching of the LED in an optocoupler or a panel display.

In the diagram we show the phototransistor driving the input of a 7414 Schmitt trigger, a device designed with a large hysteresis loop which prevents its output from responding to trash that would send a standard TTL device into convulsions. Part b of the figure represents the behavior of this chip when confronted with a noisy input signal.

Phototransistors can, of course, be connected directly to TTL, but the problem you tend to run into there is that the optos are typically out on the end of a cable somewhere, quite removed from the logic circuitry. In between the distant machine sync generator and the CPU may be SCR-controlled motors (noisy), a robot welder (noisy), a machine operator wearing silk lingerie on a cold, dry day (noisy), and possibly even a bad connector or two (catastrophic). Ah, industrial reality.

There are many ways to handle this, depending on the brutality of the installation. Shielded cable is *de rigueur*, of course, and in extreme cases it may even be necessary to provide logic at the opto site which converts the single-ended signal to a bit stream on a current loop. Once the information is converted to a well-buffered, digital signal, it is much easier to handle without degradation.

Before we abandon the subject of optical sensing techniques, we might as well make passing mention of CCD's and "computer vision."

A dense X-Y array of phototransistors can replace a vidicon tube and simplify video digitizing. But charge-coupled devices, which we mentioned in Chapter 3 in a memory context, are cheaper and also possess the interesting characteristic of light sensitivity, making them extremely useful in optical sensing systems.

Apart from surface scanners and simple optical comparators, the real focus of work in this area is on the development of useful computer vision. If this capability could be added to robots and industrial computer systems, the range and scope of their applications could far outpace present limitations. Systems already exist which can recognize and manipulate isolated objects on uncluttered backgrounds, enabling an assembly robot to pluck components off a conveyor reliably and place them on a jig. But analysis of three-dimensional, complex scenes still eludes researchers. It is becoming evident that the classic approach—processing the scene on the basis of aggregates of "pixels," or points in the array—is quite inappropriate and must be replaced by a type of processing that is more congruent with the frequency-domain activity of the human brain. We will have some fun with this idea in Chapter 9 when we discuss artificial intelligence.

To suggest some of the more specialized applications of CCD arrays, let's spend a moment designing a turntable for a stereo system, with the specific design requirement that nothing ever touches the grooved area of the record in any way. Money is no object. We require the unit to accept standard phonographic disks; new media are not allowed. This doesn't sound very "industrial," but like the polyphonic keyboard, it will suggest possibilities.

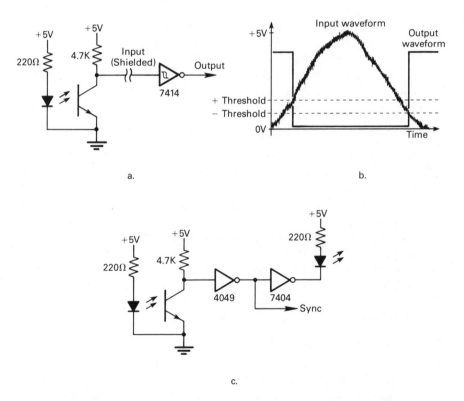

Figure 5-26   (a) If optical conditions are being remotely sensed, it is wise to incorporate a Schmitt trigger to clean up slow signal transitions and reduce noise pollution. The transfer function of these devices is suggested by (b). An alternative approach is shown in (c), together with an LED and its driver for display of the sensed condition.

As shown in Figure 5-27, we start with a scheme for physical handling of the media. Since nothing will ever be bearing down upon its surface, we can just support it by a hub no larger than the label area.

We will use a laser beam for tracking as well as extracting audio-frequency intelligence. Ignoring the details of this for a moment, note that a "head" is driven radially by a leadscrew (which can also be controlled vertically, both to dynamically accommodate warpage and allow mounting of the disk on the machine). The laser beam is generated in the mechanical drive box, and aimed at a 45-degree first-surface mirror mounted on the head.

Figure 5-27b shows the detailed action of the laser beam. After it enters the head, it is split into three separate paths. The two that are shown on opposite sides are focused appropriately, then allowed to pass through thin-film optical detectors to eventually strike the sides of the V-groove which embody the musical information in the form of time-domain surface perturbations. The beams reflect from these, and some of the returned light strikes the thin-film detectors. The signal out of each detector represents the optical heterodyning of the simple beam that had passed through it and the information-laden reflection from the record. Appropriate signal processing thus extracts the sound.

The third beam continues straight downward, striking the surface of the record at some random point. It is the job of the control system to determine its angle of reflection and continually adjust the head's position to keep it centered. The sensing is accomplished with a linear CCD array, shown in the form of an arc over the groove.

Aside from a stunning variety of engineering problems to be overcome, such a scheme should work. As the spiral groove creates error in the reflection angle, the processor detects a change in the CCD's image contour and issues correction signals to the servo drive to bring it back into line. The same system should accommodate the various nonlinearities due to warpage and poor media quality control.

That's the idea, anyway. Getting such a contraption to work reliably would require a hefty R&D budget, and the whole thing may belong in an April Fool's issue of an audiophile magazine. But it does illustrate the use of various optical techniques in a closed-loop control system. Oh yes—it would need a fourth beam (or a sonar system?) to drive the vertical correction of the head position that prevents crashes.

5.3.2 Other Sensors

However nice optoelectronic devices are, they cannot do everything. Let's enlarge our view of interfaceable sensors with a look at some other types.

The insertion/withdrawal force tester (Figure 5-17) has as its primary mission in life the acceptance or rejection of components on the basis of the force required to insert and then withdraw a tool steel pin. As shown in the referenced diagram, the force-sensing devices are called *load cells*, and are connected via charge

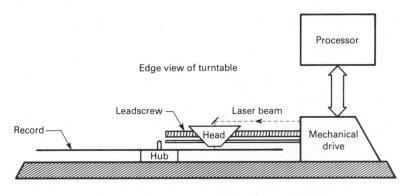

Edge view of turntable

a.

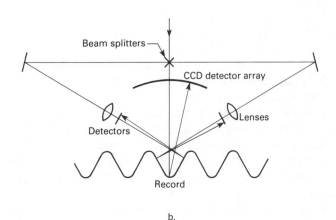

b.

Figure 5-27 This speculative laser turntable design illustrates an application of CCDs for precise position sensing. As shown in the cross-sectional diagram in (a), a head is driven radially over the record by a leadscrew. The drive signals are derived from the optical circuit in (b), wherein the angle of reflectance from the valley of a groove is sensed and used to generate a correction signal. The audio itself is derived with optical heterodyning.

amplifiers and a differential amp to the input of an analog-to-digital converter. In operation, the machine cranks out widgets which eventually pass the sensing station, where they are interrogated by the load cell assembly—mechanically slaved to the machine's reciprocating motion. At selected points during the machine cycle, as determined by the optical sync sensors we have discussed, the measurements for insertion and withdrawal force are made. The processor makes an accept/reject decision, then enters it into a queue, which advances in harmony with the part's continued travel until the latter is opposite the reject solenoid. If the decision was "thumbs down," then: wham! It's all over for the widget.

The reason for the use of two load cells in an application wherein one would seem to be sufficient involves the fact that inertial forces generate significant error, and thus have to be nulled out. The system was therefore built with a duplicate load cell assembly, whose output is subtracted from that of the primary one by the differential amplifier. The resultant signal, converted into binary, is then relatively pure measurement information.

The load cells used in this project, staggeringly expensive and precise devices made by Kiag-Swiss, respond to static force on their faces by producing a tiny proportional charge, measured in picocoulombs. Designing circuitry to accurately handle this is worth avoiding at all costs and, fortunately, the company also makes companion charge amps which produce nice bipolar output voltages on a comfortably macro scale. The circuit that takes it from there and converts it into something digital is depicted in Figure 5-28. Net performance after digitization is a resolution of 0.2 gram and a range of about 453 grams.

The chip whose input is fed by the agglomeration of minidips and discretes is a Burr-Brown ADC80, a 25-microsecond (μs), 12-bit analog-to-digital converter. Interface between it and the processor is straightforward: the processor says, "Do it!"—then the A-D says, "Got it!" and hands over the data.

A-D inputs like this can be used for all sorts of sense inputs besides acceleration or force. Let's look at a few.

The most pervasive phenomenon that is amenable to measurement without overly esoteric techniques is probably temperature. For many applications, the relatively recent development of silicon diode temperature sensors has replaced older devices such as thermocouples, resistance detectors, thermistors, thyristor temperature switches, and others. But each has its relative advantages and disadvantages.

Silicon is useful in this context because it is linear (within 1%), accurate, and stable. It is also cheap, with packaged sensors available from a variety of vendors in the $1 to $2 range, offering performance over a range of approximately −40 to +150 degrees Celsius. Compared to thermocouples, such devices are a delight to use: they need no trimming or calibration and provide healthy output swings that don't require the touchy amplification of the other's tiny change (a type K Chromel–Alumel thermocouple gives less than 10 mV of change over that range, whereas the Motorola MTS series gives 400 mV). But silicon has a limited temperature range with a usable upper limit of about 200° C, whereas thermocouples are accurate to 10 times that figure. There Ain't No Such Thing As A Free Lunch (TANSTAAFL).

Silicon is also widely used for pressure sensors—it is nearly free of the mechanical hysteresis that plagues other materials, and it is stable and rugged. It can also take advantage of proven processing techniques, enabling once-exotic sensing devices to be produced at relatively low cost (witness consumer-grade sphygmomanometers, using integrated pressure transducers such as the National

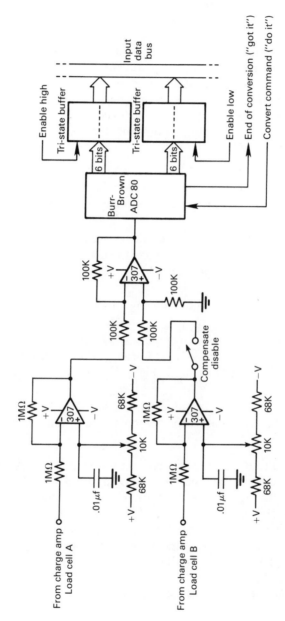

Figure 5-28 This circuit digitizes and places on the processor's data bus the difference between the signals generated by a pair of load cells with associated charge amplifiers. Handshaking with the CPU is accomplished through a pair of buffer enable lines, a convert command, and an end-of-conversion flag.

LX series). Since it does not have to be fabricated by hand, it can be small (one unit from Kulite has a diameter of 0.03 inch with a 100-psi rating) and since the ratio of its Young's modulus to its weight is even higher than that of steel, it can handle some pretty intense forces (a Foxboro/ICT device withstands gravity forces of $100,000g$ in missile applications). Ah, silicon.

There are many industrial applications for pressure sensors, especially considering the pervasiveness of pneumatic and hydraulic systems in manufacturing environments. It is a good idea to become conversant with some of these "peripheral" technologies as you begin applying microcomputers to industrial problems. There is no such thing as an isolated control system, and the range of sensing interfaces is not even limited by your imagination. There is always someone out there who will want things like:

→   Liquid-level measurement based on the heat loss of an active transducer to its surrounding medium.

→   Fluid-flow measurement based on heat transfer rate from a silicon element forming part of a bridge, in which a circuit attempts to maintain a constant temperature differential.

→   pH measurement using ion-selective field-effect transistors.

→   Sensitive acceleration measurements using solid-state cantilever beam springs which produce subtle capacitance variations on the order of 40 attofarads/g in response to forces of acceleration.

→   Chromatographic analysis based on a photolithographically fabricated spiral capillary separating column on a silicon substrate.

We could go on, and on, and on—talking about radiation measurement, sound level, electromagnetic field strength, and so on. But again, our objective here is theoretically at the heuristic level: too much detail about tangential specifics tends to obscure our intended philosophical basis. The key point of this section is just that techniques exist which allow your control system designs to augment their internal models of the world in ways that allow them to interact meaningfully with the processes to which they are wedded.

5.3.3 Analog-to-Digital Conversion

Most of the sensors discussed above, as well as those of a purely "electrical" variety (voltage, current, resistance, power, etc.) produce, after appropriate conversion and buffering, a voltage or current variation. Although this is more useful than, say, some number of picocoulombs, it still needs to be digitized to be of any value to a microcomputer.

The technology abounds with techniques for accomplishing this, most of which have become quite straightforward in the last few years as the result of integrated converters and complete multichannel data acquisition systems.

The basic tools, of course, are the A-D converters, which are available in resolutions ranging from 6 to 16 bits and a wide variety of speeds. Numerous trade-offs result in the presence of a number of performance criteria which must be juggled in the light of the intended application. These include input range, resolution, the linearity of the digital response to the input voltage, monotonicity, missing codes, quantizing error, relative and absolute accuracy, and offset error.

Some applications, such as the insertion/withdrawal tester (let's start calling that the IWT), can accommodate their analog input requirements satisfactorily with nothing more than a single ADC, as was shown in Figure 5-28. But what about a system that needs to monitor a large number of separate inputs? ADCs are expensive chips (the one in the IWT was about $80 in 1979), so it would be pleasant to avoid using one for each analog channel.

Normally, system timing requirements are such that the converter is not frantically converting during every last microsecond. Typically, it sits in an idle state until the machine's timing or a software clock calls for a reading, whereupon—bip!—it spends a few microseconds on the data conversion task and returns to its quiescent state. This state of affairs suggests the time-honored possibility of time sharing the A-D.

The general scheme shown in Figure 5-29 accomplishes just that. The "front end" is a 16-channel analog multiplexer, which responds to a 4-bit address by effectively connecting the corresponding input to its output.

The next stage is, optionally, a *sample/hold amplifier.* This device, essentially an extremely low-leakage capacitor on the input of an extremely high-impedance op amp, reduces the "measurement aperture" and allows faster net system speed. On command (from the processor), it grabs the input voltage and holds it for later conversion, effectively allowing a degree of pipelining to be added to the data acquisition hardware: the addressing of the mux and other setup activities can proceed concurrently with the actual data conversion.

With these devices feeding it, the ADC in the figure can stay busy a higher percentage of the time, fulfilling its destiny in a satisfying manner while saving the added cost of redundant conversion hardware.

Complete multichannel data acquisition systems like this are available as packaged hybrid circuits, rendering their use relatively straightforward.

5.3.4 Internal Models of the World

We fully recognize the impossibility of treating such an open-ended subject as interfacing with anything even remotely approaching thoroughness, especially within the context of a broad-spectrum assault upon the vast subject of industrial

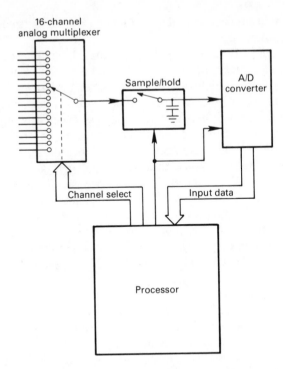

16-channel
analog multiplexer

Sample/hold

A/D
converter

Channel select

Input data

Processor

Figure 5-29   An analog-to-digital converter can be "time-shared" using analog multiplexing techniques and, optionally, a sample/hold amplifier. Schemes
such as this are frequently hybridized
and marketed as data acquisition
modules.

system design. It's a big world, imbued with complex phenomena at all levels, and
somewhere, sooner or later, just about all of them are going to be interfaced with
logic.

   To bridge the gap between sensing and our last interface category, control,
let's wax philosophical for a moment.

   We have spoken a time or two about the processor's *internal model of the
world,* that collection of numeric data and status conditions that represents, in a
fashion suitable to the system's objectives, the universe. Such a concept lends itself to a variety of graphic symbols, one of which might be Figure 5-30.

   We represent the system's environment here as a unique shape which is
sharply delineated from the universe as a whole. Presumably, this environment is
a process situation or some kind of machine that we wish to control.

   Within the system, three logical entities exist (even though they may be indistinguishable through examination of the software). One, embodied either in a
data structure or in the design of the code itself, is the ideal state of the
environment—the goal state. In simple control algorithms, this fancy concept may
be little more than a target temperature that the processor attempts to maintain
through adjustment of the current provided to a heating coil.

   The second abstract entity in the system is labeled "process," and is the set
of transformations and algorithms which the system uses in its ongoing attempt to
make the real world look like the goal state. The third is the internal model that
we have mentioned.

Again stressing that this schema is purely abstract, we can characterize the behavior of the system as follows. By periodically updating the internal model of the world with real information gleaned from sensors, the process can discern differences in the values of key parameters between the data structure thereby generated and the one that we have called the goal state. The process can then attempt to bring the two into line by issuing control outputs which are intended to modify the real world in such a way that future updates to the internal model will be more nearly aligned with the goal.

This, of course, is simply a closed-loop control system, mapped onto a higher level of abstraction. Our graphic symbol includes a link to a point outside the system's sphere of influence, representing human control or a set of instructions from a higher-level processor in a hierarchy.

With that, let's consider the last category in our rather ponderous chapter on microcomputer interfacing.

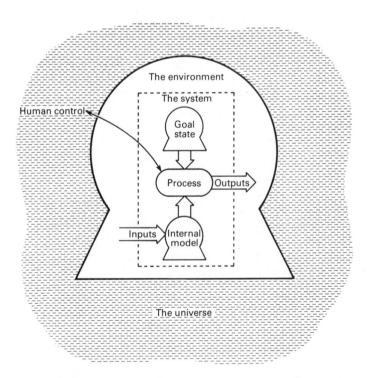

Figure 5-30   The basic objective of a control system is the continual trimming of certain output signals as a function of the perceived differences between sensed conditions and an internal goal state. In a formal sense, this takes place via an internal model which represents the significant features of the system's environment.

5.4 Control

We have already seen some of the fundamental types of "control" interfaces, in the form of simple output ports. Every now and then, characteristics of output devices are such that it is sufficient merely to connect the TTL-level lines from the computer to them and move on to the next problem.

But not very often. Let's consider next a few of the techniques that are required for most of our real-world control problems.

5.4.1 AC Interfaces

An understandably large percentage of control interface problems are directly involved with switching ac loads—motors, valves, large solenoids, lights, and other devices not commonly associated with microelectronics. This raises the problem of converting the 0- to 5-volt levels of TTL-compatible circuitry into something with a little more "oomph."

The obvious approach, albeit seemingly crude in this solid-state era, is the simple installation of a relay on the output of a transistor driver—or even on a gate if the relay's current requirements are gentle enough. One of a variety of acceptable schemes is shown in Figure 5-31.

The use of a relay possesses a modicum of attractiveness because of the obvious isolation between the logic and the load, not to mention the comfort in being able to determine its state by means of visual inspection. Something precious has been taken away from us by the solid-state revolution that began around 1960: more and more functions have been closeted in obscure bits of silicon which are in no way amenable to human senses, increasingly coarse by comparison. Gone are the days of relays that can be prodded with an index finger, of thermionic valves whose friendly glow gives reassurance of circuit function, and of logic levels that you can *feel*. Every now and then, it's nice to go back home, kick off your shoes, and relax with an electromagnetic component and a 5-cent Coke. Sigh.

Aesthetics aside, however, there are still a lot of good reasons to use relays in many applications. Often, your control system will end up receiving routine maintenance from plant electricians of the "old school," and relays have a demystifying effect upon microcomputers. Further, they are often cheaper than

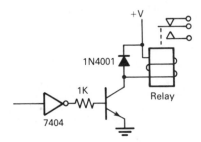

Figure 5-31 Driving a relay typically involves little more than a current amplifier and a diode to absorb the "back-EMF" that results from the collapse of the field in the coil upon deenergizing.

their newfangled counterparts—especially in high-current configurations—and they allow much more flexibility in terms of contact wiring and high-voltage logic. There is no shortage of dc isolation, either.

There are a few things to remember, however, when dealing with components that depend upon mechanical realities. Like switches, relays have bouncy contacts (although as output devices driving ac loads, who cares?). But they also require a finite period of time to change state, and while that observation is true of even the fastest Josephson junction, we are talking milliseconds with relays—logic propagation delays that you can see and hear.

Again, that makes little difference in most applications, but consider the situation represented by Figure 5-32. Here, some "condition" is shown which activates a switch connected to a processor. Suppose that the software in the processor has been taught to react to the presence of this condition by activating a relay which fires a solenoid that somehow remedies it. What actually happens?

Microprocessors are fast—much faster than relays and solenoids. The software probably detects the switch closure and reacts to it by calling a subroutine that delivers a pulse to the output bit connected to the relay driver, then returns to the normal scan. It is not difficult to imagine the following scenario: the relay starts to close, the CPU goes back to its scan loop, and surprise: there's a bit set indicating a switch closure. So it calls a subroutine that delivers a pulse to the solenoid via the relay.

Now if everything is more or less normal, the solenoid will eventually fire, the mysterious condition will be eliminated, and everybody will be happy (even if the software made a couple of passes at it, we might never know about it). But what if something is wrong with the real world, and the condition is stuck? Now the processor goes into a loop: hey, a switch closure—bang goes the solenoid—hey, a switch closure—bang goes the solenoid ... the system might just loop there forever.

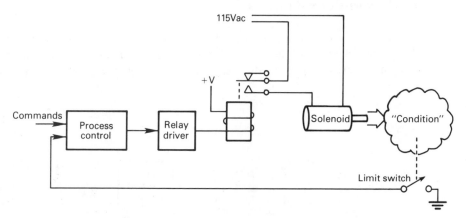

Figure 5-32 Something that has to be considered when interfacing with electromagnetic components is the fact that propagation delays are thousands of times longer than those of the processor logic itself (see the text).

There are two overlapping problems here. The first is sloppy design that ignores the time required for the relay and the solenoid to change state, allowing a status check before the control function has had time to take place. The second is the implied possibility of the system getting stuck in a loop. Depending on the type of "condition" implied by the drawing, this design ranges somewhere between careless and worthless.

So design with relays, although more or less intuitively obvious, does call for some awareness about their physical characteristics when the time comes to create some software. Also some hardware: little things like a diode to absorb the "back-EMF" that is generated when the relay is turned off must be remembered or driver failure can result.

For these reasons and others, various solid-state relays have hit the market. Basically opto-coupled triacs, these devices are intended as direct replacements for some of the more commonly used relay styles. A typical design (which can also be implemented with appropriately scaled discrete components) is shown in Figure 5-33.

The use of a zero-crossing trigger circuit assures that sharp high-voltage transients will be avoided, eliminating surges and radiation, an especially annoying problem with SCRs and triacs. The optoisolator prevents any current paths between the control logic and the load circuitry, and input and output filters encourage clean and orderly operation. Since it is all solid state, the SSR is a rugged device that has no contacts to corrode or take their time about moving, and it's easy to use. Fully modular devices of this sort are available from Monsanto, Opto 22, Crydom, Motorola, and others, and the same basic philosophy is applicable to various discrete designs. In particular, opto-coupled SCRs (such as GE's H11C2) are available in 6-pin DIPs (the same package as standard optocouplers), making such implementations relatively straightforward.

5.4.2 High-Current Drivers

Another common class of output interface problems involves circuits that allow the processor to drive heavy dc loads such as solenoids. As with the previous category, there are so many different types of configurations that a whole book on

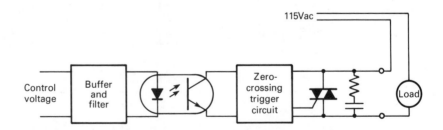

Figure 5-33 Solid-state relays allow isolated control of ac loads, but are not available in the wide range of contact configurations of their electromagnetic counterparts.

circuit design would be required to do them justice, but let's look at two fairly typical kinds of problems, the first drawn from the insertion/withdrawal tester (IWT).

As we mentioned earlier, the business end of the machine is a large solenoid, which acts upon a reject decision in the processor by breaking the offending widget from its carrier strip and hurling it into a bad-parts bin. The solenoid selected was a Ledex size 6EC, capable in a "short stroke, 10% duty cycle" mode of exerting nearly 90 pounds of force.

The circuit of Figure 5-34 shows how this is driven by the logic. When the processor generates FIRE PULSE (simply a bit on an output port), the J-K flip-flop becomes set. This has no immediate effect, but when the next SYNC C is produced by one of the three optical sync pulse generators we described in Section 5.3.1, a short pulse finds its way to the trigger input of the 555 timer, and the fun begins. Its output, amplified substantially by the discrete driver circuitry, provides a ground return path for the solenoid and it becomes enthusiastically energized, meeting the doomed widget at precisely the right time (SYNC C). After an interval determined by the time constant of the variable RC network connected to pin 7 of the 555, the driver circuit is turned off, the flip-flop is reset, and the circuit is ready for another cycle.

Perhaps the only aspect of this design that bears some emphasis is the fact that it performs some logic operations that could have been done in software, but deliberately were not. Sure, the output port bit could drive the transistors directly, eliminating a few parts, but the design convenience of being able to say, "Here comes a bad one," and then simply forgetting all about it was well worth the extra hardware. The processor has plenty of other problems.

Our second solenoid-driving example is presented both to demonstrate another current-driving technique and also to show how it is occasionally possible to "beat the specs" when the power of logic is at your command.

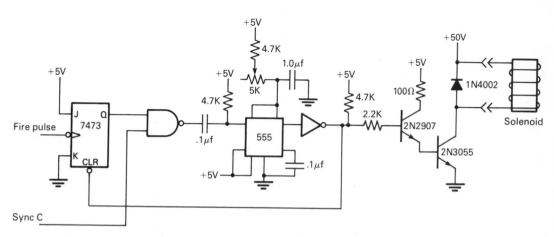

Figure 5-34 Solenoid driving circuit fires for a preset interval upon occurrence of "Sync C" only if the "Fire pulse" has taken place. After firing, circuit is inactive upon further syncs until again armed.

The problem is this: We need to mechanically intervene in a process, depending upon certain decisions reached by a controller, but we must be able to do it during an extremely narrow window in the machine's cycle. Specifically, we must energize a solenoid rapidly enough that it completes its mechanical travel within 7.5 ms, and then we must keep it energized for an arbitrarily long period of time.

This all sounds trivial, but when you start looking at solenoid specs, it becomes a bit complicated. A small enough unit driven with a high enough voltage can certainly complete its travel within the time limit, but if we keep that voltage on it for much longer than that, we will smoke it. The curves shown in Figure 5-35a express the problem, showing the solenoid's stroke vs. force and stroke vs. time curves for the current levels representative of 10% and 100% duty cycles.

This is a classic problem, normally solved with a strategically located capacitor somewhere in the solenoid driver circuit that has the effect of providing an initial current surge. But in a developmental context, it is nice to feel that you have control of all the parameters. The circuit in Figure 5-35b gets us pretty close.

Again, a pulse from the processor starts things rolling, but this time it does it by firing a one-shot, set for—you guessed it—7.5 ms. As long as its output is active, the top opto conducts, the transistor driver circuitry is on, and −20 volts is applied to the bottom of the solenoid. The top is connected to +5 volts, so for 7.5 ms, the solenoid has 25 volts across it.

When the one-shot times out, the rising edge sets the flip-flop beneath it, and the lower driver circuit is activated. It has the effect of placing 0 volts on the bottom of the solenoid, holding it energized coolly with 5 volts for an arbitrarily long period of time—months, if need be. When the processor grows weary of this condition, it issues the RELEASE pulse, resetting the flip-flop, removing the ground, releasing the solenoid, and leaving everything in a condition that is ripe for another such experience.

Both of the solenoid drivers that we have shown are of the relatively esoteric variety, as such things go, since you can make do with a lot less circuitry if you just want to stuff electrons through a coil. But our intent here is not a catalog of current driver circuit designs, but a suggestion of possibilities and perhaps even another "between-the-lines" clue concerning some of the subtle hierarchies of system design. Let's move on.

5.4.3 Stepper Motors

It seems that the bulk of machine control interface problems involve motion, not a particularly shocking notion since the prevailing image of "machine" in an industrial setting is comprised largely of moving parts. Sometimes, appropriate motion can be created by energizing a solenoid or switching on a motor, but frequently that's not enough: you find yourself needing to produce precise angular displacements or rotational rates.

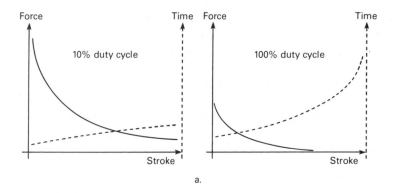

a.

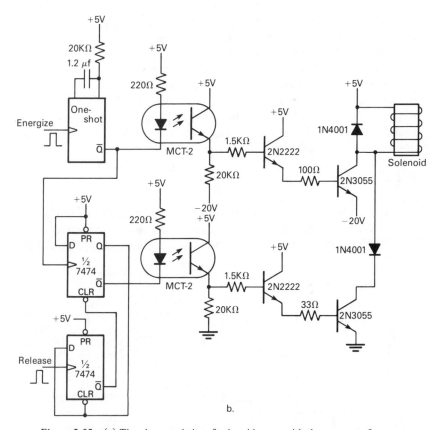

b.

Figure 5-35 (a) The characteristics of solenoids vary with the amount of current used for energizing—a high current suitable for 10% duty cycle provides faster and more forceful operation but will burn out the coil if applied indefinitely. (b) Advantage can be taken of the high current mode if the driver circuitry reduces the current for holding purposes.

These sorts of problems are addressed handily with the use of *stepper motors*, devices that move one angular increment for each digital pulse they receive. The pulses are typically in a four-phase sequence, allowing straightforward direction control by phase reversal.

We have seen one application of steppers already, in the Blanchard controller. In that situation, the problem was simplified for us by the use of a complete stepper–driver package which required only TTL-level pulses for operation. The Slo-Syn motor mounted on the Blanchard grinder was then driven by circuitry buried somewhere in a control box.

Such circuitry was a pain to design but, like almost everything else, it has since been implemented in chip form. The CY500 from Cybernetic Micro Systems exemplifies the breed, with enough internal intelligence to take all the work out of stepper control. Billed as a "stored program stepper motor controller," the device has 22 instruction types which allow everything from rate generation to execution of buffered and timed bidirectional sequences. Figure 5-36 shows a typical stepper control system based on this chip.

When the Blanchard controller was designed, no such panacea existed and it was necessary to implement the rate generation function with TTL. The box that did that handled five Blanchards; with the necessary 22-bit resolution, it ended up requiring 65 chips, 30 of which were 74193 counters. Again, something that once called for all the logic design horsepower available has been reduced to just another bus-interface exercise.

5.4.4 Analog Outputs

In our discussion of sense interfaces, we made mention of analog input devices and, needless to say, they have their complements on the output side of the system. Typically called *DACs*, digital-to-analog converters take care of all those situ-

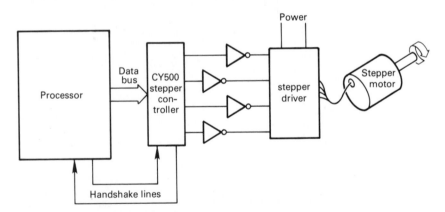

Figure 5-36 The control of stepper motors can be simplified considerably by offloading most of the work onto an integrated stepper control chip. The processor can then direct motor operation with high-level command sequences.

ations in which you find yourself with digital information that needs to be converted into a corresponding voltage (or current). These are used as components in some types of A-Ds, where they are driven by a binary counter that stops as soon as the generated analog value is equal to the input value, yielding the binary equivalent. DACs are also useful wherever it is necessary to digitally generate analog control information.

Most of the commonly discussed applications of DACs seem to be associated with such things as vector graphics systems, speech synthesis, data collection for digital signal processing, and so on, but there are some "hard" industrial uses as well. One idea, suggested by the engineering staff of Analog Devices, is shown in Figure 5-37. Here, a DAC is used in the interface circuit of an industrial scale to permit local "zero setting." In operation, an empty container is placed on the weighing platform, and a SET TARE button is pressed. This loads the digital equivalent of the container's weight into a latch, whose output is connected to the input of the DAC. Its output then becomes the analog equivalent of that, and is subtracted by a differential amplifier from the voltage produced by the load cell. Subsequent changes in the weight on the platform brought about by filling the container are then automatically adjusted for the tare weight, and the output reflects only the value of the load.

It might seem a little redundant to be doing such arithmetic processing with hardware, when there is apparently a microcomputer only a few inches away. It would, of course, be little trouble for the software to maintain the tare weight

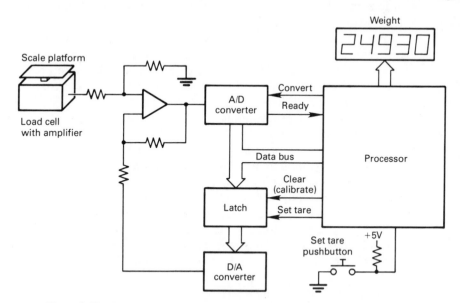

Figure 5-37   One application of ADCs and DACs is the automated tare setting on an industrial scale. Although this could be trivially done by the processor, such a setup obviates the need for CPU support, rendering it more easily interfaced without custom software.

somewhere in memory, subtracting it from the reading whenever necessary and updating it when the SET TARE button is pressed. We show it this way for one solid reason: It is very common for such items as scale interfaces to be supplied by vendors as stand-alone units, billed as "easily interfaceable with microprocessors." This allows someone to use the device without a computer, but makes no restriction on bus interfacing and the like for those who need a little more processing horsepower. A user of the first type might simply want to display the weight, or print it on a ticket. The other fellow, however, might want to make the scale a subsystem in a far larger scheme, possibly a continuous production batching system. The scale interface manufacturer will sell more boxes if he can serve both markets with equal ease.

DACs, like the less pronounceable ADCs, are available in a variety of resolutions, speeds, and degrees of accuracy, with prices ranging from a dollar or two to hundreds. Selection for a specific application must balance those trade-offs to focus on the most appropriate device. What's new?

5.5 Random Interfacing Notes

There is probably some suggestion of the significance of microcomputer interfacing in the fact that this chapter is approximately as long as the first four put together. This should not be interpreted an any obsession on the author's part: he is getting just as impatient as the reader to place this subject squarely in the past and turn to some of the more sublime matters in the world of industrial microcomputer system design. There are many.

But interfacing is such a pervasive and potentially disturbing phenomenon that it seems to monopolize much of our discussion. However clever, nay, brilliant, the concepts embodied in the architecture and software of a system, any carelessness in the creation of its real-world interface can dash the hopes of engineer and customer alike by dooming the ill-fated contraption to a turbulent existence of poor performance, human contempt, and expensive attempts at maintenance. It can be depressing.

Tantamount to the success of a system is the application of good circuit design philosophy, not to mention extremely careful construction techniques. While detailed discussion of such things is somewhat outside the scope of this book, there is one body of principles that is so crucial to the success of your projects that we must give it a few precious pages of attention.

In fact, we give it its very own chapter. It may even be the most important one of this book, for although techniques and design philosophies evolve with varying degrees of rapidity, there are a few realities that are absolutely timeless—mercilessly so.

It is to these that we now turn our thoughts.

6

The Ideal vs. the Real

6.0 INTRODUCTION

At various points herein, we have tried to make it clear that this book is not one of those that depicts reality as an ordered and predictable world of lumped constants and ideal devices. That kind of thinking not only distorts one's understanding of the truth, but also results in potentially flaky system designs and expensive relearning processes. If we recognize as we progress that we are not dealing with isolated phenomena and laboratory conditions, then perhaps the longevity of our systems will reflect our awareness.

In Section 1.1.2, we presented a snapshot of a beleaguered industrial microcomputer system gamely holding on until the inevitable end despite the ruthless assault of numerous pernicious environmental influences. In this chapter we enlarge on that, hopefully arriving at some design heuristics that can be used to minimize the pain.

More or less arbitrarily, we can identify five major categories of abuse: noise, heat, contamination, vibration, and human perversity. We devote a section to each, and tack on a couple at the end dealing with physical reality (things don't stop instantaneously) and safety (just as we don't want our machines to be hurt by human beings, we don't want human beings to be hurt by our machines).

Let's begin with noise, perhaps the most nefarious of all.

6.1 NOISE

Before considering specific types of noise problems, we need to decide just what we mean by ''noise.''

In a rigorous philosophical sense, noise of the random variety can be characterized as a phenomenon with an autocorrelation function of zero and an average spectrum that has a constant amplitude over all frequencies. The average value of a random signal is zero, and it can be observed by simply cranking the gain of an amplifier up to a high level: the hiss, snap, crackle, and pop is due primarily to the thermal motion of conduction electrons and random variations in the numbers of electrons (or holes) dancing about in the amplifying components.

Such a view of noise is not of much use to us, partly because a purely abstract concept like that relegates most of our programs and data to the same category—the coding depth of the instruction stream observable on a microprocessor bus is so great that someone wielding an FFT analyzer would be hard-pressed to come up with a meaningful interpretation aside from the spectral peak attributable to the CPU cycle rate. Wouldn't that be embarrassing?

So let's consider noise from a pragmatic viewpoint: It is a phenomenon that in some fashion interferes with the intended operation of a system, although not necessarily to the point of malfunction. Since we are taking a more practical view here, we need not restrict it to randomness. That's fortunate, since some of the more insidious types of noise have very pronounced peaks at 60 hertz. Others can be strongly correlated with various aspects of system operation.

If you sit back and contemplate, it becomes clear very quickly that noise is a diverse subject. Let's try to order our thinking somewhat by splitting it into seven categories.

6.1.1 External Noise

This is the big one, and either causes or is blamed for a huge percentage of system irregularities. The universe is abuzz with noise, ranging from the background radiation echoing the Big Bang to that big SCR motor controller on the same circuit with your system that was made back in the days when nobody bothered with zero-crossing switching. Amazing things happen.

It is a typical day in a large shipyard. A welder lowers his mask and strikes an arc on a barge hull—the communication line overhead spikes and knocks a bit or two out of a part number currently being transmitted. Purchasing thus issues a purchase order for 82 diesel engines at $1.04 apiece, instead of 82 engine mount retaining bolts.

Elsewhere in the shipyard, a security guard keys his business-band walkie-talkie and a nearby matrix printer, made by a company you know quite well, picks up and rectifies the RF in the optical linear position sensor at the end of a 3-foot cable, and loses its mind. The print head smashes into the stop at the right-hand edge of the paper and the wires, still pulsing, eat a little hole in the platen.

Two field engineers are called—one for the subtle communication problem that was only discovered after an incredulous engine vendor called to inform Purchasing of the correct price, the other for the flaky printer. When they arrive, of course, both the welder and the security guard are long gone, and there's no problem. They sadly conclude that any mechanical or electrical device with any mal-function short of a complete breakdown will function perfectly in the presence of a trained serviceman.

Some external noise problems, like those, are attributable to quirks of fate. Others are a little easier to find. Consider a machine like the IWT.

Recall from our drawing (Figure 5-17) that we have various antennas con-nected to the logic. In particular, there are three sense lines coming in from the machine sync generators, two analog lines from the load cell charge amplifiers, some miscellaneous control lines, and a solenoid driver. All these things involve wires of appropriate lengths, and are very efficient methods of picking up noise and coupling it into the logic.

During the succession of all-nighters that comprised the system's installa-tion, this fact became quite apparent. One particularly evasive failure mode finally correlated with the operation of another assembly machine over 100 feet away.

If you were hoping for a cheap device that can just be soldered in to solve problems like this, forget it. The only way to deal with environmental noise is to keep it out. This calls for careful use of shielding and grounds, proper line termi-nation, and a respect for the black magic of broad-spectrum RF phenomena. If that last point sounds completely absurd, consider the case of the worthless CRT terminals purveyed by a now-defunct business system manufacturer. The opera-tors could clear their screens and transmit garbage to the system just by stepping close to them after walking across a carpet. Everything was tried, to the point that the field engineers charged with the task were literally seeing them in their sleep—what little of it they got.

The "fix" finally involved grounded mats for the operators, recommenda-tions concerning undergarments, a replacement of the steel base with an alumi-num one, paralleling a No. 22 ground wire at the card cage with a length of braid, cutting miscellaneous "ground loop" traces on the PC boards, hanging capacitors on the communications lines, wrapping foil around the keyboard cable, and about a dozen other little patches designed to magically circumvent the noise that was unaccountably getting in there and wreaking havoc. About the time it was finally done, the customer changed computer vendors.

Proper use of shielding calls for some combination of hard-core theory and common sense. Perhaps most important, it must be connected intimately with the system ground, lest it worsen the problem by coupling more noise into the sys-tem. Further, the system and all connected hardware must share a shield that is intact at all frequencies at which noise might occur. Dc conditions don't tell the whole story. At 100 MHz, a simple piece of wire might become an effective insu-lator, an antenna element, or a matching stub.

The objective, then, is the development of a grounding system that approxi-mates a low-resistance "mesh," a not-insignificant problem. If the holes in the mesh are large enough to accommodate waves of a troublesome frequency, then instead of a mesh, you have a collection of loops. Madness.

Incidentally, the matrix printer problem presented above was solved, although not by the manufacturer's field engineers (who scratched their heads very convincingly, added a few unrelated changes, and sent a bill). Figure 6-1 shows the situation.

That long wire connecting the processor with the rotary position sensor is the culprit—especially in view of the fact that the printer's cabinet is made of structural foam, quite transparent to electromagnetic radiation. At the end of this long cable, actually three wires twisted together, is an LED and a phototransistor.

As all you radio fans out there know, we have just identified the essential elements of a receiver: an antenna, a detector, and, in the sense circuitry on the board, an amplifier. It's fairly safe to say that a sufficient dose of RF will swamp, or at least confuse, any optical sensing process that happens to be under way.

The solution is simple: shield the cable and bypass the remaining high frequencies to ground with capacitors. These measures effectively place the optical sensor and its wiring into the same ground system as the processor, eliminating the need to replace the structural foam cabinet with a metal one or line it (kluge!) with aluminum foil.

Pleasant, eh? These are the kinds of things that are much more handily done at the design stage, when there is no irate plant manager hovering behind you with his chin on your shoulder, counting the centimeters on the scope graticule as you helplessly probe about in search of a clue.

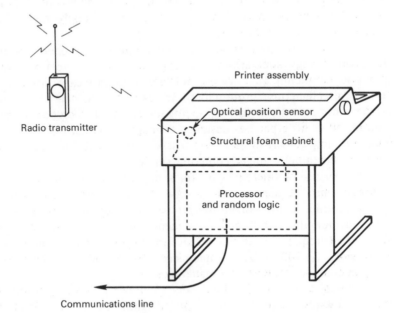

Figure 6-1   The matrix printer contains an effective radio receiver: a rectifying device on the end of a long cable. Nearby walkie-talkie transmission confused the print logic until the cable was locally shielded.

6.1.2 Power-Line Noise

Noise from the environment is insidious stuff, and it is not bashful about trying every door and window in your system looking for an entry. Since so much of it is somehow involved with ac power distribution, that much-neglected three-wire interface with the outside world is often a superhighway of noise propagation (Figure 6-2).

The basic approach here is a filter of appropriate dimensions, since there is little point in shielding a low-impedance noise conduit. A variety of line filters is on the market, rated according to current-handling capacity and filtering characteristics. A schematic of a typical one is shown in Figure 6-3. Devices of this sort are available from R-Tron and Corcom, among others.

Line filters handily absorb high-frequency noise traveling along the ac line, but it is frequently necessary to deal with glitches and transients as well. GE markets a line of metal oxide varistors (MOVs), two-terminal devices that can be connected across the line as shown in Figure 6-4, absorbing high-voltage transients.

Line surges can be caused by inductive load switching, lightning hits nearby, and even physical shock to power lines. These are commonly in the 400-volt range, but occasionally reach 5000 volts or more. Even though they are typically only a few microseconds in duration, they can do respectable damage (or at least confuse a processor). This has led to a reasonably healthy market in surge protectors, modules designed to be inserted between the line and the load which respond in picoseconds to any voltage rise greater than 10% above peak nominal, clamping it. One manufacturer (RKS) points out that their units are rated at 600,000 watts of dissipation for 100 μs. They can save the life of a system, and are definitely worth considering.

Another venerable option, justifiable in many noisy environments, is the ferroresonant constant-voltage transformer, made famous by Sola Electric. These devices typically reduce input voltage fluctuations of $\pm 15\%$ to $\pm 3\%$ at their out-

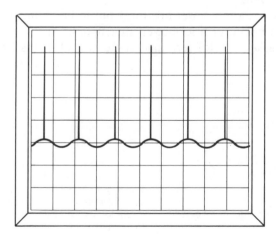

Figure 6-2 Possible appearance of the ac line, showing 60-hertz spikes from a nearby "sloppy" machine. Spikes of this magnitude can sometimes pass through a power supply and pollute the dc, or can reradiate inside the box and be picked up by cables.

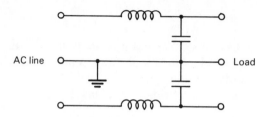

Figure 6-3 A simple but effective means of eliminating high-frequency noise riding on the power line is to install a line filter just inside the equipment enclosure. Metal-cased units of this general design are available from a number of vendors.

puts, and provide noise rejection of 120 dB (common mode) and 60 dB (transverse mode). They can be had in power ratings of 120 to 7500 volt-amperes.

While we are on the subject of the power line, we might as well mention a problem that is not usually called "noise" but is just as annoying: power interruptions.

It usually doesn't take much of a power glitch to send a microcomputer system reeling. There are three things you can do about them: prevent them from reaching the system, provide an orderly shutdown when they do, or latch the power so that they cannot come in rapid succession—a particularly damaging effect common during storms.

Preventing power interruptions from reaching the system is accomplished with a UPS—an *uninterruptible power supply.* These can be built with a dc supply, an automobile battery, and an inverter, or can be purchased in much sexier form from a number of vendors. Battery banks should be selected to provide continuous power through the longest anticipated failure, or at least long enough for personnel to take whatever steps are necessary to "back up" the doomed data.

Providing an orderly shutdown is a good idea even if you are not particularly worried about power failures, especially if there is some medium involved such as a floppy disk. It is possible to detect the imminent removal of power, generate an interrupt, and have the processor take emergency action to close a file before it dies (or at least prevent random disk writes). This requires some backup power, but may well be worth the trouble.

The third approach, latching the power off, turns out to be an inexpensive and useful solution to the problem of brief power interruptions: just turn them into long ones. There is at least one CRT terminal on the market that is virtually

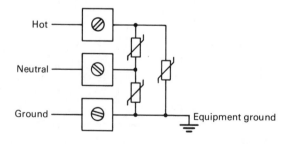

Figure 6-4 Metal-oxide varistors installed across the power line are good insurance against equipment damage from high-voltage transients.

guaranteed to smoke its horizontal driver transistors if power is removed and then immediately reapplied, and a popular matrix printer has a strong propensity for similar inexcusable behavior. The circuit in Figure 6-5 can solve that problem, latching power off and turning on an indicator light whenever an interruption of any length occurs. Power is reapplied by pushing the button.

Clearly, there's more to dealing with the power line than just hanging on an open-frame dc supply and assuming that the filter electrolytics will pick up anything that the regulator misses.

6.1.3 Crosstalk

The first two classes of noise problems that we have discussed specifically imply the introduction of electromagnetic or electrostatic disturbances from the outside world. It is satisfying to point our fingers "out there" and state with assurance that the brutal environment is the source of our problems.

Well, our other five kinds of noise are, um, generated right inside our machines—by our own designs. That's right.

For the first of these, *crosstalk*, we have to thank that peculiar characteristic of electrical behavior that causes two conductors separated by an insulator to form a capacitor. Who ever said that capacitors always have to be purchased components with two leads and a part number? You can accumulate a fair number of picofarads just by placing two wires side by side.

Figure 6-6 gives some suggestion of the areas in which this can cause problems. Here, we assume that a length of ribbon cable has been used to interconnect a processor with a peripheral—say, a keyboard interface. At some point, someone strikes a key, and a pulse arrives on the processor's interrupt line calling for a

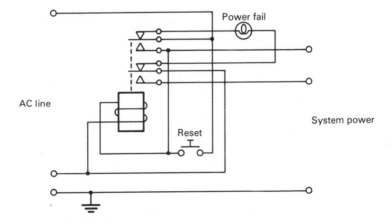

Figure 6-5 This line-interruption circuit responds to a momentary loss of power by unlatching and requiring manual reset for the reapplication of ac to the system. This protects the connected load against the erratic cycling of the line voltage often associated with violent storms.

software response. The interrupt routine is executed, the data are grabbed from the data lines, and an ACKNOWLEDGE pulse is issued to the interface. But wait: someone wired the unit such that the ACKNOWLEDGE line is right next to the DATA AVAILABLE line in the cable, as suggested by all those virtual capacitors in the drawing.

Look what happens: the capacitive coupling between the lines causes the sharp edges of the pulse, with lots of very high frequency components (well into the 100 MHz range), to appear on the adjacent wires, one of which happens to be the interrupt line. *Voilà*—the processor gets an interrupt which tells it to go service the keyboard. It grabs the data, issues the ACKNOWLEDGE pulse, and gets a keyboard interrupt. This goes on indefinitely: we have just constructed an oscillator.

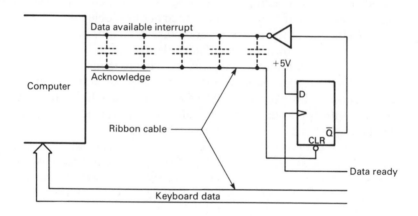

a.

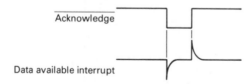

b.

Figure 6-6 In (a), the virtual capacitance of a ribbon cable carrying edge-sensitive handshake lines is shown, effectively constituting high-frequency coupling between the interrupt and the acknowledge signals. In (b), the result can be seen: issuance of an acknowledge pulse by the computer causes a spurious interrupt which looks like a keystroke. In operation, striking a single key would lock up the machine in endless digital oscillation.

Such effects occur in many places, but cables are by far the most famous (ribbon in particular). One of the author's first adventures in the world of system design was a 1972 home-brew 16-bit minicomputer, with bells and whistles galore. He planned on an 8-MHz clock.

It finally started working, albeit flakily, down at about 500 kHz. There was one huge piece of unclad perfboard called the CPU, on which there were nearly 100 ICs (including 74181 ALUs). They were all wirewrapped together and, since the designer wanted it to be as pretty as possible, all the No. 30 Kynar wire was bundled tightly in giant raceways bisecting the board horizontally and vertically. At some places, the cable carried hundreds of wires and was over an inch thick. And, of course, there was no ground plane on the board—and no decoupling. Somehow, anything beyond static operation was unreliable until the CPU was re-built from the ground up, and by then the Intel 8008 was on the scene and, well, why mess with all that random logic?

As insidious as crosstalk is, there are a number of design guidelines that can help minimize it. Parallel wire runs like those just mentioned can be kept to a minimum, although it usually isn't too much of a problem with data lines, which typically have at least a few nanoseconds to settle down before some synchroniz-ing signal declares their collective state to be valid. In cabling, it is good practice (well, necessary, actually) to make every other line ground wherever control or asynchronous signals are involved. There is also some ribbon cable on the market which carries its own ground plane—an extension of the system ground which, closely coupled to the signal lines, tends to reduce their effect upon each other.

The basis of this phenomenon, of course, is the behavior of these virtual capacitors that are formed by the proximity of the wires. The effective impedance between the lines decreases as a function of frequency, resulting in very efficient coupling at the high frequencies involved in rapid-rise-time TTL transitions.

But crosstalk is only one of the types of noise associated with the "simple" transmission of signals through wires. Let's look at the others.

6.1.4 Signal-Current Noise

Long associated with power distribution systems, the phenomenon of *IR drop* is the voltage across a conductor resulting from the flow of current through it. This is rarely a problem with logic signal lines, since current levels are only a few mils, but wires that interconnect dc power supplies with masses of circuitry are fair game.

The resistance of a length of wire is a function of its metallic composition, its length, and its cross-sectional area (kl/A, where k is a constant, l is the length, and A is the area). In the simplest case, placing long wires of insufficient gauge between a supply and a load will create such a large *IR* drop that the noise margins of the devices will be substantially reduced. This simple fact is often overlooked: when building a system, the standard approach seems to be just picking some wire that looks like it will handle the current. But the figures can be impressive:

Number 20 wire, for example, can safely carry 7.5 amperes according to the copper-wire tables. But its resistance is 10.35 ohms per 1000 feet. If we want to power a unit that is 3 feet away from its supply, we have a round-trip wire resistance of 0.062 ohm. That sounds fine until Ohm's law is applied: a 5-ampere current through this resistance yields a voltage drop of 0.31 volt, large enough to cause serious degeneration of the logic's noise immunity. This might be seen as general flakiness and sensitivity to heat and stray noise.

That is the straightforward case, and can be largely solved with bigger wire or, as in the S-100 bus, with on-card regulation. But sometimes the same phenomenon can crop up on printed circuit boards, where too many devices end up sharing the same ground line. The thing that makes that a little hard to spot is the fact that steady-state conditions may not represent enough of a current drain to highlight excessive *IR* drop in the power traces. The problem may become acute only when the totem-pole output stages of the devices switch—when charge stored in the base–emitter junction of the ON transistor makes it turn off just a little bit more slowly than the other one turns on ...

6.1.5 Collector-Current Spikes

... causing sharp current spikes which generate noise proportional to the impedance of the power circuitry. Look at Figure 6-7. Here, we see the typical totem-pole output stage of most TTL devices, accompanied by curves representing the effects on the collector current and the supply voltage brought about by logic transitions. The current requirements of the device are increased during changes of state not only because of the conduction overlap (which is worse on the 0-to-1 transition than on the other), but because of external loading and the transients brought about by charging and discharging capacitive loads. Still, the

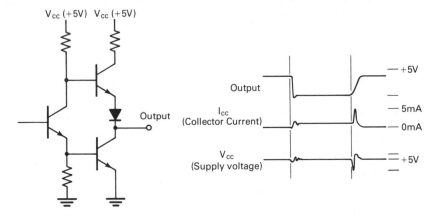

Figure 6-7   The totem-pole output of a typical TTL device is shown here along with a set of curves characterizing its switching behavior. On the 0-to-1 transition, conduction overlap causes severe current spikes which can confuse nearby logic circuitry unless enough capacitance is provided to accommodate the surge.

conduction overlap is the big problem, and the spikes generated can be so severe that they confuse nearby logic devices or even the switching device itself. The current that flows during this brief period is limited only by the total resistance of the transistors and the collector load resistor.

This universal phenomenon is the reason for the characteristic profusion of decoupling capacitors between V_{cc} and ground on all well-designed boards containing TTL circuitry. These should be well distributed across the board, physically proximate to known offenders. The capacitance that they provide must be capable of supplying the requirements of the collector-current spikes long enough to prevent a drop in the supply voltage. Although this is a large value of capacitance, it must present a very low series inductance, so that it can respond rapidly to the sharp leading edges of the transients. The most popular approach is the use of a few tantalum electrolytics in the range 10 to 50 μF, along with a number of ceramic disks between 0.001 and 0.01.

6.1.6 Transmission-Line Effects

Here we have another of the cans of vermiform organisms that seem to appear whenever we depart from our philosophical stance and delve carelessly into detailed and complex explanation of things better pursued through arcane texts strewn with graphs and typeset predominantly in Greek. If you really want to know how and why transmission-line effects are significant in the communication of digital signals, try TI's "Designing with TTL Integrated Circuits" or a book on antenna theory.

Yeah, antenna theory. The basic problem is that the frequencies involved in digital logic are such that the characteristic impedance of the wires has to be considered if you really want to do a bang-up job of noise minimization. TI suggests shooting for 100 ohms, which can be approximated with 93-ohm coax or No. 26-28 wire with thin insulation twisted about 30 turns per foot. Lines of higher impedance introduce crosstalk problems, and those of lower impedance introduce driving problems. Straight wire connections without directly associated ground returns are recommended only up to about 10 inches, and should in any case be close to a ground plane. Decoupling should of course be used at driving and receiving gates (0.1 μF), and such gates should be used for no other purpose, largely because of reflection problems, which can wreak havoc with flip-flops and one-shots.

This last point was observed during the development of the system named BEHEMOTH (the original version). The interrupt request LED driver on the front panel card was connected directly to the Q output of a 7474 via about 3 feet of wire in a big, fat ribbon cable. For some obscure reason, the flip-flop refused to stay set, despite the fact that all the logic responsible for setting up its "D" input and clocking it was fine.

Eventually, it was noted that disconnecting the wire to the LED solved the problem and, in fact, driving the line with a dedicated buffer provided a permanent solution that did not require forgoing the luxury of another blinking light. The problem arose from the reflection of the flip-flop's output transition from the other end of the wire, whereupon it reset the flip-flop via an effect called *collector commutation*. It stayed set only for the length of time required for the round trip to the input of the LED driver and back—not long enough to be of any use. On the scope, it appeared as a faint spike on an otherwise flat trace.

The key factor in all this is the length of the line. If it is so long that the propagation delay is longer than the rise or fall time of the signal (2 to 6 nanoseconds), then the whole contraption behaves like a transmission line driven by a generator with a nonlinear output impedance. Reflection of the voltage steps produces summing effects that render the state of the line indeterminate until well after the round-trip period, slowing system response and, in some cases, presenting erroneous logic levels. Also, as noted above, inappropriate choice of line driving devices can have even more blatant effects.

These progressively more esoteric sources of noise problems have collectively caused many a headache, and only sound circuit design and a good measure of intuition can offer any hope of total freedom from them. Let's back away from the black magic now, and turn to our last category, which is at once the most obvious and the most frustrating.

6.1.7 Intermittents

This class of noise problems consists of those traceable to poor connections, including cold solder joints, bad IC socket pins, worn connector fingers, broken wires in flexible cables, improperly crimped terminators, badly plated-through holes, worn insulation, connector contamination, thermal component stresses, bad grounds, and perhaps even some other things.

Noise problems from these sources can masquerade as anything from environmental interference to outright device failure. A flaky connection in concert with the vibration of a machine to which a system is mounted can look like electrical noise from the machine's rotating components, causing much barking up wrong trees and subsequent head scratching. (Eventually, someone usually notices that kicking the system, an act originally performed with an emotional, not analytical, motivation, creates a noise spike. From there on, it's a lot easier.)

Design for minimization of this class of problems is best accomplished with a continuous awareness of component quality and physical integrity. *El cheapo* PC boards (and even some better ones) have been known to contain little crystallized whiskers of conductive material joining two plated-through holes inside the glass-epoxy substrate. Only the most vigorous gouging with an X-Acto knife can eliminate them. The big problem is finding them in the first place, but fortunately they are not very common.

But other things are, such as connector intermittents. IC sockets, edge connectors, and various types of cable connectors are all fully capable of bringing a system to its knees. As a general rule, every interconnect decreases reliability—

but there's a brutal trade-off here. Connectors vastly simplify service, allowing unrestricted chip and board swapping (an effective if inelegant diagnostic tool). Repeated desolderings can introduce more problems than a PC board full of sockets. Perhaps a good rule of thumb is the use of chip sockets for expensive LSI devices and those which take a worse beating than the rest (such as interface buffers, line drivers, optos, etc.).

The very nature of intermittents and bad connections is such that we can't really say much about it, except perhaps: "look out!" You can make things easier for field service people (one of whom may be you, if things get bad enough) by using high-quality interconnects, strain reliefs on cables, and general good construction practice.

With that, let's move on to the second of our five major aspects of harsh reality.

6.2 HEAT

Thermal effects, like noise, are all-pervasive phenomena that are seldom recognized as the immutable and unforgiving limits that they are. Few, if any, semiconductor devices will function above 200° C, and most CPU chips and their families are spec'd at 0 to +70° C (with industrial versions rated at −40 to +85, and military versions at −55 to +125).

This wouldn't be particularly disturbing except for the fact that a designer's best intentions are seldom realized in the harsh environment of a factory. A box that is adequately cooled by a 4-inch fan juxtaposed with a filter might well overheat when the filter clogs and reduces the flow of cooling air to near zero, especially when the ambient in the factory reaches levels that make people perspire profusely.

This book makes no attempt to replace thermodynamics texts, but a quick discussion of heat flow might be useful in our context. We can follow this with some practical suggestions for controlling thermal problems.

6.2.1 Heat Flow in Equipment

Heat is an energy associated with the random motion of atoms and molecules in matter, and may be produced in a variety of ways. For our purposes, we can think of electron flow through semiconductor devices as the primary source, with the power dissipated being equal to I^2R. One immediate threat posed by this generated heat is the failure of the semiconductor junction, usually as the result of current densities high enough to create plasmas in the silicon. More common, however, is system degradation caused by alteration of component characteristics (especially analog circuits), effects that are doubly frustrating because they are not permanent.

Clearly, it is worth our while to remove heat from the devices and, preferably, to transfer it to the outside world in a way that does not appreciably raise the ambient temperature inside the box.

Heat, as we said, is a form of energy. *Temperature* is an expression of the amount of heat per unit volume for a given material. There are two laws of thermodynamics that underlie the processes of heat transfer in a piece of equipment (or anywhere, for that matter). First, the total heat input must equal the total heat output plus any heat energy that is either stored within the system or somehow converted to other forms (such as motion, light, electromagnetic radiation, etc.). Second, heat always flows from a hotter body to a cooler one. These laws are suggestive of those underlying the behavior of electricity, but the interactions between the three mechanisms of heat transfer render the subject a bit more complex.

The three mechanisms are *radiation*, *conduction*, and *convection*. The first is the transfer of heat between two bodies not in contact, and is seldom considered significant in the cooling of electronic equipment. The second is the flow from one part of a body to a cooler part of the same body, and the third is heat transfer brought about by mixing a hot part of a fluid (water, air) with a cooler part of the same fluid. If this last process is the result of motion brought about artificially, such as with a fan, we have *forced convection*; if it is simply the result of turbulence caused by thermally derived density differences, we have *natural convection*.

Whatever the mechanism, the flow of heat is always from the hotter regions to the cooler ones, and the extent of cooling that results in the former is a function of the temperature difference between the two. This is why ambient temperature specifications are important: the amount of heat that can be removed from a device is limited by the temperature of its environment. It is also a function of the hot body's exposed area and the thermal resistance of the transfer path connecting it with a cooler area.

This, in a nutshell, is what equipment cooling is all about: attempting to maximize the heat flow out of the circuitry through judicious manipulation of the conductivity of the transfer path, the exposed area, and the amount of temperature difference.

The basic method, of course, is *heat sinking*, enlarging the surface area of a thermally active body by bonding it to an extended surface that will increase the rate at which it can dissipate heat to the environment, thus lowering its equilibrium temperature. The analytical design of appropriate heat sinks is nontrivial, resulting in a relatively blind empirical approach in most projects. Usually, the selection process consists of rooting around in a junk box looking for a heat sink that is already drilled for the device package in question. This is common, but hardly to be recommended.

But there are so many things that can go wrong—and we're still just talking about hot devices. One of the most subtle is due to the fact that two "flat" pieces of metal—say, an extruded aluminum heat sink and the bottom of a TO-3 package—when brought into contact and fastened tightly do not touch quite as much as human senses and intuition would suggest. Ignoring differences in flatness, the total contact area is typically about 1% of the total surface area. This is well known in the connector industry, where it markedly affects contact resistance, and when it comes to heat sinking, well—consider: the thermal resistivity of air is 1200° C in./watt, and that of aluminum is 0.19° C in./watt. If 99% of the mating surface is actually comprised of air gaps, the anticipated heat flow might not occur. This is the reason for thermal joint compound, a white silicone-base jelly loaded with zinc oxide (or some newer materials), which, when properly used, can reduce the thermal resistance of an otherwise dry joint by a factor of 3.

In the design of industrial microcomputer systems, problems of elevated ambient temperatures within the enclosure are usually more severe than those of isolated devices. You will probably most often be using commercially made power supplies, hopefully already embodying good thermal design with regard to power transistor cooling. But this, together with the combined effects of perhaps a hundred chips, raises the specter of inadequate heat transfer between the inside of the enclosure and the outside world.

6.2.2 Cooling the Enclosure

Let's establish a paradigm for our cooling problem, and consider various techniques within its context. Riffle, if you will, back through the pages to Chapter 1, and let your eyes rest once again on the mournful sight of Figure 1-2. Within the sweltering confines of that battered Hoffman 20H16CLP gasketed NEMA 4 enclosure struggles a microcomputer system, its power supply, and some random interface logic. Its future is uncertain. The filter is clogged with airborne crud, and there is virtually no airflow through the box.

Use of a fan is the most obvious approach to the cooling problem, but it introduces some problems that are unacceptable in a harsh industrial setting. However well it works on the bench, usually one or both of the following take place in the field: the filters become clogged within a few days or even hours and the box overheats, or the maintenance people get so sick of cleaning filters that they leave them out, filling the enclosure with all sorts of nasty stuff. Over a rather short period of time, said stuff can cause connector degradation, heat concentration around power semis, and possible circuit malfunction due to either conductive or corrosive content in the foreign matter.

So it becomes desirable to isolate the contents of the box from the outside air—which was the point of the gasketed enclosure until the designer suddenly realized that it was going to need active cooling and tore into it with a saber saw.

There are other options. If the thermal design is clever enough and the dissipation requirements are not too severe, it may be possible to couple the primary heat sources tightly to the walls of the enclosure, letting its extended metal surface (perhaps augmented with an external heat sink) transfer heat to the environment. This is very tricky, however. Among other things, it is difficult to predict the amount of respect that factory personnel will have for your environmental requirements in the vicinity of the finned heat sink—someone might find it a perfect place to hang a jacket. And even if that doesn't happen, you might find that the primary airflow pattern around the enclosure is laminar, greatly decreasing the efficiency of the thermal conduction between the heat sink and the fluid medium which you are counting on to convect the heat out of your life. In a laminar, rather than turbulent, flow situation, the fluid medium moves parallel to the surface with no velocity components normal to the flow direction. At the surface, the velocity of the fluid is zero, and the only applicable mechanism of heat transfer is conduction into the stagnant fluid film and then through the successive stream-line layers. The turbulent flow that is preferred is rife with eddy currents and continually brings the cooler ambient air into contact with the bounding surface, producing a steeper temperature gradient and better cooling.

Ideally, of course, the heating of the air along the surface of the enclosure would cause a decrease in density, resulting in vertical airflow as it rises and becomes replaced by cooler air from below. But the uncertainty concerning the unit's thermal environment makes this risky, since variations in ambient temperature, airflow, surface contamination, and obstruction are completely unpredictable.

Well, that was no help. How about a *Peltier effect* device? Now available from a number of manufacturers, thermoelectric devices are fabricated from bismuth telluride in the form of oriented polycrystalline ingots interfaced with ceramic plates. They are essentially solid-state heat pumps, and can be had in a variety of physical configurations, including packaged air conditioners with ratings from 240 to 1100 Btu/hr in the cooling mode. (They will heat, too, with their inefficiency adding to performance in that mode, resulting in a corresponding range 600 to 3200 Btu/hr.)

The advantage of these units is that you can actively transfer heat to the outside world ... but of course they need some dc, and that, in all likelihood, will increase the requirements on the system's supply. TANSTAAFL.

Somewhat less elegant, but probably cheaper, is a liquid cooling system (as absurd as that sounds in the context of a single-board computer and some support logic). Shown in its basic form in Figure 6-8, this scheme transfers the heat of the electronics to a fluid such as water by circulating internal air through a heat exchanger. If the airflow path is cleverly contrived, then stagnant, high-thermal-resistance pockets can be eliminated and a large percentage of the generated heat can be dumped into the water. A pump carries that to another exchanger in the outside world, where it is transferred to the environment.

This scheme is identical in principle to the solid-state air conditioner, but is easier to implement on a budget. The heat exchangers can be automotive heater cores, and the remaining components can be of similar sophistication.

If you have always thought of logic as being fairly low-key in terms of its power requirements, remember that most applications permit open-air circuitry. In an industrial setting, this usually is not the case. A group of PC boards that would run quite happily without cooling if just resting on your desk run into real problems when sealed within a closed box. Because of thermodynamic realities, the temperature will rise and rise until the differential between it and the outside world is sufficient to cause an amount of heat flow through the various resistances

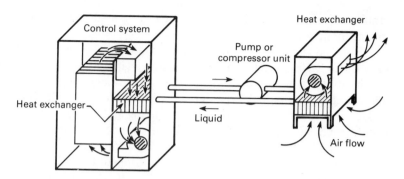

Figure 6-8   A liquid cooling system may be justified in some industrial applications where the environmental contamination is too great to allow airflow through the system enclosure.

that results in thermal equilibrium. It will finally stabilize at this point, which is well above ambient ...

... but by then it may be dead.

6.3 CONTAMINATION

In the preceding section we saw that fear of contamination is a primary motivation behind the sealing of the system enclosure, an act that creates severe thermal problems. It's a healthy fear.

Shortly after the IWT was placed into service, an anguished phone call from the customer brought a field engineer running: the machine had abruptly died and refused to respond in any consistent way to the ministrations of their technicians. It didn't take too much analysis to find the problem. The clamps of the enclosure's door had not been tightened, and numerous tiny metal chips—tailings from the process—had unaccountably worked their way into the box, whereupon they nestled snugly among the pins of the wirewrapped interface card, the CPU's edge connector, and the discrete components of the open-frame dc supply. The customer was fortunate: after careful cleaning with compressed air, the box resumed normal operation.

One of the IDAC/15 terminals fared less well: metal chips invaded and found one of the Self-Scan Displays. Recall that there is +250 volts floating around on those. All it took was one little spark.

But metal chips and other conductive contaminants are not the only problem. The perspiration that rolled off the glistening skin of an unfortunate technician charged with the task of repairing a production batching system located in a Venezuelan jungle caused problems many weeks after his visit. The combined effects of the mildly corrosive salty fluid that dripped on the PC boards, the nearly 100% relative humidity, and the omnipresent heat caused enough electrolysis to significantly change the characteristics of some analog circuitry, and the system became inoperative.

Here's a classic. An environmental control and energy management system based on an S-100 processor was installed in a large university in Kentucky. A few months after the nightmare of installation and initial debugging was complete, the unit started becoming flakier, and flakier, and flakier—responding to a kick with a stack crash and often needing vigorous shaking and a curse to run properly. Investigation revealed that the PC board edge connectors, mounted on a backplane parallel to and a few inches above the floor, were nearly full of a sticky glop that had somehow collected on the boards and dripped down. It was a whitish, gooey substance that was invading the not-exactly-airtight connections and producing approximately 100 intermittents per board.

It turned out to be a mixture of detergent and floor wax. That's right: every time the janitor mopped, waxed, and buffed the linoleum floor, the fan drew tiny airborne droplets into the computer and sprayed them over the card cage. It was not long before there were some respectable accumulations of the material, which remained in a liquid state while on the warm boards but became a paste as it puddled in the connectors. It took a couple of grumbling hours of scrubbing with warm water and a brush to clean the system and restore operation—and, oh yes, not only was the airflow redirected but recommendations were made to the custodial staff concerning the care of the computer room.

The industry abounds with stories, many no doubt even stranger than this, which indicate the range of effects attributable to environmental contaminants. Even such common substances as Coke and coffee can wreak havoc. All we can really say from a design standpoint is that there's a lot of dirt out there, and even tiny openings in the enclosure can admit significant amounts. This severely affects cooling, as we have discussed, but also limits front panel component choices to those which allow the maintenance of something approaching a hermetic seal. In the IDAC terminals, this was a particularly annoying problem since one of the necessary panel devices was a badge reader.

6.4 VIBRATION

Physical vibration isn't as much of a problem as it used to be, when the widespread use of tubes (remember tubes?) caused the effect called microphonics to be of concern. But it still deserves some attention, especially if installation requirements call for the mounting of your system on the frame of a machine.

The biggest problem here is the behavior of connectors when they are subjected to cyclic inertial forces. Something that feels like a mild hum to the touch may actually be, upon stroboscopic examination, a series of sharp mechanical transients along an axis parallel to the circuit boards. Without shock absorption and physical restraint, they can work right out of their connectors or develop intermittencies that are nearly impossible to locate.

Another vibration-induced problem manifests itself in wires, especially those that connect one physical body to another which is undergoing a different mode or amplitude of vibration. Even though it is probably invisible, the result is constant flexion, which can quickly fatigue the conductor material and lead to maddening glitches and eventual hard failure. This effect can be minimized with the use of stranded wire, careful strain relief, and stress distribution.

Perhaps we should mention here that a major source of "vibration" is the mechanical shock associated with shipping. Heavy components, such as large electrolytics, can be ripped from their mountings, and delicate components, such as crystals, can be subtly damaged. There is an inexpensive indicator on the market, called Shockwatch, which can be attached to a system prior to shipping. If it experiences an acceleration above a certain limit, it changes color and strengthens your case against the carrier.

6.5 HUMAN PERVERSITY (AND INCOMPETENCE)

This is a big one, and beautifully exemplifies the situation suggested by the title of this chapter.

There is a temptation, when developing a system, to treat it lovingly and design it in a way that appeals to your finest sensibilities. Creation of a well-integrated hardware and software package calls for a balanced synthesis of art and engineering, and it is impossible to remain emotionally unaffected. There is something of yourself in that system flickering to life on the bench.

This is all wonderful, of course, and greatly increases the probability of a high-quality product. But there's a catch. The people who are going to use it are probably not you, and very likely will fail to recognize the brilliance of your command parsing scheme or the elegance of that vector generation hardware. They may not recognize the subtle power of the central data structure, and may even resent the system as another attempt by management to keep an eye on them—or gradually phase them out. They may want to damage it.

Also, the technicians who take care of field service may not have the respect for the hardware that you have. The folks who ship it or install it may see the machine as just another steel box. A scratch on your carefully finished front panel means nothing to them, and in time, it might even be seen as a good place for graffiti.

In 1979, a foreman at a certain foundry in America's heartland gave the order to clean up the shop. One employee was responsible for the area around a large assembly machine, which happened to include a microprocessor-based controller.

"Sure thing, boss," he said, and set to work. When nobody was looking, he took a pair of boltcutters and, ahem, "unlocked" the system's enclosure. He then grabbed a firehose, which was coiled on a nearby wall, and used it to "clean" the inside of the box.

It happened that the controller was connected by both parallel and serial communication lines to a PDP-11/40, and contained high-voltage switching circuitry for motor control. Sure enough, the computer was damaged severely, a motor was burned out, and the controller was effectively destroyed. Net repair cost: $160,000.

"Hey, man," he said when questioned, "the dude wanted a clean machine." The union threatened a walkout if the employee was punished in any way.

It's sometimes a little difficult to deal with this kind of thinking. After all, somebody stayed up nights gazing at the wall to formulate the control algorithm in that box, then spent weeks designing and building it. Debugging probably ruined a few evenings, but it was a personal creation—a monument to microprocessor technology and human ingenuity. But in one vicious splash of water and sparks, it was gone, reduced to charred scrap. The incident lived on as a knee-slapper over beer in the neighborhood tavern ... and as a bitter memory for one disillusioned engineer.

Of course, willful vandalism is not the only human problem we face—we can always invoke the JOKE program if things get bad. Physical abuse of the machine can also be attributed to the simple difference in finesse between the system designer and some of the users out there. Coke spill—so what? Armload of steel? Just whack the switch with the end of an 80-pound bar. Need a place to set some boxes? There's a horizontal surface on the control system.

This is at least as severe a problem as noise, heat, vibration, and contamination all combined. Understand, we're not even talking about malicious damage here—just the rough treatment received from people who don't think the way you do about the hardware. You respect it.

As painful as it is, your testing procedure must somehow accommodate this.

Before leaving this entertaining but somewhat sobering bit of rumination, we should probably talk about incompetence. It is all over the place.

One system that comes to mind was installed and brought to life, and the customer was given a set of written procedures which accommodated various emergencies—power failures and the like. One morning, after a storm, the system was in its "huh?" mode, and it was necessary to reboot the program from disk.

The operator dragged out the manual, found the cold start procedure, and proceeded to perform it. Essentially, the task was to push the RESET button, push the START button, then wait for the symbol ">" to appear on the screen, indicating that the operating system was loaded. At that point, the line word "RUN" was to be typed, and the system would do the rest.

About two hours later, he called the system designer, desperation in his voice. "I've been waiting all morning," he moaned, "and the thing still hasn't said 'seven.'"

"Seven?"

"Yeah, it says in the manual that you hit RESET and START, then wait for a seven before telling it to RUN."

Good grief.

This reflects back to our opening comments: the folks who end up using your system are not likely to understand it, and any "human interface" schemes must take this into account. A command structure that maps perfectly on the 8086 instruction set probably doesn't make much intuitive sense to someone who thinks a computer is a giant thing with spinning reels of tape behind a local politician in a campaign commercial. The system must be literally "idiot-proof," able to accommodate absurd entry sequences. To most operators, the sequence of steps necessary to make the system do its job is totally arbitrary and must be learned by rote.

Somewhere in association with this, we should probably mention mechanical factors. A famous one is the use of identical connectors to carry sense, communication, and solenoid drive lines to and from a machine control system. The connectors were physically interchangeable, but not logically: all it took was one error on the part of the plant electrician and numerous blown chips needed replacement. You may have no trouble associating labeled plugs with correspondingly labeled sockets, but you are not the only one who is ever going to touch the machine. Key 'em, at least—better still, use different styles and numbers of pins for each interface connector.

Grim, eh? Human factors are perhaps the most frustrating part of this business, for hominids are capable of supremely illogical and irrational behavior when confronted with a computer—or even a box with a blinking light or two.

6.6 NEWTONIAN PHYSICS, AND OTHER REALITIES

In any purported compendium of differences between reality and that ideal state of affairs often dubbed the "textbook world," one must include certain aspects of the environment which are so universal that they should be considered absolute.

Perhaps the most visible of these is inertia, expressed by the observation that things don't start or stop instantaneously. This affects the design of both hardware and software, for our relatively slow human perceptual apparatus tempts us to believe, at least in a visceral sense, that devices such as solenoids, relays, and motors respond very quickly to the electromotive force of actuation. But from the standpoint of a software testing loop, there are actually long acceleration curves.

This can be seen in a practical sense if you consider the design of a linear positioning apparatus such as that found in a multiplatter disk drive. The function of the hardware is simply the movement of the head assembly from one concentric track to another, as quickly as possible. But if it is told to go "flat out," then told to stop the moment it reaches its target, overshoot is guaranteed, which then has to be accommodated with alternating forward and reverse commands to correct the progressively smaller errors until, at last, the heads are positioned properly. The correct approach is the issuance of deceleration commands sometime before the target, in a fashion perfectly analogous to the dynamic behavior of an automobile and its driver when a light turns red.

Other manifestations of this difference between real and ideal are more of an electrical nature. Circuits of any sort take a finite time to change state, for reasons of circuit capacitance, device propagation delays, loading, and so on. Strange things can happen during the transition periods.

A classic in this category is the circuit that converts a single clock signal to a series of successive states, useful for timing synchronous systems (Figure 6-9). The typical approach to this problem is the use of a counter permuted to an appropriate range, followed by a decoder that produces a 1-of-n output for a $\log_2 n$ bit input. To make this work properly, it is necessary to gate the decoder with one phase of the clock. Why?

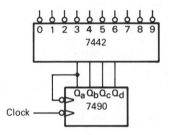

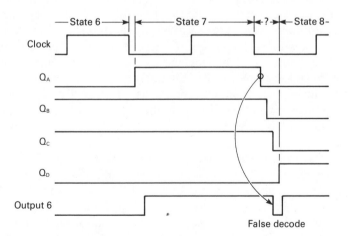

Figure 6-9 The various unequal propagation delays found in ripple counters and asynchronous decoders can have disastrous effects upon a logic design if attention is not paid to timing.

Consider the old standbys—the 7490 ripple decade counter and the 7442 BCD-to-decimal decoder. The counter consists of four flip-flops which propagate an input pulse in the form of state changes; the decoder is just an array of gates that accomplishes the function noted above. But when the counter counts, the collective state changes are in no way simultaneous. For a change from 7 (0111) to 8 (1000), for example, the low-order bit changes first, yielding a 6, then the next bit, yielding a 4, then the next, yielding a 0, and finally the last, yielding an 8. Each transitional state lasts for about 10 to 15 ns.

This is bad enough, but there are various unequal propagation delays through the 7442 (which are responsible for the additional delay of the false "6" state shown in Figure 6-9). The combination used as a state generator produces the desired state transitions with all sorts of trash and glitches on the lines at times other than those expected.

We could go into some of the other textbook classics, such as circuit calculations yielding resistor values of 250 ohms (ever try to buy a 250-ohm resistor, other than a power type?). But things like this, and like the scarcity of lumped constants, become fairly obvious after one or two bouts with real hardware. Let's instead touch briefly on safety and then get into some software.

6.7 SAFETY

It is not terribly likely that the systems with which you involve yourself will present shock hazards reminiscent of good old 110-volt relay logic. From that standpoint, microcomputers are little threat to human organisms.

But the nature of the business about which we speak involves an unknowable variety of interface possibilities, many of which are quite capable of shocking, crushing, pinching, burning, striking, and otherwise insulting human beings who may be in the wrong place at the wrong time. Hydraulic cylinders, ovens, and machine tools all offer their own interesting possibilities when driven "open loop" by a processor gone berserk.

We have discussed the effects of noise and other influences enough to know that there is nothing sacred about the orderly execution of a program. What happens if a power glitch or noise spike sends the code into a loop which either takes something out of range or operates it at an inappropriate time?

Good question. But it is at the system designer that a lawsuit will be first directed when someone gets hurt, and if you are controlling a dangerous process, it is well to give the problem some serious thought.

OSHA requirements state that application or removal of a controller's ac power shall not, in itself, cause any movement of connected machinery. That's a good start. It is good to add interlocks that stop machine operation when a human being strays within striking distance (see the opto schemes in Chapter 5), and a big red STOP button should be within easy reach of an operator wherever he may be. In addition, some systems might justify the use of a watchdog circuit, which essentially keeps an eye on the processor's health by requiring a periodic "I'm OK" signal to maintain machine operation. A disruptive glitch that sends the CPU off chasing geese is then unlikely to leave life-threatening conditions alive in the plant.

Of course, there is always something that can go wrong, and determination of all possibilities and development of techniques to accommodate them is the responsibility of the system designer. That's a tall order, but that is how a jury would see it. It sometimes helps to keep that in mind when the complexity of the fail-safes gets out of hand. Just remember one of the corollaries to Murphy's Law: A transistor protected by a fast-acting fuse will protect the fuse by blowing first.

With that, we come to the end of our brief insight into reality. Let us now delve into something a little more magical.

Part II

7

To Teach a Machine

At some point we must consider how these exquisitely contrived hardware edifices called microcomputers can be coaxed into operation. Throughout Part I we made numerous fleeting references to "software," never really pausing long enough to consider it at more than a subterranean level. We had enough to worry about.

But software, really, is the point of all this. The basic motivation that spawned the industry was the desire for universal, general-purpose hardware that would effectively offload a larger portion of the system design effort onto the chip vendors. With standard circuitry applicable to a wide range of projects, designers could spend more time defining the task and less doing battle with device timing and other classical manifestations of hardware design.

That was quite reasonable when the complexity of a microprocessor application rarely exceeded 1000 lines of code. In the early 1970s, people without access to the crude development systems and rudimentary assemblers of the day didn't feel too left out—they just hand-coded their programs and manually entered them in binary or hex.

But memory became cheaper and cheaper. Soon, the availability of RAM and ROM in 16K and 64K chunks made feasible some substantially more complex microprocessor applications, and program size grew and grew. Fat listings, once only seen around big DP shops, began cluttering the desks of harried microprocessor software designers. Programs of over 100,000 lines appeared, and the complexity and sophistication of development tools grew in erratic spurts to help accommodate the ever-increasing flexibility of the hardware. But the burden remained on the programmer, for the tools couldn't keep up. What was needed was another evolutionary development like the microprocessor which would allow engineering costs to be shared, in part, with the hardware vendors.

But what actually appeared was a confused jumble of nonstandardized high-level languages which further alienated all but the *cognoscenti*. Communication between programmmers and management, programmers and customers, and even programmers and programmers became more and more erratic as the myth of machine-independent portability was shattered on the rocky shores of conflicting philosophies. Documentation ranged from inconsistent to nonexistent, and the general inefficiency of software development continued to increase as a function of project complexity—for programming was still a human task and could not take advantage of production techniques that made the hardware revolution possible.

"What's to be done?" was a standard refrain. There were no established testing techniques, nobody was really sure how software should be specified, and on top of these and other problems there was an ominous shortage of programmers. The lack of good software management techniques ensured the proliferation of "hotshots" who insisted on doing things their own way—keeping the field notoriously unmanageable and attracting more of the same. The software design business became exclusive, frightening newcomers with its implied emphasis on wizardry and further alienating them with its profusion of obscure concepts and arcane notational schemes.

Nobody came out ahead during these turbulent years, with the possible exception of the chip vendors, who strode profitably through various recessions with nary a glitch in their exponential growth curves. Programmers grayed prematurely or turned to a life of reckless hedonism, managers found themselves helplessly watching production schedules turn yellow and crumble before their very eyes, and customers—those trusting souls whose money kept the whole industry alive—found themselves with giant aggregates of undocumented in-line code, incomprehensible even to the original programmer. "You want a revision? Uh, gee, I'm kinda busy these days. ...."

What's it all about, anyway? What is this thing called software that maketh grown men to tremble and sets, lo, their pants afloat? What, indeed, *is* to be done?

7.1 SYMBOLIC THOUGHT

Perhaps the best way to approach questions like that is to begin at the top, with the human thought processes, which, after suitable distillation, become programs.

What is intelligence?

That is a question that has inspired works of Great Thought for centuries, furrowing the learned brows of philosophers, psychologists, theologians, and neurophysicists as they have sought, each in their own fashion, to find The Answer. The question has remained well outside the domain of technology until recently, but we'll get to that later.

One of the exalted hallmarks of intelligence, at least as we human being see it, is the ability to manipulate symbols—to represent the universe internally in symbolic form and communicate with each other through linguistic representations of the resulting internal model. We have already espoused the "internal model of the world" concept herein, in the context of a control system: add a few dimensions and some unquantizable phenomena such as consciousness, and you have, more or less, a mind.

Our symbolic processes are potentiated by the fundamental operating characteristic of the brain—association. Unlike computers, which represent data internally via some structure of absolute and indirect pointers, brains represent information in an associative fashion (note the use of the words "data" and "information"). A new item—be it sensory input, an abstract insight, or an acquired fact—is somehow integrated with related information in an exquisite associative web that allows us access to it via any of a number of pathways. We touched on this in the Introduction, with a brief aria on the general subject of learning.

The mechanism by which all this magic takes place has been widely speculated upon, but there seems to be a gap between investigations at the "device" level (neuron and synapse behavior) and those at the "system" level (cognitive modeling, etc.). One of the more colorful ideas is the hologram metaphor, which is based on the relatively well-founded assumption that memories are stored as a distributed pattern across a structure, rather than as the contents of isolated cells as in a typical computer system. This is supported by observations concerning our levels of perception: you remember a scene not as an array of "pixels," but as a fully dimensional echo of the original perception whose detail in various areas is a function of the depth with which they are associated with other memories. (Walking into someone's computer room and glancing about will leave you, not with a precise image of the room in balanced detail, but with a general "gestalt," highlighted in spots by more complex memories of items that you recognized or that particularly sparked your interest.) It has also been observed experimentally that widely variant memories are stored in the same areas of the brain and, in fact, portions of said areas can be surgically excised without grossly affecting the memories—just their texture or resolution. You cannot operate to remove an isolated bad memory, even with medical finesse that would render today's medieval by comparison.

This is all very interesting, because holograms (from the Greek *holos*, meaning "whole") exhibit similar behavior. It is well known that a small piece of a hologram can reproduce the entire original scene, but with coarser resolution. It is less well known, but equally wonderful, that one can create holograms with "reference beams" that are not simply floods of laser light: they can be complex images as well. The same holographic plate can record numerous images, each stored with a different character of reference beam, and each recoverable with subsequent illumination by its particular one. This, *voilá*, is a somewhat primitive associative memory. Presumably (if we can stretch the metaphor somewhat), one could carefully contrive, or even nest, the reference beams such that each one elicits not a single image, but a set of images of various intensities. Perhaps some of those could then serve as reference beams to other holograms, allowing exploration of intertwining associative pathways.

The key to this is the idea that memory, and thus the basis for human thought, is a phenomenon better considered from the standpoint of frequency-domain wavefronts and interference patterns than from the old idea of storage cells that somehow embody the sum of our experience, which is then supposedly processed by the neural equivalent of a computer.

But what has this got to do with software?

This: when you attempt to teach a machine (which is what programming is all about, although it is still pretty primitive), you are actually mapping a conceptualization of an idea onto a set of system resources via some kind of structured language. Let's take that a little more gently.

Consider Figure 7-1. Here we have a model of the modeling process—a meta-model, if you will, in which we represent the act of doing just what we are doing now when we talk about this model. Starting with a cloudy, abstract, deeply associative representation of a concept (which is, itself, a symbolic structure, although at a supra-verbal level), person X engages in a mapping process, a projection of that relatively ineffable structure onto the somewhat more tangible fabric of language. The concept becomes a succession of verbal symbols, chosen not

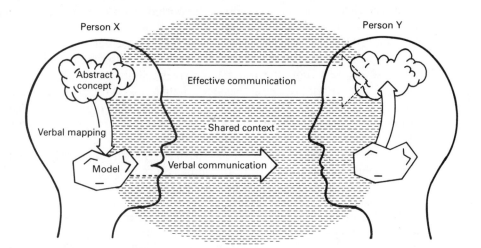

Figure 7-1 Communication between human beings involves not only a sequence of words, but an extensive shared context and a symbolic mapping process which adds coding depth to the linguistic data. The net effect is a substantially higher degree of abstraction in the communication process than would be suggested by the words alone.

only for their representation of the concept but also for their appropriateness to the communication task—which depends very much on person Y.

Presumably, the fact that X and Y are talking at this level suggests that they have constructed a mutually acceptable context which provides a framework for their ongoing communication. Utterances that pass between them link associatively with this framework to reduce the need for explicit description of every idea, thus vastly increasing the bandwidth of their information channel by taking advantage of association to make their words more efficient. Low-level structures are created, agreed upon, and then used in the communication of subsequent, more complex structures. Eventually, a single word becomes capable of expressing a highly abstract image which might otherwise require painstaking representation. The coding depth increases geometrically. (Could infinite coding depth be distinguished from random noise? Hmmmm.)

In the drawing, then, we see that the distillation of the concept that is produced by person X is linked with this shared context. X invokes linguistic mechanisms to convey this symbolic model, and in the process, actually within the model itself, suggests to Y the type of mapping that yielded it. Y then has enough information to project the incoming structure onto his own set of high-level constructs, regenerating, in the context of his experience, the concept originally spawned by X.

It is an elegant and graceful event: things unspeakable have been funneled through the relatively narrow channel of speech. The participants have encoded their thoughts in a spatiotemporal fashion that requires only the establishment of a context and the exchange of tangible models.

Have we not done just that? Do you not "feel" the idea with a depth that would be impossible to express to someone else without the equivalent preparation of a contextual basis and subsequent transmittal of highly charged verbal mappings?

This highlights a crucial truth: language has to be considered as only one part of a much more elaborate communication process, one in which the knowledge bases and instantaneous states of mind of the participants are at least as much responsible for the interpretation of verbal utterances as are the lexical elements from which those utterances are formed. Throughout the progression of a conversation (or even the one-way evolution of the written word), the internal states of the participants continually change to represent the modified reality that is the result of the communication: concepts are assembled and then associatively integrated with their ever-evolving internal models of the world.

But really, what has all this got to do with software? Consider Figure 7-2. Here, instead of attempting to transmit an idea from one person to another, we are attempting to implement it in a computer. It starts out similarly: X performs a verbal, or quasi-verbal (depending on the need for written high-level specs) mapping which represents in a more symbolic form the original concept, itself perhaps something as abstract as a totally unstructured vision of a particularly alluring microcomputer application.

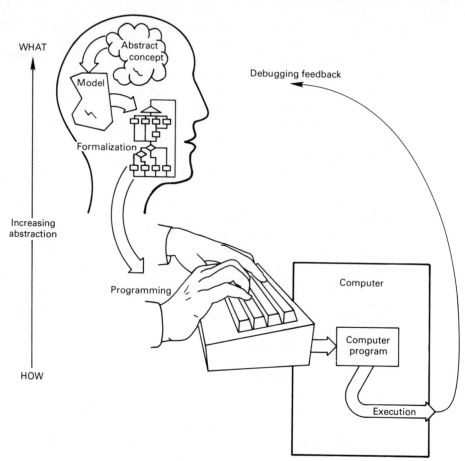

Figure 7-2 Replacing "person Y" of Figure 7-1 with a computer places a much greater burden on X, for he must carry the modeling process much further— producing an explicit, unambiguous description of the concept.

But now, things start to look very different. Typical computers (ignoring the science of artificial intelligence for now) find rambling verbal descriptions a tad useless. First, any "shared context" has to be at a very inflexible, rigorous level, and it must be strictly on the computer's terms. Second, the resources presented by the computer for the communication of ideas are so semantically rigid that they demand strict formalization of the process before they can begin to assimilate it meaningfully. The programmer, therefore, is required to go through some intermediate steps before he expresses the idea to the system: starting with the verbalization of the idea, he progressively strips it of ambiguities until it is an explicit, step-by-step description of precisely HOW the computer is to accomplish the task. Once this program—for that's what it has become—is written down, it can be handed to the computer in some appropriate fashion. (We'll talk about that in a minute.)

The computer, now equipped with an absolute procedure, can begin performing the sequence of instructions in order to carry out the programmer's wishes. In the diagram, we indicate a feedback loop which makes possible the inevitable debugging: "Ah, let's see, it crashes. I wonder what could be wrong?"

Even without considering clever tools like high-level languages, we can see that there is a fundamental difference between this communication process and the interhuman one we described previously. The bandwidth of the information channel is extremely narrow, there is little or no shared context, and X cannot reasonably assume that the computer is capable of deducing his concept by internally projecting the communicated model upward into some abstract conceptual space. On the contrary, the computer just says, "duh," and begins blindly following simple instructions.

When the complexity of computer applications grows from trivial control loops to sophisticated information-management systems, this rather agonizing process begins to place an unreasonable burden upon X. He is required, among other things, to maintain in his head a complete image of the program's intended operation, so deviations can be made apparent via the feedback loop, analyzed, and acted upon. He is also faced with some rather formidable hurdles in problem expression, for as the concept becomes more complex, the interactions between various parts of the program become so convoluted and intertwined that the human mind is, alas, boggled.

A reasonable approach to the problem is the creation of some tools that effectively insulate poor X from some of the computer's low-level realities. If he can express the problem to the machine at some level that is closer to his original conception, the chance for error can be reduced somewhat.

Part of the reasoning behind this is that, as we have suggested, human beings are associative creatures whose memories are not on address buses and whose processors do not suffer from the von Neumann bottleneck that makes computers so pathetic. We tend to represent complex arrays internally in a high-level fashion that allows us to extract their meaning without having to manipulate bits. A ham radio operator with some years of experience on CW, for example, does not decode each Morse character that he hears; he inhales entire words and phrases as individual elements. A "dahdididah dahdah dah didahdit" is a transmitter, not an X-M-T-R. Similarly, speech is conscious only at a high level: saying the word "speech" is hardly accomplished by deliberately letting fly with an unvoiced sibilant, then doing a labial plosive followed by a hundred or so milliseconds' worth of voicing with a constricted forward tongue hump, at last wrapping the whole thing up with a carefully contrived unvoiced tongue-to-gum-ridge stop that immediately turns into a fricative on the hard palate. Phew! Sure is hard to talk—let's just look at each other for a while.

What this all boils down to is that people are much more comfortable thinking WHAT, not HOW.

But computers are just the opposite. They have no idea WHAT, they just inhale the HOW and go to work. There must be a way to span this vast and awkward
gulf.

7.2 MOVING ALONG THE WHAT–HOW SPECTRUM

As we mentioned at the beginning of this chapter, people were content in the Old
Days to talk to computers on their own machine level. Perhaps the machines were
enough of a novelty back then that nobody was terribly upset about the philosophical implications of this; and if they were, it would not have done any good anyway. There were no assemblers, interpreters, compilers, operating systems, editors, or simulators. Just machines. (Actually, if you delve into prehistory, you
find that really ancient computers were programmed by hardwiring, but that's just
too grim to think about.)

Figure 7-3 shows, more or less arbitrarily, various degrees of synchrony
between computers and human beings. This model is often used in the AI field to
express the relative sophistication of a system, specifically with regard to the level
of expression necessary to implement human wishes in the form of machine
behavior.

Down at the murky HOW end of the spectrum, we see hardware and microcode, which get you so close to the machine you can feel the bits between your
toes. One very small step up from that is *machine language* (in an 8080 or Z-80,
"11000011" means "jump to the address specified by the next two bytes), and
one somewhat larger step up from that is *assembler* (which lets you say "JP" to
get the same point across). An assembler is a program that runs on the machine,
accepting "assembler source code" as input data and producing "object code"
(machine language) as output data. Here, there is still a one-to-one correspondence between a written instruction and the one the machine executes, but an encouraging effect begins to occur, as suggested by the horizontal arrow: you can
"extend" the language just slightly by identifying commonly used structures and

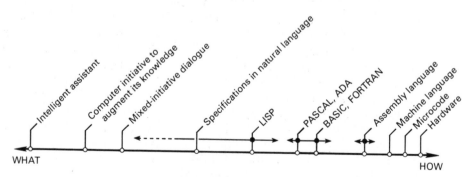

Figure 7-3   The WHAT–HOW spectrum illustrates the range in levels of expression necessary to communicate with a system.

calling them "macros." If for some reason you note that your program frequently likes to rotate the accumulator three times to the right and then increment it, you can inform the assembler that henceforth the new instruction called "GEORGE" should be thus interpreted.

This helps a little. If nothing else, an assembler lets humans express instructions in the form of mnemonics and addresses in the form of named labels, totally eliminating the need to remember numbers. That relieves us of one of the things that we are not very good at.

Sooner or later, a deft assembler programmer notices that there are some larger structures that he uses over and over, and that his life is simplified by defining them, labeling them, and putting them in a library for future use. That way, it is unnecessary to write a floating-point divide routine every time the need arises; he just fetches it from the library and inserts it into the program at the appropriate spot, interfacing with it as necessary.

As this library grows, it becomes possible to write whole programs with nothing but a scattering of these handy structures, a few macros, and some logic to handle the flow of control. From that idea, it's but a short step to a compiler: if you build the entire mess into a package that obviates the need to invoke any assembler-level operations, then suddenly you have a *high-level language* (HLL). That is a fairly significant advance, since now it is not even necessary to care how the machine accomplishes the management of a DO-WHILE or other structure—it just does it somehow. (But you still have to tell it to do the DO-WHILE.)

HLLs come in three flavors. First, there are *compilers*, programs that accept a high-level program in the form of a text file, chew on it a while (fetching appropriate code from a library), then create an output file in a form that is executable by the machine. It is important to note that the compiler is just a program to which your program is just data. When your high-level statement says IF RATE = MAXLIMIT THEN DECREASE, the compiler invokes a parser to determine what kind of instruction it is, what kinds of data are involved, and what library routines need to be put to use in order to produce meaningful object code. If you handed the same text line directly to the computer, it would barf.

Second, there are *interpreters*, which remain resident in the computer as the program actually executes. Instead of translating the high-level code into low-level code and going away, interpreters bite off one line at a time, figure out what it means, perform it, then go on to the next one. Execution is obviously much slower, but they offer the advantage of on-line debugging, for unlike compilers, interpreters let you sit down and just play with the program until you like it.

In Figure 7-3, interpreters and compilers are lumped together as one group. Aside from marked operational differences, they are about the same—the distinction lies primarily in the amount of source text bitten off at any given time for conversion into machine code. They both operate on the general principle of inhalation of essentially undifferentiated strings of ASCII, detection of delimiters (spaces and CRLFs), identification of important labels and reserved words, determination of an appropriate object code segment or subroutine call, and finally, execution (in the interpretive case) or addition of the new statement(s) to an output file.

And then, there is the third flavor—*LISP* (as well as FORTH, and a few other "threaded" languages). Closer to the WHAT end of the spectrum even in its basic form, LISP offers the primary feature of extensibility. Instead of being based on a set of sharply defined structures that are predominantly oriented toward the manipulation of numeric data, LISP makes a valiant attempt to provide the programmer with facilities for dealing with information at a symbolic level. Sound familiar? As we move farther from the HOW end of the spectrum, we find it easier and easier to communicate with the system without having to accept the brutal language restrictions that characterize even the most elegant of conventional interpreters and compilers.

The characteristic of extensibility is interesting, and is the reason for the horizontal arrow that just sort of trails off at some undefined region in the WHAT–HOW spectrum. In essence, LISP is an interpreter—but with a twist. There is no differentiation whatever between functions and data, allowing new functions to be built from old ones. This sounds suspiciously like macros and library routines, but here entire programs are just function definitions. Representation of a concept like PLUS is no different from representation of a concept like 3: there is nothing unique about PLUS except its meaning in certain contexts.

This is all a trifle obscure without a lot more explanation, so let's put it off until Chapter 9. What is important here is an understanding of Figure 7-3 in the light of the shared context we created in the preceding section.

So continuing along the line, we find it becoming progressively more brainlike until we reach the WHAT end. "What would you like me to do, Mr. X?"

"Well, I'd like you to control the downfeed rate on that Blanchard over there by keeping an eye on the power level of the grinding head motor. Try to keep it within limits that maximize the life of the diamond bits while making the best use of machine time."

"OK, boss. Am I appropriately interfaced?"

"Yup. Go to it."

7.3 ONE MORE GRAPHIC SYMBOL

The only problem with the WHAT–HOW spectrum as an all-encompassing symbolization of software "levels" is that it shows no distinction between system and applications programming. So let's try Figure 7-4.

First, one caveat: as we noted in the Introduction, there is rarely such a thing as an absolutely clear distinction. This made things tough enough in Figure 7-3, but it assures us that Figure 7-4 will be the kind of stuff of which arguments are made.

In this sketch we have taken the WHAT–HOW spectrum and expanded it into a two-dimensional space. The new dimension is SYSTEM-APPLICATION, allowing us to spatially observe differences in the purpose of a given piece of software.

At the lower right, we again see machine code and its more palatable equivalent, assembler. We give assembler the benefit of the doubt and show it a little bit removed from the W–H baseline (after all, we could say that the ability to communicate with the machine in mnemonics and macros places us closer to the world of computer applications).

Somewhat closer to WHAT are operating systems—entities that we discuss in Chapter 8. They can be thought of as cushioning blankets of code wrapped around the machine, insulating the programmer from the details of file management, I/O device control, and memory allocation. Instead of thinking of the line printer as a handshaking parallel interface on port 54, one might just refer to it as "LP." A device driver deep in the bowels of the operating system (OS) recognizes the significance of this and deals with the machine-level trivia whenever necessary. In a similar fashion, the OS takes care of files—if the text of this book had to be explicitly placed on disk at user-defined track and sector addresses, the author would probably be using a typewriter instead. But via the OS, he can create, copy, edit, print, format, and kill text files, and a homebrew utility called COUNT even allows a display of the word count for each chapter—enabling

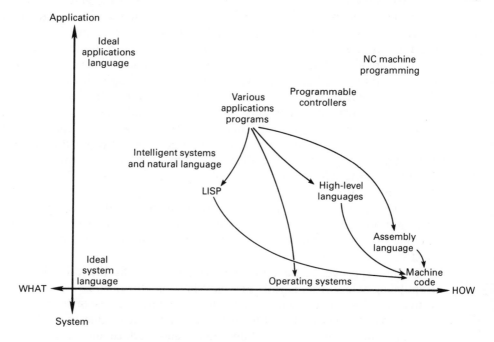

Figure 7-4 The WHAT–HOW spectrum can be expanded into a two-dimensional space which also represents the purpose of a piece of software in the system-application range.

informed, rather than naive, panic at the proximity of the deadline. Ain't technology wonderful?

Just about all useful systems have either an OS or something that looks a lot like one, so it might be worthwhile to imagine adding a third dimension to the figure which represents resource depth or at least something that suggests an undercurrent of interconnectedness. For example, the aforementioned OS via which these chapters make it from brain to paper supports, among other facilities, a word processor. Now the WP is really an application program, so it should have a healthy Y-axis position in the diagram, but it would be lost without the OS, through which it deals with disk files, the terminal, and the printer. It would just sit there in memory and quietly format imaginary text, giving the system the general appearance of catatonia.

Nevertheless, we can see in the upper right the usefulness of this extension to the original model. Numerically controlled machines represent a class of programming that is about as HOW as you can get, but is thoroughly rooted in the application. Programming here reflects the specific requirements of the machine, eliminating generality which would only complicate the task. The user can just tell an X-Y table to move right 35 thousandths—he doesn't have to think about subroutines, testing loops, and all that other computer stuff. To him, the "machine" is the mechanical system, not a board full of chips.

We show various applications programs—those in the general spirit of microprocessor technology—somewhat removed from the very top of the drawing. This suggests room for improvement in "user friendliness," that quality of a program which determines the ease with which someone can use it and get to know it. There are a lot of systems out there which can suddenly confront the operator with something like "ERROR CODE 19," generally somewhat more mystifying than: "For some reason, I just got confused. Could you try that again?"

The applications programs are shown linked via the three language groups—assembler, high-level languages, and LISP—to the operating system which ultimately provides the machine interface. This is confusing, for in many (probably most) small, dedicated microcomputer systems, there is no resident OS. In the IDAC terminals, for example, there is just a pair of ROMs with about 3.5K of machine code.

This can be resolved by thinking in terms of the software development itself, not the actual "target" system—which is usually stripped down to a minimum configuration, since the flexibility of an operating system would be wasted.

The items over at the left are deliberately vague, for the technology is yet far from being able to claim, with a straight face, intelligence. Even AI work is rife with disclaimers, so let's leave speculations along these lines for later.

It may be clear from all this that one of the primary objectives of software is the ability to accept one level of description and produce another. A human says "A = SQRT(X)"—and the BASIC interpreter says 10111001 101000110 00001011 11110100 ... and eventually assigns to A a binary representation of a decimal number equal to the square root of the present value of X. Even in the whimsical scenario that concluded Section 7.2, in which Mr. X chatted with the machine about Blanchard control, nothing could occur until the human's natural language was parsed, contextually analyzed, considered in the light of both common and specialized knowledge, symbolized, interpreted, and converted into a sequence of machine code instructions at the 1 and 0 level. Pretty awesome, eh? The sheer magnitude of the processing task that the above represents somehow suggests that reaching the WHAT end is going to involve some new forms of computer hardware.

The value of this intellectual exercise extends beyond the convenience of having a mutually comprehensible model as a backdrop for upcoming chitchat about software. This industry is famous for human confusion of "levels."

Take the person who went to a business school and took COBOL programming courses, for example. Now, COBOL is much closer to the APPLICATION level than most other high-level languages, allowing the programmer to sort files, tabulate columnar data, and do other business-oriented data processing with a minimum of machine awareness. To this end, it is a relatively abstract language, and our hypothetical business programmer probably thinks of the computer as the big box with printers and disk drives at the other end of his time-sharing terminal's cable or phone line. He might quite reasonably think that the computer just "knows" what he means when he says, "SET TAX-RATE TO 5."

He has been taught, of course, that the machine uses 1's and 0's to think, but his understanding of everything below COBOL's level is probably cloudy. When he encounters a microprocessor, he might be shocked—not only at the capability that it represents at such a low cost, but also at the staggering amount of machine-level detail that is apparent upon leafing through an assembler-level program.

He has always been isolated from the reality of the machine by a body of sophisticated software which carries on its own shoulders much of the level-conversion effort that is required. But the system doesn't think in text. It knows nothing about payroll.

It knows nothing about Blanchards.

7.4 HUMAN PROBLEMS

Presumably, all this modeling and symbolizing brings to light some of the problems with the software business. We have seen that computers and humans are at opposite ends of a conceptual spectrum: the former are speedy and precise idiots working in the time domain, and the latter are sluggish and error-prone *illuminati* working in the frequency domain. It is a wonder that they can communicate at all.

Consider our Blanchard example. The programmer chews on the idea for a while, then spits out a well-masticated chunk of code that supposedly tells a microprocessor how to interact with its I/O devices to save somebody money on diamond tooling costs. The computer, of course, has no idea what it's doing; it has no conception of the task. It does have an internal model of the world, but it is so primitive compared to the programmer's that it could be unplugged, carried to the summit of Mt. Everest, plugged in again (!), and if it could still work at that temperature, resume execution of the control loop with quiet assurance that everything was quite OK. It wouldn't say, "Um, where's the grinder? What am I doing on this mountain? Hello?"

Of course, this awesome ignorance may be reassuring to people who shudder at neo-Luddite imagery of servitude to omniscient machine masters, blinking with cold efficiency in every corner. (They needn't worry until Chapter 9.)

No, the real problem with all this is the spanning of that vast conceptual gulf between human and machine to which we have been alluding from various angles since we began. It is somewhat disrespectful to charge a human brain with the task of mapping an abstract concept onto a precise sequential pattern of machine-level instructions, and as we have noted, the technology has spawned a variety of software tools and techniques that purport in assorted fashions to somehow ameliorate this onerous burden. It has to be done. The computer population is increasing at a much higher rate than the human one, and if programmer efficiency is not somehow increased to match, everyone in the world is going to have to learn to write code just to keep up.

As we noted near the beginning of this chapter, expectations of another (r)evolutionary development comparable to the microprocessor haven't yet materialized: programming is only vestigially amenable to production techniques that can make it keep pace with hardware development. Difficulties exist at almost every level in the conversion of a task from WHAT to HOW. Formal problem expression is complex and usually overlooked. Management commonly leaves such things to the software group, preferring to criticize the end result rather than create unambiguous task definitions in the first place. It is uncertain whether software design is an art or a discipline. Modules are rarely documented well enough to make them useful to someone else. Implementations of languages don't quite fit other machines. Testing a finished piece of code is a chancy business, with exhaustive analysis of all possible states nearly impossible. Field updates cost a fortune.

And few people even have a clear idea of what constitutes good software. The standard definition is "a program that works." But this overlooks most of the above, for even a giant mangled kluge finally beaten into hesitant operation after months of hair-pulling, cursing, and debugging can be said to "work."

So let's turn our attention now to some of the tools, now that we have a contextual framework (aha!) into which they can all be fit. We will look at operating systems and languages on varying levels, touch on AI, then discuss all that in the light of a generalized design project. At that point it will be obvious that we are missing a key, so we'll spend a chapter on software engineering, then map everything we have discussed onto a development system and the much-overlooked task of implementation.

By that time we will have achieved a reasonable balance between hardware and software. We can then solder up some systems, write some code, and make 'em work.

8

Software Tools

8.0 INTRODUCTION

Bolivar Shagnasty, hotshot programmer, snored quietly at his desk in the software development area. Listings and notes spilled to the floor. The wastebasket was overflowing, as was the ashtray. An empty coffee cup stood amid dozens of overlapping stains on a copy of Yourdon's *Structured Design*.

As his head lolled on his chest, Bolivar Shagnasty dreamed. Bedecked in military finery and ribbons representing every theater of war, he led his Crackerjack Coding Corps into its finest hour. The massive amorphous task stood forbiddingly before them.

Suddenly, he gave the order. The CCC hoisted aloft the ultimate weapon and brought it crashing against the task. With a cacaphonous roar, the great mass splintered into a pile of smoldering disorganized rubble on the ground.

"OK, boys. Code it!"

The corps set to work with a variety of sophisticated tools, restructuring the rubble into a system. Where the original had been amorphous, the new construction was crystalline and structurally flawless. Where the original had been forbidding in its massive impartiality, the edifice that rose before them was elegantly modular. Shagnasty climbed to the summit and surveyed the scene. "Run it!"

"Bolivar? Bolivar, we have a problem." A firm hand was shaking his shoulder. He awoke with a start.

"Huh? Did it work?"

"Well ... it's kinda hard to tell."

Kinda hard to tell, eh? Could the fundamental conception of the problem have been obscured by detail? Could, perhaps, the exuberant use of sophisticated tools have contributed somewhat to a muddling of human judgment?

This absurdity is a fitting start for a discussion of software tools, those thoroughly essential layers of sugar coating that sweeten the machine and make it palatable to human beings. There is a temptation in any technological milieu to engage in a love affair with the instruments and exotic techniques which have been created as intermediaries between the person and the task. This is especially true in the world of computers, for the facilities are indeed seductive.

But remember one of our main points of Chapter 7: "The primary objective of software is the ability to accept one level of description and produce another." What are we trying to do when we write a program? We are attempting to take an idea spawned at the WHAT end of the spectrum and map it onto the HOW. That's all. That's the whole point.

In Chapter 7 we spoke of software entities that make that conversion—but there is a slight problem: the really abstract ones do not yet exist. We are thus still left with a mapping task—a somewhat less strenuous one, to be sure, but a mapping task nevertheless. How are we to accomplish this?

Aha. We must create internal tools and disciplines which are analogous to the software tools that we don't have! It is to this end that the much-touted techniques of structured design exist: until we can tell the computer WHAT, we must be responsible ourselves for a healthy share of that downward plunge through successively lower levels of abstraction. Only when we have formalized the task to the point at which a high-level language can digest it can we at last turn our backs and let the system complete the generation of HOW.

As time goes on, this should get easier. The software tools will grow in abstraction, obviating the need for us to do their job painstakingly. In the meantime, we can separate the process into two parts: *tools* and *methods*. The former are fairly well defined pieces of code that insulate us from machine details; the latter are techniques and heuristics that help us use them most efficiently.

But we have still left something out, something that should be kept in mind as we wander off into a discussion of facilities.

8.0.1 *Art and Engineering*

The fact that our "external" tools don't yet span the whole WHAT–HOW spectrum indeed suggests the need for "internal" tools—or methods—to fill the vast gap. But the highly abstract nature of the task implies that simple crank-turning application of these methods is not enough: we are not applying a set of formulas to the design of a Butterworth filter; we're attempting to formalize abstract thought. We can't just plug some variables into a discipline, push a button, and write down the results.

We need creativity. We need the intuitive insight of an artist to guide the structured discipline of an engineer. Our tools are as nothing compared to our

minds, and without the ability to bring deep cognitive processes to bear on a
task—to take advantage of our associative ability—we are little better than the
machines we purport to teach.

It goes against the grain of some engineering managers to suggest that
software design (or any type, for that matter) is at its heart an artistic endeavor.
An intuitive mind is hard to manage—it recoils at scheduling and deadlines, it
balks at "structured walkthroughs," it prefers to generate documentation at a
philosophical, not formal, level. But a totally rational mind is without *vision*, and it
is vision that synthesizes widely divergent phenomena into a single transcendent
concept. It is vision that makes the whole more than the sum of its parts.

Clearly, maximum brilliance is attainable only with both: intuitive vision
to see beyond the input data and engineering discipline to take the grand idea and
make it work. Art without engineering is dreaming; engineering without art is
calculating.

So now that we all know where we're coming from, let's talk about the tools.

8.1 FEELING THE BITS BETWEEN YOUR TOES

To illustrate some of the effects brought about by the layering of progressively
more abstract pieces of code, let's create another graphic symbol to hang our
words upon—Figure 8-1. Think of it, if you will, as a vertical cross section of a
computer landscape.

The mass of logic is shown resting on something called "APPLICATION,"
presumably some kind of industrial control task. We can assume that the interface
works and is appropriate to our objectives.

The problem with the system in this complex form, of course, is that it is
about as unfriendly as a machine can get. As is suggested by the craggy and forbidding terrain, a human being attempting to deal with it would find himself
working at this level:

```
11110101
11011011
00000000
00010111
11010010
01000100
00000101
11110001
11010011
00000001
11001001
```

It is obviously not very pleasant. However elegant this little console output
subroutine might be, it is incomprehensible to a person without painstaking
decoding.

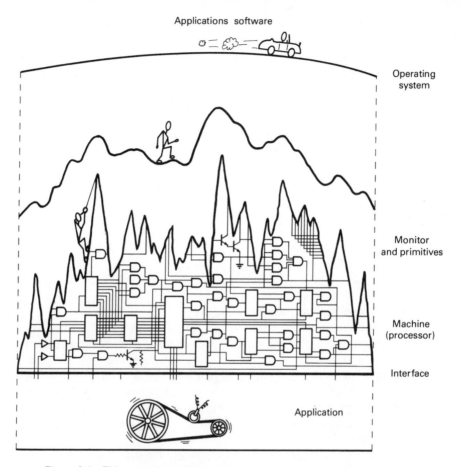

Figure 8-1   This somewhat whimsical view of a system suggests the cloaking of the craggy and forbidding machine-level terrain with relatively abstract software that smooths the programming task for humans. Many levels of code may lie between a high-level applications program and the application itself.

8.1.1 Adding a Conformal Coating

But that, unfortunately, is how our forefathers had to do it. Fingertips calloused by the front panel switch registers, minds conditioned to think in binary—the scope of a programming project was inevitably limited by the level of description necessary.

It didn't take long, though, for someone to wise up and toggle in a program that would read a little keyboard and convert the numbers into binary instructions. The code could then be keyed in using either hexadecimal or octal, making the system substantially more pleasant to use. This became the heart of the software entity now commonly known as a *monitor*.

A monitor does little more than make the machine's internal details accessible via a system console device (such as a CRT). It typically allows hex display of memory contents, modification of said contents, examination of registers and data on ports, and so on. In Figure 8-1 it is shown as a thick conformal coating on the craggy surface of the processor, smoothing it somewhat. It has had no effect on the apparent structure of the system and the "cushioning" leaves a lot to be desired, but at least some of the more hard-core aspects of the basic processor have been rendered a little less objectionable.

Using hex representation, by the way, the subroutine shown in binary on p.172 would be

```
F5
DB  00
17
D2  44  05
F1
D3  01
C9
```

Well, it's an improvement anyway.

But the problem with this as anything but an intermediate step in the "bootstrapping" of a computer is that it still requires the user to be intimately familiar with the machine's details. You couldn't take that "F5 DB 00 17 ... " and make it work on anything but an 8080-compatible machine. Of course, you may not want to. But you might get tired of looking up all those hexadecimal op-codes everytime you feel like writing a program. So before we generalize everything and discuss machine-independent coding, let's take a very quick look at a simple assembler.

In the early days of hobby computers (mid-1970s), few people could afford disk drives, and even fewer kit systems supported them. The ones that did exist outside the "professional" market were expensive and horrible.

During that grim period, one of the ways people managed to get programs written was through a ROM board containing a monitor, a simple text editor, and a rudimentary assembler. The subroutine represented above and on p.172 would be written like this:

```
PRINT:  PUSH  PSW
        IN    STATUS
        RAL
        JNC   PRINT+1
        POP   PSW
        OUT   CRT
        RET
```

This makes a little more sense, once you get used to it. If you need, for example, to rotate the accumulator left, you can say "RAL" instead of 17 hex or

00010111 binary. The texture of the machine is identical, however: the assembler provides no isolation from the processor architecture. In the context of our graphic symbol, it can be thought of as a thickening of the "conformal coating" we added in the last section (although a more sophisticated assembler to be considered later will be at a higher level).

The program we have been using as an example, by the way, is a simple console output subroutine written in 8080 code. To display a character on the CRT, you just place it in the accumulator and CALL PRINT. The routine first saves the character by pushing it onto the stack, then checks the UAR/T's status byte for the bit indicating that it is not currently busy (TBE—see Figure 5-2). If it is busy, the routine just spins in a loop, waiting for it not to be.

Then as soon as the UAR/T is free, the program pops the character off the stack and transmits it simply by doing an output instruction. The UAR/T takes it from there, and control returns to the instruction following the call. The PRINT function is shown as a flowchart in Figure 8-2. It should be clear from examination of the logic that the computer is not allowed to get ahead of the serial port, since it is forced to wait each time for the completion of the last transmission.

There are a lot of ways around the inefficiency that is suggested by that observation (interrupts, polling, etc.), but we're not here to talk about them—we are here to make some sense out of software tools.

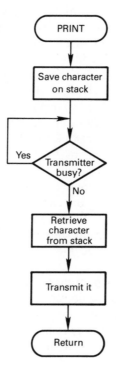

Figure 8-2 Flowchart of simple "print" routine (see the text).

Before we leave this simple assembler, note that for all its usefulness in freeing us of the need to remember opcodes and deal with the values of labels (such as PRINT, STATUS, and CRT), it still requires intimate familiarity with machine details. This is not always bad, however, for the ability to exercise tight and predictable program control over the processor is a frequent requirement in a real-time control task. But many applications are needlessly complicated by the use of assembler-level code, especially if it's a relatively crude assembler like this one.

8.2 OPERATING SYSTEMS

So far in this discussion, the system has remained at a rather useless level. One would hardly attempt to develop sophisticated software on a system providing only a monitor, editor, and simple assembler. For one thing, some sort of random-access file storage facility such as a disk is something of a necessity, and it would be nice if blocks of text and code could be passed about among system devices with a minimum of difficulty.

Typically, the creation of a program involves numerous sessions with a text editor, during which sheaves of smudged handwritten code are entered into the system as a source file. This file begins as data in system RAM, then finds its way to disk as dictated by memory-size limitations or termination of edit sessions. At some point the file of source text is compiled or assembled, yielding an object file. This probably doesn't work, so the source is edited again, producing a new text file and leaving the old one on disk as a backup. This iterative process continues until "all" of the bugs are fixed, and is frequently complicated by the need to develop many subtasks individually and then link them together.

Storage of data on disk raises a variety of interesting system problems. Its location must be duly noted somewhere so that it can be retrieved, and some means must be provided to allow new files to occupy spaces left by the deletion of old files, even if they were shorter. It must be possible for the human being to access the data, so a directory has to be maintained that cross-references operator-specified file names with track and sector addresses. Facility must be provided to create, delete, transfer, print, edit, compile, and do various other things to them, and all this must be coordinated with the current system configuration.

Clearly, doing this by hand would be a pain. That is why we have *operating systems*—probably the most important, underestimated, misunderstood, cursed, and vigorously marketed software facilities in the vast panoply of microcomputer resources.

Operating systems have two primary purposes, and fall into two major classes as a function of their relative dedication to them. The first is characterized by the system between the author and the final copy of this manuscript: it is called a *disk* operating system. The second is at home in timing-sensitive control tasks and is known as a *real-time* operating system.

Both types have the general effect of filling the craggy surface of our cross-sectional computer landscape with a cushioning layer of code which ideally renders the machine's details completely insignificant to the user, as suggested by the upper portion of Figure 8-1. UNIX, for example, a sophisticated disk operating system developed by Bell Labs and later implemented on many different kinds of computers, provides the user with few clues concerning the identity of the processor buried within it. The number of registers, the I/O port assignments, and even the amount of RAM are no longer important to the task of operating and programming the machine. (For high-level languages supported by the OS offer similar insulation; instead of using the assembler-level console routine, you can just say "PRINT.")

The net effect of this is the creation of a "virtual machine," far more abstract than the agglomeration of silicon devices on which the whole thing is based. The operating system *is* the system, and it not only frees the user from trivia but manages the various hardware and software resources in a fashion appropriate to the tasks at hand.

8.2.1 A Real Disk Operating System

As a fitting introduction to the often idealized subject of operating systems, let's look at a real one in some detail. OS philosophy is one of those things that tends to draw theorists out of the woodwork, but if we dedicated these pages to an amalgamation of all they have to say, you might be left with a lot of provocative concepts and very little idea of what really exists. So let's restrict our discussion somewhat, focusing not on theory but on a real system that is typical of those you will encounter in a development environment.

The kinds of resources available on a computer determine the sophistication of the operating system required. A large system with fast disks, some "swapping" RAM, and a dozen or so high-speed channels calls for a multitasking system that can take advantage of all that potential for concurrent execution of numerous application programs. But a micro sporting a pair of floppies begins to look a bit ludicrous when pitted against a multiuser environment: it's not that there's anything intrinsically limiting about the CPU, it's just that the resources are not large and fast enough to provide the data transfer and storage rates required for such operation. We'll see some exceptions to this when we talk about real-time operating systems, but they're in a different world.

It has thus followed that the proliferation of microprocessors in the home, small business, and software development contexts has created a need for an appropriately scaled class of operating systems. Perhaps the most prominent one that evolved from this technological period is CP/M.

It is CDOS (a close cousin of CP/M) that graces the author's computer system as he sets the Hazeltine CRT aflame with a relentless barrage of verbiage. Over a megabyte of text—plus that much again in backup files—pours from his bruised fingertips onto an ever-deepening stack of minifloppies. Something has to

keep chapters from disappearing or melting into other chapters, handle various facilities like the text editor, and allow the growth of the book in a fashion graceful
enough to justify the banishment of the typewriter into a murky corner of the office.

So what's a CDOS? The acronym is fairly obvious: Cromemco Disk Operating System. It occupies the top 11K, approximately, of RAM space in the Z-80-
based system dubbed BEHEMOTH, which we discussed in Section 4.2.3.

As was mentioned at that time, the machine is based on four minifloppy disk
drives, interfaced with the system via a controller card called the 4FDC. On this
card, there is a 1K ROM that contains a monitor, much like the one described a
few moments ago. But in addition to the standard hex operations, there is a command called, simply, "B," for "Boot."

When the B key is struck, the disk springs to life and, under direction of a
short routine in the ROM, loads about 128 bytes of code from its outermost track
into system RAM. Control is transferred to this "bootstrap loader," which then
searches the directory for a file called CDOS.COM and brings it into high
memory. Once that process is complete, control is transferred again, and the
operating system is alive.

Before going any further, let's look at Figure 8-3, which shows the general
organization of both the disk (on the left) and system RAM. The data on the disk,
of course, are actually located on a number of concentric tracks which are seg-
mented into 128-byte chunks called sectors. On the Cromemco 5-inch double-
sided diskettes, there are 40 tracks on each side with 18 sectors per track. There
are thus 1440 sectors, for a total capacity of 184,320 bytes. (There is somewhat
less in practice, because the formatting of the disk imposes a certain overhead.
Actual file capacity is 180K.)

It is therefore a trifle misleading to represent disk organization as a contiguous block of data, especially since the sectors on each track are numerically scrambled in a way that statistically minimizes the rotational "latency" time during
sequential accesses. Sequential tracks also alternate from side to side of the
media, since a head-change operation is purely electronic and requires no stepper
motor action. But the operating system straightens it all out, and at some level of
system operation, it is fair to view the disk in this linear fashion.

The "beginning" of the disk is sector 1 of track 0, the outermost track. This
is the first of 52 sequential 128-byte records collectively known as the system
area. The first one, as noted above, is the bootstrap loader, which is responsible
for finding the CDOS file and placing it into RAM. The remaining 51 comprise
the file directory.

The directory is the key to the whole system. The remaining space on the
disk is available for user files, each of which may be of any size between 1K and
the physical limit of the media. They may be created, updated, and deleted—and
without something to keep track of their various physical positions and statuses,
there would be anarchy. CDOS obviously cannot maintain the directory information in RAM, for disks need to be swappable (the author's system hosts well over

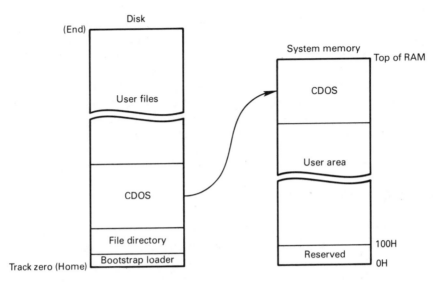

Figure 8-3 General organization of the disk (left) and RAM (right) in a CDOS-based system. Upon booting, the CDOS image is loaded into the high end of system memory, and control is transferred to it.

a hundred) and a power failure would effectively destroy the stored data by rendering them inaccessible.

The directory allows, by default, 64 entries. Special procedures may be undertaken via a system facility called STAT, which can allocate more for those applications that require a lot of short files, but 64 on a minifloppy is usually quite sufficient. The system, incidentally, supports 8-inch floppies and 10.5-megabyte hard disks with equal ease, and the various numerical parameters of which we have spoken are adjusted to suit: the hard disk defaults to 512 directory entries.

Anyway, each entry in the directory consists of a file name of up to eight characters, a file type (extension) of three characters, a byte that increments once for each 16K of the file size, a record count, a "cluster allocation map" which defines the locations of the data scattered across the disk, and a byte indicating the number of the next record. This information is sufficient to allow CDOS to search through the directory for a user-specified file name and access the file in the desired fashion.

CDOS handles the allocation of space on the disk in a way that allows new files to occupy randomly distributed available sectors. If this were not the case, deletion of a 3K file somewhere in the middle would leave a 3K space forever inaccessible unless a new file of that size or smaller were created to make use of it.

The system provides the programmer with a number of "system calls," basically a set of documented subroutines (detailed in Section 8.2.3) which allow all of the various operations necessary to interact with the disk. There are also some system calls for I/O and miscellaneous operations. Here we see one of the elegant features of an operating system that we heralded in the philosophical overview a few pages back: the user of the machine need not be aware of file structure, I/O device assignments, or timing details. The content of the individual directory entries is useless information to the system user unless he happens to be writing some new utilities which must exist within the cushioning layer of CDOS.

The system calls, themselves, represent a relatively intimate relationship with the machine, and even these are unfamiliar to most users. If BEHEMOTH were involved only with writing, for example, there would never be any reason to even think about creating programs that independently access the disk; all of that has been provided by the facilities of the operating system. So let's step up a level and see how it behaves from the console's point of view.

8.2.2 Using CDOS

Once the system has been "booted," it issues a sign-on message announcing the version number of the software, and then displays a prompt:

A.

This indicates to the user that drive A (the first of the four) is the current drive, meaning that file references to it need not be preceded by a drive specifier ("C:DEMO.TXT" to refer to a file called DEMO.TXT on drive C, for example).

As the user types commands on the terminal, certain control characters have special meaning. A control-P, for example, causes all subsequent console input or output also to appear on the printer. A control-U deletes an entered line; a DELETE or BACKSPACE wipes out the last character typed. (These keys wear out quickly; perhaps that's why there are two of them that perform the same function?) A control-S can be typed at any time, and has the effect of temporarily halting any console output from the machine. Striking control-S key again allows it to resume.

CDOS provides two classes of commands: *intrinsic* and *extrinsic*. The former are functions that are built into the RAM-resident CDOS code itself, and simply execute as soon as they are typed. The latter are executable files located on disk with an extension (or file type) of "COM," such as XFER.COM or EDIT.COM. When one of these commands is typed at the console, the system finds the appropriate filename in the directory, loads the file in from disk starting at location 100 (hex) in RAM, and then jumps to it. Location 100 of RAM is the start of the "user area" shown in the right-hand part of Figure 8-3.

In the discussion that follows, the term "file-ref" should be understood to represent a file reference of the general form

[x:]filename[.ext]

where "x" is the optional drive specifier mentioned above, "filename" is an ASCII name of up to eight characters, and "ext" is an optional filename extension. The last item deserves added comment—certain standard three-character codes have evolved which inform the system or the user what type of file is referenced by the name. Cromemco's standards are listed below, followed by a shorter group that the author has found to be useful in his attempt to organize a few thousand files:

ATO	LISP autoload module
BAK	Editor backup file
BAS	BASIC LISTed source file (also LIS or BAL)
CMD	Batch command file
COB	COBOL source file
COM	Executable command program (extrinsic)
FOR	FORTRAN source file
HEX	Intel hex format object file
LIB	Source library
LSP	LISP program
PRN	Printer or listing file
REL	Relocatable object file
SAV	BASIC SAVEd source file (also BAS)
Z80	Assembler source file

. . .

ADV	Advertising copy
DSK	Text file identifying a disk
FIL	Filler manuscript
FRM	Business forms
INC	Incidental submission
LTR	Letter
Mnn	Manuscript number 'nn'
MST	Format specification master file
MXX	Manuscript unspecified
OUT	Outline
QRY	Query letter
TXT	Random text

We take the time to detail this because a cleverly organized filenaming scheme can prevent confusion and aid at least a little in the cumbersome task of maintaining a large number of floppies. Things are bad enough without duplicate or inconsistent filenames all over the place.

Before proceeding, we should note another interesting aspect of disk filenames. Occasionally, it is worthwhile to do something to more than one file at a time—perhaps erase everything whose name begins with "GARBAGE." If the disk contained files named GARBAGE1.TXT, GARBAGE1.BAK, GARBAGE2.TXT, and GARBAGE2.BAK, then all four could be erased in a single command by simply saying

ERA   GARBAGE?.*

If, instead, the user wished to eliminate all BAK files from the disk, he could say

ERA   *.BAK

This scheme of "ambiguous file reference" can often eliminate considerable tedium. Incidentally, CDOS recognizes the power of "*.*" used in that context, and offers opportunity for a change of heart before clearing the entire disk.

So it can be seen that some of an operating system's desirability can be associated with the flexibility of file references. Let's look next at some commands, keeping in mind that "file-ref" represents any combination of the above possibilities.

There are seven intrinsic commands in the author's current version of CDOS (2.17):

ATRIB   Establishes or changes the allowable access modes of a file. The format is

ATRIB file-ref [+] [p...]

The optional "+" indicates that the following attributes are to be added to the ones already describing the file, and "p..." indicates any combination of the letters E, R, and W. An E protects the file from erasure, an R makes it unreadable, and a W prevents the system from writing to it. Various permutations of these options allow the user to reduce the chance of unauthorized access to an important file.

BYE   Returns control to the monitor. This is essentially an "un-boot," and allows some of the features of the ROM code to be used for diagnostic or testing purposes. Control is returned to CDOS by typing "B" to re-boot, or "G0" which causes it to JUMP directly to location zero— a CDOS entry point.

DIR   Displays (or prints, of course) the file directory for any disk. Simply typing "DIR" will reference the current drive; typing "DIR B:" will reference drive B. The directory listing shows the names of all the files, the length of each one in "K," the total amount of disk space used, and the amount remaining. It is also possible to attach any kind of file reference to the command to produce information about a particular subset of files. For example,

DIR D:*.LSP

would display only the names of the LISP source files on drive D.

ERA   Erases file(s) from a disk directory. We already saw an example of this in our description of ambiguous file references.

REN   Allows files to be renamed. The format is

REN new-file-ref=old-file-ref

and simply has the effect of associating a new filename with the specified file. The software makes sure that no file of the new name already exists, and issues an error message if one does.

SAVE   Causes part of the user area to be saved on the disk. This is valuable if a new executable program has been created by the user and it is currently residing in RAM. Issuing the command

SAVE file-ref n

will create a new file of 256*n bytes with the code that begins at location 100 in memory.

TYPE   Displays or prints an ASCII file. The system makes no attempt to format it; the text is simply passed from the specified file to the console. It is possible to direct the system to type a COM file, but the output is sheer garbage since object code has no ASCII significance.

That takes care of the functions that are built into CDOS itself. But there are many other things that one needs to do with files if a system is to be of much use. These take the form of executable files, which are invoked simply by typing their filename (without the COM extension). Earlier, for example, we mentioned COUNT, a simple word-count utility that was created by the author to keep track of progress on vast monolithic projects such as this one. The assembler source file was called COUNT.Z80, and there was a COUNT.BAK left over from the last edit session which produced the program. After the assembler was invoked, a relocatable object file called COUNT.REL appeared on the disk, but that wasn't enough. It needed to be linked with some library routines that handle disk access (located in ASMLIB.REL). After that link was performed, a file called COUNT.COM existed. From that moment on, simply typing the word COUNT followed by a text file name would cause the machine to wander off through the chapter and eventually display the total number of words (up to 65,535). Incidentally, an article subsequently written about that program (gotta milk these things for all they're worth, you know ...) bore the file name COUNT.M62—but that's another story.

By the way, all this detail about CDOS is not as superfluous as it may look: this operating system is conceptually identical to the majority of those running on microcomputer development systems out there, with differences of detail much more significant than differences of philosophy. Intel's ISIS, for example, sports a largely different set of commands, but upon close examination is revealed to be about the same—a little nicer here, a little clumsier there, perhaps, but not altogether alien. The most noteworthy operating system that cannot be compared comfortably to those of this class is probably UNIX, and we shall hit some of its high points a little later on.

Anyway, let's look at a few of these executable programs which, although not an integral part of CDOS, are necessary if the system is to do anything practical. Many of the facilities to be described are supplied by Cromemco with the system, others are either homebrew or purchased separately. That's another valuable characteristic of a flexible operating system like this: all the necessary "hooks" exist for the user to add capabilities that were not anticipated by the vendor. Within the category of extrinsic commands are many things other than basic utilities: languages such as FORTRAN, LISP, BASIC, and ASMB, business programs such as payroll and inventory, word processing facilities, special-purpose operations—they can all be built into the system and made to appear as if part of one integrated package. The author's little COUNT program is as easy to use as Cromemco's TYPE command: they both appear to the user to be part of the same overall structure. Here are some of the standards:

@   This is the batch command, which for some reason acquired the
             sobriquet "@." It is wonderfully useful, for it allows CDOS to
             accept a sequence of commands from a file instead of the console.
             This makes it possible to just type something like

                                        @ GO

and wander off to dinner, confident in the knowledge that the system is busily processing away on some task without the need for constant operator intervention on the "do this, do that" level.

The commands that are stored in the batch file are identical to those that would be keyed in at the console, but with additional opportunities for flexibility. First, the batch facility allows parameter passing: up to 10 codes of the form ^n, where n is a number from 0 to 9, may be used in the batch file in place of such variables as file specifiers, and then the command line (such as the @ GO above) can have appended to it a corresponding number of parameters. These replace the embedded codes in numerical sequence. This all sounds convoluted, but allows some interesting possibilities—especially in light of a second feature: nesting. A batch file can contain batch commands. Those batch files can contain batch commands. Those files can contain batch commands ... to a nesting depth of 128.

One of the author's relatively trivial applications of this facility is the PRINT.CMD batch file, which causes a text file to become formatted on the Diablo printer. Cromemco's word-processing software does all the dirty work, but the home-brew printer driver leaves the ribbon carriage energized (it gets hot), and there is usually some interest in the word count when an article or chapter is formatted. The PRINT batch file looks like this:

                        FORMAT ^1
                        RIBDROP
                        COUNT ^1

Hardly elegant. But when the command line

@ PRINT B:CHAP8.M55

is typed, the following events take place. The batch facility is loaded from disk and heralds its presence with a sign-on message. Then the command

FORMAT B:CHAP8.M55

appears on the screen—the filename was passed as parameter #1 and replaced the ˆ1 code in the batch file. The result is just as if the command had been typed in—the text formatter program, it-self a COM file, is loaded from disk into the user area, then it be-gins reading the chapter file from drive B and prints it in standard manuscript form on the Diablo printer. This goes on for a while, then upon completion the command

RIBDROP

appears on the CRT. This command is another homebrew utility, and simply outputs a signal to the printer that drops the ribbon. Finally,

COUNT B:CHAP8.M55

appears on the screen and the system just sits there for a while as the chapter file is read in its entirety and the words are counted. Finally, the total is displayed and the CDOS prompt returns.

None of that particularly justifies the full horsepower of the batch facility, but it just represents that much more freedom from the tedium of repetition and the chance of forgetting something.

Before leaving batch, two other things should be noted. First, for you recursion fans out there, the code ˆ0 is a special case that allows the batch file to refer to itself. Second, a special batch file named STARTUP can be created which will automati-cally be executed by the system upon boot—taking care of such details as printer initialization, clearing the screen, loading an ap-plication program, or whatever. It is possible with this feature to create a "turnkey" system that never allows the user to interact directly with CDOS, thereby preventing file foul-ups and similar disasters.

DUMP   This command displays the contents of any file by 128-byte records, in hex and ASCII. To invoke it, one simply types

DUMP file-ref

whereupon the entire file is piped to the console in a format that allows detailed examination of its contents. A portion of a dumped text file is shown in Figure 8-4a; that of a dumped Z-80 object file in Figure 8-4b.

INIT   This program allows the user to initialize, or format, floppy diskettes. The media are typically purchased in blank form, as just 5.25-inch disks of magnetic-oxide-coated flexible base material. There is a small index hole, located about a centimeter from the edge of the center hole of the disk, which allows the hardware to

```
        Record 4
0200  61 62 6F 75 74 2E 0D 0A-20 42 75 74 20 73 6F 66    about... But sof
0210  74 77 61 72 65 2C 20 72-65 61 6C 6C 79 2C 20 69    tware, really, i
0220  73 20 74 68 65 20 70 6F-69 6E 74 20 6F 66 20 61    s the point of a
0230  6C 6C 20 74 68 69 73 2E-20 20 54 68 65 20 62 61    ll this.  The ba
0240  73 69 63 20 6D 6F 74 69-76 61 74 69 6F 6E 20 74    sic motivation t
0250  68 61 74 0D 0A 73 70 61-77 6E 65 64 20 74 68 65    hat..spawned the
0260  20 69 6E 64 75 73 74 72-79 20 77 61 73 20 74 68     industry was th
0270  65 20 64 65 73 69 72 65-20 66 6F 72 20 75 6E 69    e desire for uni

        Record 5
0280  76 65 72 73 61 6C 2C 20-67 65 6E 65 72 61 6C 20    versal, general
0290  70 75 72 70 6F 73 65 20-68 61 72 64 77 61 72 65    purpose hardware
02A0  0D 0A 74 68 61 74 20 77-6F 75 6C 64 20 65 66 66    ..that would eff
02B0  65 63 74 69 76 65 6C 79-20 6F 66 66 6C 6F 61 64    ectively offload
02C0  20 61 20 6C 61 72 67 65-72 20 70 6F 72 74 69 6F     a larger portio
02D0  6E 20 6F 66 20 74 68 65-20 73 79 73 74 65 6D 20    n of the system
02E0  64 65 73 69 67 6E 20 65-66 66 6F 72 74 20 6F 6E    design effort on
02F0  0D 0A 74 68 65 20 63 68-69 70 20 76 65 6E 64 6F    ..the chip vendo
```

a. Dumped text file

```
        Record 2
0100  0E 19 28 06 11 74 1B CD-EF 15 CD D1 19 CD F4 15    ..(..t.Mo.MQ.Mt.
0110  0E 82 11 34 01 CD 05 00-0E 19 CD 05 00 32 D8 1F    ...4.M....M..2X.
0120  21 67 23 22 25 23 C9 11-30 02 CD EF 15 C3 34 01    !g#"%#I.0.Mo.C4.
0130  0D 0A 0D 0A 49 6E 63 6F-6D 70 61 74 69 62 6C 65    ....Incompatible
0140  20 77 69 74 68 20 74 68-69 73 20 43 44 4F 53 0D     with this CDOS.
0150  0A 24 0E 07 CD 05 00 32-F6 02 21 CA 02 11 5D 00    .$..M..2v.!J..].
0160  97 CD 16 03 D8 11 6D 00-A7 28 19 3D 28 26 3D 28    .M..X.m.'(.=(&=(
0170  38 3E 02 CD 37 03 38 49-0F 0F 0F 47 3A F6 02 E6    8>.M7.8I...G:v.f

        Record 3
0180  DF B0 18 36 3E 08 CD 37-03 38 36 47 3A F6 02 E6    _0.6>.M7.86G:v.f
0190  F8 B0 18 26 3E 04 21 E3-02 CD 16 03 38 23 0F 0F    x0.&>.!c.M..8#..
01A0  47 3A F6 02 E6 3F B0 18-11 3E 04 CD 37 03 38 11    G:v.f?0..>.M7.8.
01B0  07 07 07 47 3A F6 02 E6-E7 B0 5F 0E 08 CD 05 00    ...G:v.fg0_..M..
01C0  C9 11 F7 02 CD EF 15 C3-34 01 43 4F 4E 3A 20 00    I.w.Mo.C4.CON: .
01D0  50 52 54 3A 20 01 52 44-52 3A 20 02 50 55 4E 3A    PRT: .RDR: .PUN:
01E0  20 03 00 50 41 52 3A 20-00 53 45 52 3A 20 01 54     ..PAR: .SER: .T
01F0  59 50 3A 20 02 00 00 0D-0A 44 65 76 69 63 65 20    YP: .....Device
```

b. Dumped object file

Figure 8-4 Two invocations of the CDOS DUMP utility illustrate the appearance of files on disk. In (a), two 128-byte sectors of a text file from this book are shown; in (b), an executable object file is displayed. In each case, the body is shown in hexadecimal and the corresponding ASCII is at the right (with '.' for each nonprintable character).

recognize the start of each track, but that's not enough. A pattern must be written on each track to supply the system with sector boundaries. This scheme is called "soft sectoring," and is contrasted with "hard sectoring," in which there is a little hole indicating each angular boundary.

In addition to formatting new diskettes, the INIT program allows correction of certain kinds of soft errors (those that are not the result of physical media damage) which may have been caused by use of the disk on a misaligned drive.

The program allows numerous options, enabling creation of single- or double-sided diskettes, initialization of selected tracks only, and choice of minifloppies, maxifloppies, or hard disks.

STAT

This facility allows the user to display or change many of the operating system's parameters. In its simplest form, it produces a printout on the console which is a complete summary of the machine's current configuration. A sample STAT display is shown in Figure 8-5.

This is useful, but STAT also allows setting system date and time (which are then maintained by a 60-hertz interrupt-driven real-time clock), producing alphabetical directory listings and labeling disks. The labeling facility, invoked with the command

STAT/L [d:]

(where d: is an optional drive specifier), takes care of writing a user-specified name and date to the disk which can later be read by certain system calls. This enables application programs, for example, to meaningfully issue the message, "You have the wrong disk in drive B" if the operator errs in that fashion.

XFER

Here is one we couldn't live without. It transfers files from one device to another, where the devices are most commonly, but not necessarily, disks. It is possible, for example, to treat the console as a file in any of the facilities we have described by simply using "CON:" as the file reference. Similarly, other system I/O devices have standard names, new ones can be created, and there is even a dummy (DUM:) that is affectionately known as the bit bucket.

Anyway, XFER allows numerous variations on the basic theme of file transfer. The command

XFER D:=B:CHAP8.M55

for example, would create a backup copy of this chapter (currently on drive B) on drive D. By substituting ambiguous file references, a number of files can be copied at once. Also, by stringing together a number of files after the equal sign, separated

```
A.stat/as
STAT (System Status) version 02.09    Friday, September 25, 1981  18:35:42

SYSTEM MEMORY:                        DEVICE CONFIGURATION:
Operating system version      02.17  CON: = Console 0
Total system memory           64 K   PRT: = Printer 0    (PAR:)
Operating system size         12 K   RDR: = Reader  0
User memory size              52 K   PUN: = Punch   0

DISK MEMORY:                          DISK CONFIGURATION:
Disk label                  DISK58   Cluster size                  1 K
Date on disk              09-01-80   Sector size                   128
Total disk space            173 K    Total directory entries       64
Disk space used by directory  2 K    Directory entries used        34
Disk space used by files    142 K    Directory entries left        30
Disk space left              29 K

DRIVE:      Double sided, Single density
DISKETTE:   Double sided, Single density

1200      com    1K      Cont      mst    1K      Init      com    10K
19200     com    1K      Count     com    2K      Link      com    8K
@         com    3K      Cred981   adv    10K     Ltrhead   mst    1K
Article   mst    1K      Debug     com    8K      Print     cmd    1K
Asmb      com    12K     Diablo    com    1K      Ribdrop   com    1K
Asmlib    rel    2K      Dump      com    2K      Stat      com    8K
Cdos      com    12K     Edit      com    7K      Unix      com    1K
Cdosgen   com    28K     Fig       mst    1K      Wwhead    mst    1K
Chapter   mst    1K      Format    com    8K      Xfer      com    4K
Chat      com    1K      Honey     com    1K      Xmitopen  com    2K
Clear     com    1K      Ident     381    1K
***   32 Files,   33 Entries,   142K Displayed,    29K Left ***
```

Figure 8-5   A system status (STAT) display of drive A on the author's computer system.

by commas, one can concatenate them into one large file. If we had enough disk space (ah, for a hard disk!), we could generate a single file of a little over a megabyte that would be this entire book:

XFER/A E:BOOK.M55=E:INTRO.M55,
E:CHAP1.M55,...,E:CHAP15.M55

The "/A" after XFER is one of the option switches we will describe momentarily, and simply allows concatenation of ASCII files by removing the end-of-file mark (1AH) from all but the last one. The hard disk would be drive E, and the file names all share the M55 extension because that is the manuscript number.

There are a number of selectable options associated with XFER, each specified by appending a slash and the appropriate character to the command. These are:

A   ASCII file concatenation
C   Compare files without transfer
F   Filter out illegal ASCII characters

I   Ignore end-of-file mark
R   Read-protected file is to be transferred
S   Strip rubouts and nulls from ASCII file
T   Tabs are expanded from control-I to eight spaces
V   Verify correctness after transfer
Z   Suppress display of statistics after completion

These can be chained together as necessary to achieve the desired effects.

MEMTEST   Hardware problems are relatively rare with modern RAM, but when they occur, they can be absolutely maddening. There is no way to predict the effect of isolated cell failure, and some of the symptoms are so obscure that a bad memory can live in the system for months before somebody finally gets frustrated enough to do some serious digging.

Testing RAM exhaustively is an art, and has spawned a variety of exotic techniques that purport to allow a faster test than would be obtained by trying all possible bit combinations in each chip. With 64K RAM, that would indeed be a time-consuming process, because even at 1 million tests per second the attempt would require 3.17×10^{19708} gigayears.

The universe hasn't been around long enough for a complete memory test.

CDOS MEMTEST, therefore, uses a shortcut. Testing a 16K block at a time, it starts by shifting a single one through a field of zeros. A byte with a one-bit set is written into the highest location of the block; the next location down gets the byte shifted one position; and so on. Every nine bytes, the shift direction is reversed, resulting in a zigzag pattern of 1's in RAM. Each byte is checked after it is written and again after the entire pattern has been loaded. Then the test is repeated for each of the 18 possible ways that the pattern can be written.

Once this "peak test" is complete, the program performs a "valley test," which is precisely the inverse—zeros shift through a field of 1's. Following that, a "delay test" is performed in which the ability of the RAM to hold a pattern for at least 6 seconds is checked, then there is an "M1 test" which ascertains that the block can be read during the proper bus timing intervals. Finally, the process is wrapped up with a "bank select test" which makes sure that the memory appears in the appropriate banks (expansion of system RAM past the Z-80 address range of 64K is accomplished by selecting any of eight 64K banks via an I/O port).

At the completion of this test, which was deliberately contrived to subject the memory to the most stressful types of patterns that can be easily generated, a map of the block is displayed with any errors indicated.

CDOSGEN

When a new version of CDOS arrives from Cromemco or when the system configuration changes (new disk drives, more RAM, etc.), it is necessary to modify CDOS so that it can recognize and take advantage of the system's current hardware facilities. The program called CDOSGEN accomplishes this, and leads the user by the hand through a number of questions that establish the system's details. For each disk drive, it asks whether it is large, small, or hard, whether it is single or double sided, whether it is single or double density, and whether it is the last one in the system. It automatically sizes memory, offers the possibility of custom function key decoding for those folks who have sexy CRT terminals, and offers optional naming of the resulting CDOS file and the bootstrap loader. This is the origin of that CDOS.COM file which we mentioned some time ago in the context of booting the system.

At the completion of CDOSGEN execution, certain important addresses are displayed for interested users. These include the starting address of CDOS, the location of the "vector table" which allows access to all the I/O drivers, and the last address used by the operating system.

EDIT

This facility is so important to operation of the system for just about any purpose that it is somewhat tempting to get quite carried away and give it a dedicated section.

The text editor is the means by which new files are created. When the user creates a new assembler source file for a process-control software development task, he begins by typing

EDIT file-ref

If the named file already exists, it becomes the "input file;" otherwise, a new file is created by that name. In the former case, the file is read into the text buffer and operated on by the editor. In the latter case, new text is created in the buffer. At the end of the edit session (or when the buffer is full), the text is written to the output file, and the input file (if present) is renamed to carry the extension BAK.

The general structure of this is suggested by Figure 8-6. It is important to note that the input file can be far larger than the available buffer space in RAM. As editing progresses, blocks of text are shuffled in and out as required.

The editing process itself is accomplished with 25 command types which allow unrestricted manipulation of the text. It is all

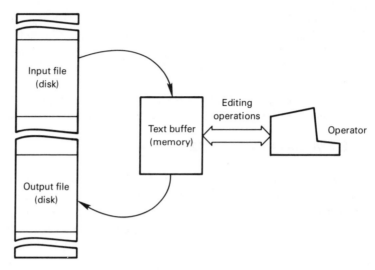

Figure 8-6   The text editor moves data into the text buffer from an input file, makes them available for extensive manipulation, then writes them to an output file.

based on an invisible "pointer" which can be moved forward or backward in the text buffer by any number of characters, words, or lines. Text can be deleted in those units, and at any point the "insert" mode can be entered, allowing new text to be added. There is a command to set the pointer to the beginning or end of text, another to set tabs, another to type a given number of lines on the console, and another to display a page at a time. There is a "Find" command which allows the pointer to be positioned at the nth occurrence of a specified character string, and a "Substitute" command which allows one string to be replaced by another.

The editor provides a "save buffer," which can be used to perform cut-and-paste operations or to bring a commonly used string into the buffer with a single keystroke whenever necessary. And of course there are a number of commands which allow disk I/O, from backing up work in progress to bringing in side files.

It will even execute commands conditionally (change all asterisks that begin a line to semicolons, for example, but leave other asterisks alone), and allow any command or group of commands to be treated as a macro.

Some editors (such as Cromemco's SCREEN) allow the user to see the actual effect on the text file of each command, representing the pointer as an easily visible marker. These are generally much less confusing for the beginner than a string editor such as EDIT.

All of these extrinsic commands we have described are, as noted, just COM files on disk. They are thus indistinguishable from homebrew utilities such as COUNT and RIBDROP, as well as from higher-level facilities such as the assembler (ASMB.COM), LISP (LISP.COM), and a data base management system (you guessed it: DBMS.COM).

Let's do one last thing with CDOS before wrapping up the subject of operating systems with some comments that tie all this into the rest of the industry.

8.2.3 A Summary of CDOS System Calls

As we have said, and shall see graphically in Figure 8-7, the operating system is a layered collection of software that cloaks the hardware in a cushioning blanket of abstraction, creating a virtual machine. Sitting at the CRT terminal, pounding away into the night with one eye glancing furtively now and then at the calendar, the author can easily forget what kind of processor is underneath all that. Who cares?

That's wonderful for the end user, of course, but there are times when it is necessary to write some code for the system itself. We're not talking about programs that will be implemented on a single-board target system in a factory—we are referring to programs that will run on the development system itself, in the context of CDOS.

If these programs are nontrivial, they will probably need to access system resources other than the CRT at some point. In particular, they may need to build new disk files or access old ones, processes that can involve quite a lot of code if they are written in their entirety from scratch.

The top layer of CDOS, about which we have been talking for quite a while, is the user interface. The next layer down consists of system calls. These are calls to the operating system that initiate certain standard functions involving the machine's I/O devices, most notably the disks. Through them, the programmer can be assured of code that is independent of system size and configuration, since the effects of the system calls are mediated by the system description evoked by CDOSGEN. So instead of writing a routine that goes out and physically accesses the disk (which is guaranteed to become obsolete when the hardware changes), one can just make use of a built-in CDOS facility. This style of programming also assures that only one change must be made when a new I/O device is added.

Actually performing a system call is accomplished by loading the C register of the Z-80 with the call's number and setting up other registers with any required entry parameters. Then, a subroutine CALL to location 5 executes the operation, following which CDOS returns to the user program that initiated it. (Location 5 in RAM is set up by CDOS as a JUMP to the appropriate entry point for the calls, eliminating the need for the user program to accommodate changes brought about by operating system revisions.)

Primarily for the purpose of communicating the general range of capabilities provided by a robust microcomputer operating system, we provide below a complete list of CDOS system calls, with a short description of each. The information of primary importance here is between the lines: the specifics of CDOS are of much less concern than the suggestion of flexibility inherent in this kind of structure and the demonstration of the machinations necessary to interact effectively with file storage devices.

The system calls are shown in ascending order of the function number that should be placed in the C register:

01   READ CONSOLE (with echo)—Waits for a character from the system console and returns it in the A register. In most cases, the character is echoed to the CRT, although certain control characters have special system meaning and are accepted as such (like control-P, which toggles the printer on and off).

02   WRITE CONSOLE—Writes a single character, passed in the E register, to the console (as soon as it is ready to receive it). Depending on the status of certain flags, it may also go to the printer. If the character is a tab (control-I or ASCII 09), it will be displayed and printed as the number of spaces necessary to leave the cursor at one of the tab stops (which are columns 1, 9, 17, ..., 73).

03   READ READER—Reads one character from a paper tape or card reader, returning it in A. It checks the "pause" status flag (toggled with control-S) so that one can clear a jam. Cromemco systems don't come with readers or punches, as they are a bit *passé* these days—but a good use of this call and the next one is the system-level interface of a modem, allowing transmission or reception of a file over the phone with a minimum of programming inconvenience.

04   WRITE PUNCH—Punches one character (from register E) on paper tape or card (or outputs it to the appropriate port—see call 03). All of the actual device drivers are accessible for modification.

05   WRITE LIST—Prints a single character on the printer (passed in E). As with the WRITE CONSOLE call, CDOS will wait for the device to be ready. It does not, however, expand tabs, and will quite willingly send along garbage characters which may have no meaning to the printer.

07   GET I/O BYTE—Places into A a special byte that CDOS sets aside for the specification of additional I/O devices such as extra consoles.

08   SET I/O BYTE—In conjunction with call 07, allows the user to modify the I/O BYTE as desired. It is passed in E.

09   PRINT BUFFERED LINE—Displays (and optionally prints) a string of ASCII characters terminated with a dollar sign. The starting address of the string is passed in the DE register pair.

0A   INPUT BUFFERED LINE—Reads an input line from the console, and loads it into memory starting at the address in DE. The first byte of this specified buffer contains the buffer's length, and on return the second byte will contain the number of bytes actually entered. The entered line of text begins at the third byte. Earlier we mentioned certain control keys that have effect during console interaction with CDOS, and this is where those manifest themselves. They are:

CTRL-C	Abort and go back to CDOS.
CTRL-E	Physical CRLF with no logical effect.
CTRL-P	Toggle the console/printer link.
CTRL-R	Repeat line as typed so far.
CTRL-U	Delete entered line and reset buffer.
CTRL-X	Delete previous character and echo it.
Rubout	Delete previous character and back up cursor.
DEL	Same as Rubout.
Underline	Same as Rubout.
Backspace	Same as Rubout.

0B   TEST CONSOLE READY—Checks the console to see if a character has been typed, returning FF in the A register if so, and 00 otherwise.

0C   DESELECT CURRENT DISK—Outputs a value to the disk control port which shuts off the drive motors and deselects the current disk.

0D   RESET CDOS AND SELECT DRIVE A—Initializes CDOS, logs off all disks, and selects drive A as the current drive. Accessing the other disks will log them on again.

0E   SELECT DISK DRIVE—Selects as the current disk the one whose number is passed in E. (A = 0, B = 1, etc.)

0F   OPEN DISK FILE—The DE register pair points to an FCB (File Control Block—see system call 86) which specifies a file name. The disk is accessed, and the system returns FF in A if the file is not found, or the directory block number if it is found. DE is set to the directory entry, which has by this time been transferred to memory.

10   CLOSE DISK FILE—DE again points to an FCB, and the named file is closed. Then the disk's directory is updated to represent the new status of the file: the FCB containing a new "cluster allocation map" is written to disk. The returned parameters are the same as on the previous call. Closing a file is necessary if any changes have been made to it, because the process allows the effect of those changes on the file's disk addresses to be accessible to other programs. (In a program such as COUNT, no changes are made and the file does not have to be "closed.")

11   SEARCH DIRECTORY FOR FILENAME—This one simply scans the directory on the specified disk for the filename in the FCB referenced by DE. If the filename happens to contain some "?" characters, it has been specified ambiguously and the function call will find the first occurrence. Interestingly, this function and the next one (12) will fetch the directory entry even if it has been erased: erasure of a file affects only a single byte in the directory. This enables sneaky programmers to un-erase recently wiped files, as long as no other write operations have trodden on their space.

12   FIND NEXT DIRECTORY ENTRY—This is always executed after function call 11, and is identical in behavior except for the fact that the

next, not the first, directory entry satisfying the FCB's name is obtained. In this fashion, ambiguous references allow access of multiple files; also, system organization requires a directory entry for each 16K of a file's size (although all but the first are invisible). These phantom entries are called "extents" and are handled via this function call.

13   DELETE FILE—The specified file is deleted from the disk directory. If ambiguous reference is involved, any number of directory entries may be tagged with the delete character, and that number is returned in A.

14   READ NEXT RECORD—Through this call, files are read into the RAM disk buffer. The record number is noted in the FCB (the information comprising the file control block was noted in Section 8.2.1). A code is returned in A which indicates read completed, end of file, or read attempted on an unwritten cluster (a cluster is 1K of disk space, consisting of eight sectors).

15   WRITE NEXT RECORD—This is the opposite of function call 14: data are written from a buffer in RAM to disk. In similar fashion, status is returned indicating write completed, entry error, out of disk space, or out of directory space.

16   CREATE FILE—The file specified in the FCB is created. A returned value indicates the directory block number or a directory full condition. This operation writes no data to the user area; it just builds a directory entry.

17   RENAME FILE—One or more files are given new names, specified in the second half of an FCB in which the first half specifies the file reference.

18   GET DISK LOG-IN VECTOR—This call returns a map (in A) which shows the disks that are logged in.

19   GET CURRENT DISK—The number of the current drive is returned in A.

1A   SET DISK BUFFER—This allows any area in RAM to be used as the disk buffer, as determined by DE. The "normal" location is 80H, and is set whenever a program is loaded.

1B   GET DISK CLUSTER ALLOCATION MAP—Used by STAT, this call returns information about the space on a specified disk. The BC register pair points to a bit map indicating allocated clusters, DE reveals the capacity of the disk in clusters, and A contains the number of sectors (records) per cluster (8).

80   READ CONSOLE (no echo)—Same as system call 1, except that the character entered is not displayed on the CRT.

81   GET USER-REGISTER POINTER—Reserved for system expansion into a multiuser environment.

82   SET USER CONTROL-C ABORT—In normal operation, typing Control-C on the console will abort a user program and return control to CDOS. In some applications, this might be disastrous (imagine losing everything after five hours of editing because of an erroneous keystroke!). This call allows redefinition of the address to which the system

will jump when the key is struck (it is normally location 0).

83   READ LOGICAL BLOCK—This call allows a program to read data from disk at a specified physical address. No regard is given to file structure; it just moves the head to the appropriate track, waits for the sector, and gets the data. It is possible with this call to specify not only the disk and block number, but also whether the read should be in an interleaved or noninterleaved form. Recall our earlier mention of scrambled sector numbering to allow faster sequential addressing—this is called interleaving and on the 5-inch disks is simply accomplished by thinking of every fifth sector as the next in sequence.

84   WRITE LOGICAL BLOCK—Just the opposite of call 83.

86   FORMAT NAME TO FILE CONTROL BLOCK—This builds an FCB. HL points to the start of the input line (which is a file-ref per our earlier discussion) and DE points to the location in RAM where the FCB is to be constructed.

87   UPDATE DIRECTORY ENTRY—This allows modification of a directory entry by finding it using call 17 or 18, changing it as required, and storing it back on disk.

88   LINK TO NEW PROGRAM—When it is desirable for one user program to call another from disk, this call is used to get it and begin execution.

89   MULTIPLY INTEGERS—This is one of the miscellaneous convenience calls, and provides a 16-bit multiply function. The operands are passed in HL and DE and the result is returned in DE.

8A   DIVIDE INTEGERS—Provides a 16-bit divide of the form HL = HL/DE with DE = remainder.

8B   HOME DRIVE HEAD—Returns the head of the specified drive to track 0 (closest to the outer edge of the disk).

8C   EJECT DISKETTE—If the system happens to be equipped with Persci 8-inch drives, this call will physically eject the disks.

8D   GET CDOS VERSION AND RELEASE NUMBERS—So that the user may know the software revision level or so that a program can determine its compatibility with CDOS, this call returns the version number in B and the release number in C.

8E   SET SPECIAL CRT FUNCTION—Modern intelligent CRT terminals provide a healthy array of special features, such as clear to end-of-screen, address cursor, blink characters, and so on. This call accepts any of a long list of input parameters and pipes them to the CRT in order to perform the special features.

8F   SET CALENDAR DATE—This allows the user to set a date (which may then be maintained by a real-time-clock program or equivalent hardware). The values comprising the date are stored in CDOS and may be accessed via system call 90 for addition to listings or enhancement of sign-on procedures.

90   READ CALENDAR DATE—Retrieves data stored with function call 8F. If no date has been stored, it will yield 00/00/00.

91	SET TIME OF DAY—Similar to call 8F, but with time data. The system makes no checks for its plausibility, and it must, of course, be maintained after initialization by appropriate software or hardware.
92	READ TIME OF DAY—Reads system clock.
94	SET FILE ATTRIBUTES—Attaches E, W, and R attributes to a directory entry for erase, write, or read protect. This operation was described under ATRIB in our listing of the CDOS intrinsic commands.
95	READ DISK LABEL—The STAT facility allows labeling a disk with a name, which actually becomes a dummy directory entry. This call allows that label to be read and used by an application program.
96	TURN DRIVE MOTORS OFF—Many programs, such as EDIT, may execute for hours without need of the disk drives. To save wear and tear on the heads and media, this call is provided to issue a hardware command to shut off the drive motors.
97	SET BOTTOM OF CDOS IN RAM—Normally, all RAM space below CDOS is available to user programs. But there are situations in which it is desirable to associate a custom I/O driver or similar utility with the system permanently, and prevent its destruction by rampant text files or assembler symbol tables. This call allows creation of a protected zone in which such things (as the author's Diablo driver software) may be safely kept.

An exhaustive list of system calls is hardly recreational reading, but now it all pays off. Consider Figure 8-7, the layer cake with golf tees stuck into it. This representation of an operating system wouldn't make much sense if we had not just explored a real one in considerable detail. The bottom layer, interfaced appropriately with I/O devices and disks, is the computer—whatever it may be. In this particular case, it is a Z-80 in the form of a Cromemco Z-2D, but that doesn't really matter.

The next layer is a collection of device drivers—something we have not talked about, but so simple that a couple of sentences will do it. The drivers are just short subroutines, accessed via a vector table (that's a list of addresses—we computer folks like buzzwords), which take care of primitive I/O operations. They have no intelligence and are not designed to be easy to use: their sole purpose is machine-level interface with the physical devices connected to the system.

But on top of these are the system calls, those formalized routines that we just sketched. At this level, the details of disk access and timing, the trivia of waiting for a UAR/T to finish sending a byte, and other such nuisances, are removed from the programmer's vision. You want to find a directory entry? No problem, just point to the FCB and use function call 11. This level is key insulation between machine details and the human mind.

And finally, the "top" layer: the user interface. When you sit down at the CRT and begin chatting with the operating system, you don't even have to think at the system call level. You type

DIR B:

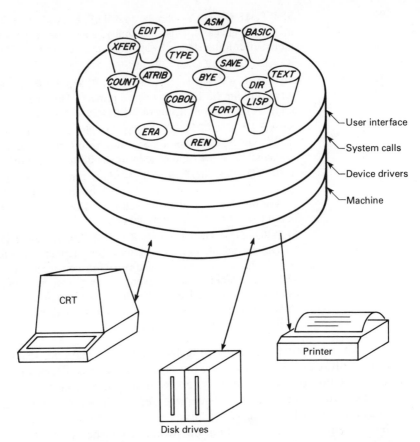

Figure 8-7   This layered view of the CDOS operating system suggests the activities that take place below the user's awareness, primarily involved with the details of file and device handling.

and presto! The directory of the disk in drive B is neatly displayed, with the size of each file in kilobytes, the number of entries, the amount of space used, and the amount remaining. You want to transfer a file? A program called XFER is brought in by CDOS. It accepts the parameters in your command and pulls the strings on the system calls until the job is done. As the system user, you are completely isolated from the machine level—which is the way it should be.

Incidentally, in Figure 8-7 we show the intrinsic commands on the surface of CDOS and the extrinsic ones as external entities that have been poked in, just to emphasize their conceptual difference as we work our way through these various levels of abstraction.

Let's tie this into the rest of the world and then move along to a look at some languages.

8.2.4 CDOS vs. the Others

All of that apparent proselytizing should not be interpreted as an ad for CDOS. In line with our avowed real-world orientation, it was decided that a purely philosophical approach to the subject of operating systems would be just too much to take: the field is rife with delightful concepts that map only clumsily on reality.

The operating system we just described is fairly typical of those commonly implemented on 8-bit systems. There are many different flavors, but there is little point in attempting a comparison. Instead, let's use that as a basis for a short overview of features that become desirable as the complexity of either the processor or the environment increases.

First, almost any application can benefit from a measure of multitasking. The CPU is sufficiently fast from the viewpoint of human users that clever "resource management" can cause it to appear as more than one processor. This has obvious value in situations where more than one user is involved, but even in dedicated applications it can speed things up: assemblies and program compilations can proceed concurrently with edit sessions, a chapter could be formatting on the printer while the author works on another, and so on.

An elemental form of this principle may be seen in the *spooler*, a relatively simple facility that can be grafted onto a conventional operating system without the all-out conversion that would be required for true multitasking. The spooler simply allows a file to be printed as a background task while normal system operation progresses in the foreground.

But in a true multitasking system, a fascinating variety of problems suddenly becomes evident. Whether intended for a single user or many, the system is inevitably faced with the need for intertask communication. Here, the problem is not with physical constraints (as it is with the printer), but instead with interference of more subtle forms. Consider: There are two tasks running, A and B. They are connected via a protocol technique that prevents one of them from grabbing a piece of data that is currently in use by the other. In our scenario, there are two data items of interest, P and Q.

A has P, and needs Q before P can be released. B has Q, and needs P before Q can be released. The system, unless very carefully designed for such rare but crippling eventualities, locks up and requires manual intervention before it can be restarted. The condition is frequently called *the deadly embrace*, and is justifiably famous.

The author has seen a microcomputer-based business system, built around 1975, that incorporated no facility for restricting data access between concurrent tasks. It was garbage: a customer file being updated by an operator at one terminal could produce gross errors if a billing or other program were being run from another terminal. The user had to schedule operations around the machine's proclivities—negating the main selling point of the nearly $30,000 system.

Obviously, upgrading a system to a multitasking environment requires more than simple "context switching" and memory segmentation. But what if multiple processors are used?

This naturally relieves CPU allocation problems, but some care is still required in organizing file access. The operating system for a multiprocessor must very cleverly minimize contention, and also needs to provide some inter-CPU communication scheme. This is commonly done with "mailboxes" in RAM and a rigid protocol.

Clearly, there is no panacea, but the widespread need for multitasking capability has spawned some very capable operating systems. Perhaps the most eminent among them is UNIX (together with its various offspring), which arose from Bell Labs after nearly a decade of refinement.

UNIX, unlike "conventional" operating systems, allows machine users almost unrestricted freedom in customizing their environments. Philosophically, at least, UNIX is friendly—although at first, some of the command formats tend to appear counterintuitive. Like CDOS, it provides device-independent I/O, but it also allows software entities called *filters* to be inserted in data paths, producing the general effect of modularity. Instead of writing a program to perform some custom data manipulation task, the system user can often produce a *shell* of system-provided filters (interconnected by *pipes*) which accomplishes the job without the need for any coding effort. This intrinsic modularity naturally affects software design, and one of the major selling points of UNIX is its enforcement of good structure on the basis of its very nature.

The elegance of UNIX has spawned some non-Bell systems that boast varying degrees of equivalence, most notably XENIX, IDRIS, CROMIX, and Uni-FLEX. It is reasonably safe to assume that the architectural flexibility of this approach will continue to have a greater and greater effect on system design, more effectively utilizing hardware resources and offering users higher levels of system abstraction.

Some hint of the power of UNIX-like operating systems can be gleaned from an examination of the directory structure. In CDOS, the file directory is akin to a simple list of the file folders in a cabinet: there is a direct 1:1 correspondence between the directory entries and the files themselves. Further, each diskette in a system is a self-contained entity, and there is no convenient way to cross the interdisk boundaries.

But Cromemco's CROMIX is a different story, as is suggested by the sample file system drawn in Figure 8-8. The structure, far from being a linear array, is an open-ended hierarchy.

It is often described as an inverted tree, with the *root* directory at the top. At this level, a command to list the directory entries would produce a display of all the names in the top row.

But of those, only one is actually a file (CROMIX.SYS). The others are themselves directories. ETC, for example, is a directory of seven files: MOTD

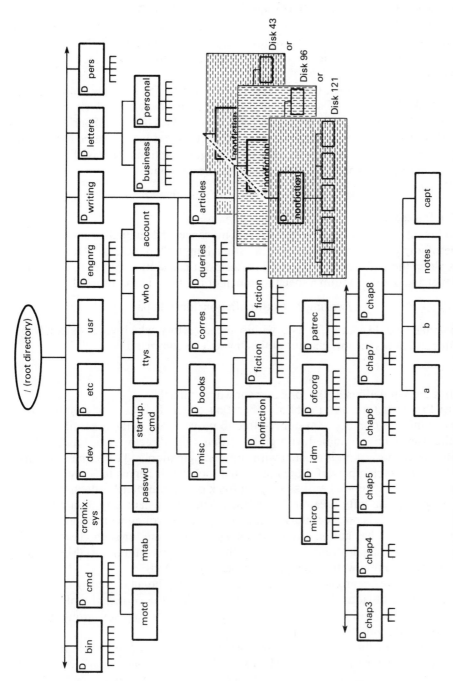

Figure 8-8 The CROMIX operating system (derived from Bell Laboratories' UNIX) allows a hierarchical file structure that can extend in three dimensions: breadth, depth, and additional disks. Entries in directories can be directories themselves.

(Message of the Day), MTAB (Mount Table), PASSWD (user information), STARTUP.CMD (same idea as CDOS STARTUP.CMD), TTYS (terminal information), WHO (list of users currently logged on), and ACCOUNT (the file displayed by WHO). The fact that ETC is itself a directory is indicated by a D in the diagram and on the display.

But this can go much deeper. The root directory, for example, contains an entry called WRITING. It is a directory with five entries, all of which are directories (MISC, BOOKS, CORRES, QUERIES, and ARTICLES). Each is also a directory of directories such that, finally, every actual text file is located at the end of a path that explicitly defines its nature. The figure captions for this chapter, for example, could be "typed" on the console via the command

ty /writing/books/nonfiction/idm/chap8/capt

The full file specification is known as a path name.

This, of course, would get to be a real nuisance; the full path name, since it can be of any depth, might represent a rather lengthy typing task. But since the nodes along the path are themselves directories, it is a simple matter to leave the root level and operate within a specific branch of the tree. The command

d /writing/books/nonfiction/idm/chap8

for example, would place the user within the directory which consists solely of text files and notes for this chapter of this book. The command

ty /capt

then has the same effect as the long one shown previously.

Such a structure has a sublime liberating effect. It is as if one system becomes an infinite number of systems—free to move about in the great file hierarchy, the user is never saddled with great complex file directories of totally unrelated items. If the system is created with some imagination and care, the structure itself becomes such a powerful organizational tool that it begins to serve the user at a much higher level—as well as keeping track of appointments, notes, things to do, and so on. The lofty dream of a tireless machine assistant is quite a bit more reasonable when it is not necessary to constantly fiddle around with system details (the author never has gotten around to using the machine for such classic "personal computer" applications as handling the checkbook—until CROMIX came along, things like that were much faster and easier by hand).

In a system hosting a multimegabyte hard disk, the applicability of such a tree structure is obvious. With all that space available, the complexity of the system can grow to a useful level. But if the disks are minifloppies, wouldn't the sheer overhead render it all a bit pointless?

No. Much of the elegance of this system centers around its device independence, some of which is manifested by its behavior in a floppy system. As is suggested in the drawing, branches of the tree can be interchanged just by "mounting" and "unmounting" diskettes. This adds a third dimension to the structure, for now it becomes dynamic and modular with unlimited depth in a third axis as well as in the other two. And "networking" with other systems simply adds new nodes to the tree.

The CROMIX system incorporates features that allow freedom to move within this structure with a minimum of operational inconvenience. I/O can be redirected to other files or devices, the "shell" that we mentioned earlier is user-configurable for customized environments, and up to six users can be on the system at any given time (given the appropriate amount of RAM—64K each plus 64K for the house). Like "big" systems, it provides password access and file protection at all levels and supports system date and time.

Execution is fast. Unlike CDOS, the directories are all resident in RAM when the disks are "mounted" on the system, and bank selection and context switching are used instead of "swapping" on disk. Multiple tasks thus appear to be simultaneous, although at the CPU level, of course, they are not. But a flexible operating system like this actually becomes "the machine" from the user's point of view, further supporting our concept of insulation from the system details. Somewhere down there, beneath the user interfaces, the hierarchical structure, the system calls, the resource manager, and the device drivers—there is a lonely $20 chip, frantically servicing multiple levels of interrupts and concurrent processes, wistfully, perhaps, remembering the days of long boring hours in a tight little keyboard wait loop. "Did anybody hit a key yet? No? Well, then, did anybody hit a key yet? No? I see. Um, did anybody hit a key. ..."

With a bit of horsing around, files created under CDOS can be supported and/or converted to the CROMIX environment. The system sports a CDOS emulator that runs as a subtask, masquerading as the other operating system. But it is reasonably safe to conclude that in time, tree structure and all of its inherent potential for power and flexibility will take over the micro world and further blur the already vague distinctions among micros, minis, and maxis.

Now that we have established a realistic system environment that is somewhat more substantial than the unaided CPU, let's turn our attention to an overview of a couple of languages.

8.3 PROGRAMMING LANGUAGES

In Chapter 7 we spoke extensively of the philosophies underlying computer programming, and in the first part of this chapter we attempted to express the structure and behavior of an operating system. It remains now for us to look at the languages which effect that much-touted conversion from one level of abstraction to another.

It is important to note, first, that languages—high level or low—do nothing to extend the potential of the computer. That is already defined by the instruction set and the hardware architecture. But they do allow vast increases in the efficiency of a programmer's time by allowing problem expression in a form that is much less arcane than explicit definition of machine-level operations. As the ratio of the number of machine instructions to a line of corresponding source code increases, so does the *level* of the language. Thus, assembler, with a 1:1 correspondence, is considered a low-level language, and complex math applications languages such as APL are considered very high level.

But this is a bit misleading. We tend to confuse the "level" of a language with some sort of value judgment about its power, and nothing could be further from the truth. PL/M, for example, is officially lower level than BASIC if the object/source code ratio is taken as an absolute. But PL/M accommodates features that channel the programmer into a much more structured design than does BASIC, which encourages sloppy and amorphous programming. So is BASIC better than PL/M? Is any one of them better than any other? One might find FORTH to be nearly optimum for machine control, COBOL to be just fine for an accounting job, and PASCAL to be preferred for general-purpose development.

Something else: one of the popular marketing buzzwords in the language business is "portability." Presumably, the fact that a high-level language allows the source code to be machine-independent would suggest the code could be carried from processor to processor, working equally well on each as long as the language translator was implemented. This is fine in theory.

The problem here is that portability can exist only if languages are truly standard, and that doesn't seem to happen very often. It is not at all uncommon for a language to be developed and hustled by a system manufacturer—who of course wants to highlight the particular features of his machine. The language implementation, therefore, is probably a dialect or "enhancement" of a standard that incorporates all sorts of delightful features ... but is incompatible with other languages of the same name. There must be a hundred BASICs by now, and the market offers a number of variants on PL/I, LISP, FORTH, C, PASCAL, and most others. Some of the languages, of course, are not products of a specific company but instead emerge from a standards committee; but by the time they do, they are squabbled over, compromised, and may border on obsolescence.

And then, there are the problems with machine-level accessibility. A nice insulating blanket of language abstraction is indeed a delight, until you have to bang an output port every 213 μs or respond within a dozen machine cycles to an interrupt. If the machine is a vague and distant entity, it becomes a bit difficult to accomplish tasks like this.

Thus, there must be a break in this insulation somewhere which allows access to the lower levels. Lower than the system calls, lower than the device drivers: you must be able to read port 14, rotate it left, and issue an acknowledge pulse on bit 7 of port 22. This is very hard to do in COBOL.

Cleverly implemented languages accommodate this by allowing relocatable modules written in other languages to be linked with their own object code. In the Cromemco system, for example, a single application program can have a language parser written in LISP, a number cruncher composed in FORTRAN, a real-time executive executed in assembler, and a few miscellaneous library routines thrown in for good measure. This enables one to take advantage of different languages' particular areas of competence but to turn to others when the competence begins to wear a little thin.

Incidentally, one of the problems with PASCAL and BASIC is that in most implementations, great effort must be expended to assemble a system in a modular fashion. As we shall see in Chapter 11, there is a lot to be said for this approach, and it is good to choose a repertoire of software tools that are somewhat accommodating.

The final point in this general overview is that the increase in programmer efficiency that we noted is subject to a trade-off in memory requirements and execution speed. There is a hefty overhead associated with a language implementation, and it is definitely felt in the robustness of the system resources that become necessary to support the performance of an apparently simple task. But programmer time is expensive, at every level from system specification to field maintenance, and it is often worthwhile to spend a little more on the machine and let it host some high-level tools. (As time goes on, hardware gets cheaper and brains get more expensive. It's a strange business.)

The remainder of this chapter would be hopeless if we even attempted to meaningfully compare all of the popular languages. This is not a programming book. What we will do is look at a powerful assembler and then contrast that with an implementation of PL/I. This should suggest the general flavor of the level trade-off without committing us to an overly ambitious stab at comprehensive coverage of a turbulent field.

8.3.1 A Macro Assembler

By definition, the most flexible language available is the assembler. Unconstrained by high-level constructs chosen by the language designer and cast in concrete, the assembler can express absolutely anything that is executable on the machine. It allows code optimization to whatever degree the programmer wishes for those situations in which space is at a premium; it provides predictable timing; it can be single-stepped in a fashion that makes some sense in the context of a source listing. It may be a pain to use, but it certainly is flexible. It is like offering somebody 26 English letters as components of a book versus offering a specific set of canned paragraphs with certain words left as plug-in parameters.

As we have said, the assembler itself is a program that accepts a "source file" as input data. It scans through it line by line, ignoring anything that begins with a semicolon (that indicates a comment) and keeping track of labels. It typically accomplishes the assembly in two passes through the file—one to identify all the labels and macros and the other to resolve all the references to them with the definitions that were established the first time.

As an example, let's take the COUNT program we have mentioned a few times and examine its innards. The combined source file and assembler output is shown as Listing 8-1.

The first line,

NAME   COUNT

establishes a module name for the program. After assembly, when it is time to link this together with other modules to build an executable program, the name COUNT is used as a reference. Actually, we could get away without it, because the system will automatically assign the first six characters of the filename in the absence of the NAME pseudo-op.

```
CROMEMCO CDOS Z80 ASSEMBLER version 02.15                        PAGE 0001
WORD COUNT UTILITY PROGRAM

                      0002         NAME    COUNT
                      0003 ;
                      0004 ;This program wanders through a text file and accumulates the total number
                      0005 ;of words (up to 65,535). A "word" is any group of one or more characters
                      0006 ;delimited by a space, a hyphen, or a carriage return.
                      0007 ;Execution is invoked with the standard CDOS command format:
                      0008 ;           COUNT file-ref
                      0009 ;The assembled file COUNT.REL must be linked with ASMLIB to provide the
                      0010 ;disk access and binary-decimal routines.  These are characterized as foll
                      0011 ;
                      0012 ;     FNAME - Sets up extended FCB from an FCB.
                      0013 ;     ZOPN  - Opens an existing file.
                      0014 ;     ZIOER - Prints file error message from ZOPN.
                      0015 ;     ABORT - Prints a message and exits to CDOS.
                      0016 ;     GCHAR - Inhales one character from the file.
                      0017 ;     BINDB - Converts binary number to decimal string.
                      0018 ;
                      0019 ;HISTORY:       Created 10/3/80 (SKR)
                      0020 ;               Edited for MS62 1/1/81 (SKR)
                      0021 ;               Edited for MS55 1/7/81 (SKR)
                      0022 ;
                      0023         EXTRN   FNAME,ZOPN,ZIOER,ABORT,GCHAR,BINDB
                      0024 ;
0000' 3A5D00          0025 START:  LD      A,(5DH)         ;If the first byte of the filename
0003' FE20            0026         CP      ' '             ;  is blank, print error message
0005' CA6500'         0027         JP      Z,ERROUT
0008' 97              0028         SUB     A               ;Set up file control block.
0009' 215C00          0029         LD      HL,5CH
000C' 119600'         0030         LD      DE,IXFCB
000F' CD0000#         0031         CALL    FNAME
0012' CD0000#         0032         CALL    ZOPN
0015' CC0000#         0033         CALL    Z,ZIOER
                      0034 ;
0018' 210000          0035         LD      HL,0            ;Initialize word count
001B' 229400'         0036         LD      (COUNT),HL
001E' AF              0037         XOR     A
001F' 329300'         0038         LD      (LAST),A        ;  (and the LAST status bit)
                      0039 ;
0022' 119600'         0040 LOOP:   LD      DE,IXFCB        ;Get a character from the file
0025' CD0000#         0041         CALL    GCHAR
0028' FE1A            0042         CP      1AH             ;If it is EOF, quit
002A' 2829            0043         JR      Z,EOF
002C' FE20            0044         CP      SPACE           ;If space, hyphen, or CR, we are between w
002E' 280C            0045         JR      Z,YUP
0030' FE2D            0046         CP      DASH
0032' 2808            0047         JR      Z,YUP
0034' FE0D            0048         CP      CR
0036' 2804            0049         JR      Z,YUP
0038' 0600            0050         LD      B,WORD          ;Otherwise, we're in a word
003A' 1802            0051         JR      TEST            ;Set B=1 and go test
003C' 0601            0052 YUP:    LD      B,INTRVL        ;Set B=0 and test
003E' 3A9300'         0053 TEST:   LD      A,(LAST)        ;Get the last status
0041' B8              0054         CP      A,B             ;Same?
0042' 28DE            0055         JR      Z,LOOP          ;If so, go get another character
0044' FE01            0056         CP      A,INTRVL        ;(Only count word -> interval transitions,
0046' 2007            0057         JR      NZ,SAVE         ;  not the opposite.)
0048' 2A9400'         0058         LD      HL,(COUNT)      ;Else, increment the word count
```

Listing 8-1

Ah yes, new term: A *pseudo-op* is any one of about 30 special instructions which can be embedded in the source file to control the assembler's behavior. They have no equivalents in the instruction set of the processor, and their effects are entirely defined by the assembler; they are, in fact, the embodiment of its power. Macros, conditionals, includes—all are invoked by pseudo-ops.

The listing for COUNT consists of two parts. The original source file, which would be redundant herein, simply consists of everything to the right of (and not including) the line numbers. The file (COUNT.Z80) was created at the terminal with the text editor, in exactly the same fashion as a file of manuscript text. When this file is handed to the assembler, however, it is appropriately masticated and then appears as shown in Listing 8-1. To the extreme left are the relative addresses of the code, followed by the machine instructions themselves, followed at last by the line numbers.

```
CROMEMCO CDOS Z80 ASSEMBLER version 02.15                    PAGE 0002
WORD COUNT UTILITY PROGRAM

004B' 23            0059          INC     HL
004C' 229400'       0060          LD      (COUNT),HL
004F' 78            0061 SAVE:    LD      A,B             ;  then save the new status
0050' 329300'       0062          LD      (LAST),A
0053' 18CD          0063          JR      LOOP            ;    and loop.
                    0064 ;
0055' 218C00'       0065 EOF:     LD      HL,STRING       ;Translate binary count into decimal string
0058' ED4B9400'     0066          LD      BC,(COUNT)
005C' CD0000#       0067          CALL    BINDB
005F' 217F00'       0068          LD      HL,EOFMSG       ;Print it
0062' CD0000#       0069          CALL    ABORT
                    0070 ;
0065' 216B00'       0071 ERROUT:  LD      HL,ERRMSG       ;Print an error message
0068' CD6300#       0072          CALL    ABORT
                    0073 ;
006B' 53706563      0074 ERRMSG:  DEFB    'Specification Error',13
      69666963
      6174696F
      6E204572
      726F720D
007F' 576F7264      0075 EOFMSG:  DEFB    'Word count = '
      20636F75
      6E74203D
      20
008C' 30303030      0076 STRING:  DEFB    '000000',13
      30300D
                    0077 ;
0093' 00            0078 LAST:    DEFB    0               ;Word/interval flag
0094' 0000          0079 COUNT:   DEFW    0               ;Binary word count
                    0080 ;
0096' 00            0081 IXFCB:   DEFB    0
0097' (0022)        0082          DEFS    34
00B9' BE00'         0083          DEFW    IBUFF,0
      0000
00BD' 04            0084          DEFB    4
00BE' (0200)        0085 IBUFF:   DEFS    80H*4           ;Disk buffer
                    0086 ;
      (0020)        0087 SPACE:   EQU     20H
      (002D)        0088 DASH:    EQU     2DH
      (000D)        0089 CR:      EQU     0DH
      (0000)        0090 WORD:    EQU     0
      (0001)        0091 INTRVL:  EQU     1
                    0092 ;
02BE' (0000')       0093          END     START

Errors              0

Program Length  02BE (702)
```

Listing 8-1 (Cont.)

But we stray. Back to the top of the listing: You may have noticed that the line number of the NAME pseudo-op is 0002. That is because line 0001 was a TITLE pseudo-op, which read

TITLE   WORD COUNT UTILITY PROGRAM

The assembler responded to that by titling the top of every page in the listing accordingly, whereupon a print of the pseudo-op itself became somewhat superfluous.

Lines 0003 through 0022 are *comments*, as decreed by the leading semicolons. They explain the purpose of the program, describe how it is used, and offer tips concerning its operation which may not be at all obvious from the raw listing. They also include HISTORY information, which shows the most recent changes.

One of the keys to good programming style is the liberal use of comments. However lucid and self-evident a piece of code may be when it is written, it is likely to be an obfuscation a few months later when it becomes necessary to patch it. The problem is severe enough when you are dealing with your own code, but things can get downright grim should you find yourself staring at an uncommented listing produced by someone who is no longer with the company, especially if that someone was one of those egomaniacal programmers who deliberately generate obscure code to convince the world of their unfathomable brilliance. Comments allow a reflection of the WHAT level on the complex and counterintuitive texture of the HOW.

Among the comments in Listing 8-1 are brief definitions of six external routines that are needed to make this program work. COUNT itself is just a module that must be linked with these library routines (located in the file ASMLIB.REL) to produce an executable package. They actually map very closely upon some of the system calls, but have been implemented at a slightly higher level to further relieve the programmer of details. The disk I/O routines include all the logic necessary to set up the buffers and character pointers, allowing a straightforward sequence of subroutine calls to handle the details of opening a file and getting data from it.

Line 0023 is another pseudo-op, and tells the assembler that the names of those six routines are *external*. It would otherwise panic and produce an error message when it encounters each reference to one of those undefined labels. The EXTRN simply tags the six labels with a flag indicating that their addresses will be resolved when the LINK takes place. This can be seen in the object code of lines 0031 to 0033 and a few others: the address in the CALL instruction is left as 0000 and followed by a # character.

In the library routines themselves, the same labels have been listed in an ENTRY psuedo-op, which tells the linker that they are publicly accessible by modules that treat them as external. The EXTRN-ENTRY scheme simply provides a facility for connecting modules together, allowing a large task to be broken up into several understandable subtasks.

Line 0025 is the actual start of the program:

> START: LD   A,(5DH)

The statement's label is START, a name by which it can be referenced from any other point in the program. There is actually no reason for it to ever pass this way again, for once the file is open and things are initialized, the program simply reads the text file (file-ref) and exits. But it's standard practice to label the first instruction anyway, and in this particular assembler that name is used as the operand of the END statement (line 0093) to tell the linker where execution should begin.

The instruction itself is a simple data transfer, as indicated by the mnemonic LD, for "load." The destination is A (the accumulator) and the source is (5DH), which is interpreted as the content of memory location 5D hex. If the parentheses had not been there, the value 5DH (0101 1101) would have been loaded into the accumulator.

But memory location 5DH, in CDOS, happens to be where the first byte of a filename resides when someone types a command on the console which includes a file-ref. The logic that begins with line 0025 is simply checking to see if one was really included in the command, and an error message will be displayed if not.

This is easily done: now that the first byte of the filename area has been loaded into the accumulator, we can trivially test it to see if it is something other than a space. How?

> CP   ' '

The CP is the mnemonic for "compare," and simply subtracts the operand (which may be another register, the content of a memory location, or an explicitly defined value such as ' '—which is a space, or 20H) from the accumulator. Although it avoids actually changing the data in the accumulator, it does set certain flags which indicate the results of the subtraction, and through these, subsequent conditional tests can determine whether they were equal (the result was zero), whether the operand was greater than the accumulator (the result was negative), and so on. In this particular case we are interested only in equality, so the test is

> JP   Z,ERROUT

or "Jump if zero to the error routine." ERROUT is the label of the code beginning at line 0071 which points to the error message and returns to CDOS via a library routine that takes care of printing it.

Once it is established that there is at least something masquerading as a filename out there, the logic sets about building an extended file control block. This is rather trivially accomplished by calling the FNAME routine after subtracting A from itself to clear it, pointing to the system FCB with HL, and pointing to the extended FCB with DE (an XFCB is basically an FCB, but it includes character pointers to keep track of the current position in the file). The program then calls ZOPN, which either opens the existing file or returns with the zero flag set and HL pointing to an error message ("File not found"). The instruction in line 0033 tests the zero flag and returns to CDOS via an error display routine if it is set.

Those nine lines of code just took care of checking for erroneous or missing file reference and opening the file if everything was OK. The blank comment line at 0034 indicates a logical break, since the next step is initialization of the word count logic itself.

This deserves a bit of idle commentary. We are not really writing a manual of software style, but there are some things that have such sweeping effects on the programming process that they deserve mention anyway. One of them is initialization.

It may seem obvious, but forgetting to initialize some important variable has probably been responsible for at least 1 million hours of debugging time since the dawn of computing. There have probably been another million or so lost to redundant initialization: it takes forever to get to 100 if you set a counter to zero, increment it, check to see if it has reached 100, and then jump back to the initialization if it hasn't.

When a program is intended to integrate with others in a modular fashion, establishment of calling parameters and other initial conditions is crucial. In a monolithic program of your own devising, it is occasionally safe—well, sort of—to make certain assumptions about the conditions of variables that are not explicitly set to a needed starting value. "Oh, no problem, the E register was cleared back over here someplace. ... " But somebody might decide to use the E register for something, and then go quite berserk trying to figure out why an innocuous change precipitated apparently unrelated problems in a different part of the program. Avoiding this may occasionally call for instructions that seem redundant, but it helps. It is also good practice to avoid leaving trash in the wake of a subroutine that could cause confusion to others.

Anyway, lines 0035 to 0038 of the COUNT program are involved with variable initialization. The value called COUNT (yeah, same name—interesting coincidence) is a two-byte location somewhere in RAM, and it is handily set to zero by first loading the HL register pair with zero and then loading COUNT with HL. LAST is a single byte, and is similarly cleared via the accumulator. Line 0037 is a stylistic carryover from the days when memory was expensive: instead of an easily understood "LD A,0" it performs an exclusive-OR operation between the accumulator and itself, which has the effect of clearing it. Saves a byte, you know—AF versus 3E 00. Line 0028 accomplishes exactly the same thing by subtracting A from itself.

These little *faux pas* could have been trivially edited out of this program before its immortalization in print, but we left them in to illustrate a point. Obscure code is a pain. If someone were just handed this listing and directed to add a patch, and that someone did not happen to be intuitively aware of the fact that exclusive-ORing something with itself results in zero, then that statement might give pause. It is comparable to sparse use of comments, placing an unreasonable burden on anyone who might happen by with a need to understand the code. In these days of expensive people and cheap memory, it makes more sense to write straightforward and understandable programs; it pays off when debugging time rolls around ... and around ... and around.

There is little point in an exhaustive study of this little program. It just initializes a counter and a flag which indicates whether the last byte was part of a word or part of an interword interval, then begins reading the specified file character by character while counting the number of times the LAST flag changes from "word" to "interval." Spaces, carriage returns, and hyphens are treated as delimiters, and the program is not smart enough to recognize and ignore embedded formatter commands such as "@B = 1" (which results in a blank line in the printed output). But they are of little consequence.

When the loop which does this encounters a 1AH (control-Z) from the text file, it knows that it has reached the end. The two-byte location known as COUNT at this time contains the number of words in binary, so it is passed to a library routine called BINDB, together with a pointer to a string area (defined in line 0076). After execution, the ASCII decimal equivalent of the 16-bit value resides in STRING, with blanks replacing the leading zeros. It happens that this immediately follows the character sequence "Word count = " in RAM, so that the whole combination, addressed as EOFMSG, is printed by ABORT and control returns to the operating system.

We should make a few other random observations about the listing. First, note that the character strings have been assembled into a sequence of ASCII values: ERRMSG, for example (line 0074), has become a list of 20 hex bytes starting at address 006B'. (The apostrophe after the address means that it is relative to the actual starting address of the program that will later be assigned by the linker.) Enclosing the ASCII string in single quotes in a DEFB (DEFine Byte) pseuso-op tells the assembler to consider it a "literal."

Another form of the DEFB statement is shown in line 0078. Here, it is used to set aside a single byte in memory for use as a flag, in the process initializing it to zero. The explicit initialization about which we agonized a few moments ago is thus technically unnecessary, but (1) it is good practice, and (2) data areas are normally set aside with a DEFS (DEFine Storage) psuedo-op which does not itself preset the data to any particular value.

DEFW is just like a DEFS, but it defines a 16-bit word, not an 8-bit byte.

Lines 0087 to 0091 are called "equates," as indicated by the EQU pseudo-op. These reserve no memory, but instead establish an equivalence between a label and a value. SPACE, for example, is defined as 20H in line 0087. Note that line 0044,

CP   SPACE

is assembled into address 002C as FE 20. The FE means "compare A with the byte following this instruction," and the 20 is the hexadecimal equivalent of the ASCII code for a space. It all ties together very neatly.

That, more or less, is a typical assembler program. It is written for the Z-80, and is guaranteed not to work on anything else. To be understood fully, it requires intimate knowlege not only of the assembler's proclivities but also of the architecture of the Z-80 CPU. Execution is fast and proceeds precisely as the step-by-step sequence dictates, but it probably took a lot longer to code and debug than the equivalent written in an appropriate high-level language.

The trade-offs are there, and they are brutal. It is tempting when viewing it from one standpoint to say, "Ugh, why would anybody want to write programs at that level?" But there are times when it is the only reliable approach—at least for a machine-interface module that can plug into a system written in something at a higher level.

Actually, a robust assembler like the one we are describing contains a few features that take some of the sting out of working at the extreme HOW end of the spectrum. Some of the pseudo-ops offer some very powerful functions.

The *macro* facility, for example, allows new instructions to be created that not only permit the use of a single name in lieu of a block of code but also accept a string of parameters which modify the behavior of the code actually generated. Consider the following macro definition:

DISPLA:	MACRO	#MSG
	CALL	SSDCLR
	LD	HL,#MSG
	CALL	SSDISP
	MEND	

Further, consider the following message statements:

MESG1:	DEFM	'Self-Test Mode'
MESG2:	DEFM	'Lamp Test'
MESG3:	DEFM	'Receive Message Overflow'
MESG4:	DEFM	'Terminal Off'
MESG5:	DEFM	'Terminal On'
MESG6:	DEFM	'Badge Data:'
MESG7:	DEFM	'Transmit Error'
MESG8:	DEFM	'Receive Error'

(These are the internally generated messages that can appear on the self-scan displays of an IDAC/15 data collection terminal; all other operator communication comes from the host system.)

The code bearing the label DISPLA is called a *macro definition*. The three instructions between MACRO and MEND simply call a subroutine that clears the top display, point to an ASCII string in a "define message" pseudo-op, then call a subroutine that takes care of the mechanics involved with its actual output to the display hardware.

None of this is particularly frightening, but it is a mild programming nuisance. If the need to display messages arises frequently, the normal response would be to simply make an equivalent subroutine (DSPMSG, for example) out of those instructions instead of repeating the sequence each time, but look at this: Each use of the subroutine would still require the programmer to pass the message number.

```
LD        HL,MESG3
CALL      DSPMSG
```

It would be much nicer if the programmer could just say

```
DISPLA       MESG3
```

Here is what happens. The macro definition is assembled early in the program—before any references to it are made. The first line establishes the macro's label and any parameters that it can accept, in this case the message label. The next three instructions take care of getting it displayed, but note that the value with which HL is loaded is the same as the *formal parameter* (prefixed by the # sign) in the macro definition's first line. Finally, MEND informs the assembler that the definition is complete. Now, whenever "DISPLA MESGn" appears somewhere in the program, it is replaced by the body of the macro, but with MESGn plugged into the LD instruction which sets up the pointer to one of the messages. The result is easier programming and cleaner code, since the assembler can be directed to suppress the printing of "macro expansions."

This is actually rather provocative. With macros, program readability can be vastly improved by defining complex blocks of code and packaging them as single instructions. These can even be collected in a library and made available to a number of programmers, who at some point might find that entire programs can be comprised of nothing but macro calls. Is this not bordering on a high-level language? At what point does the macro assembler become a compiler? Is there any limit to the "level," as determined by the source/object code ratio?

Actually, this phenomenon would be sorely self-limiting if it were not for a few features we haven't mentioned yet. Macro definitions can be nested indefinitely and their calls can be nested to eight levels. Also, there is an interesting feature called INCLUDE which allows blocks of source code to be pulled in from other files at any point in a source program. And third, there is a wondrous facility called *conditional assembly.*

Conditionals, otherwise known as IF statements, are another of those relatively esoteric features that are sadly underutilized by the majority of programmers. Most commonly seen in association with macros, IF statements allow certain pieces of code to be assembled or not, depending on the satisfaction of certain conditions. It must be understood that the assembler does not look within the program's logic—the meaning of the code itself is unknown to it. It looks instead at the values of data items maintained within the assembler itself: counters, Boolean expressions, labels, and so on. An example will help.

```
BUFSAV:      MACRO       #REG,#POINTR,#BUFFER,#DIR
        ;
             LD          HL,(#POINTR)
             LD          (HL),#REG
        ;
             IF          #DIR EQ 1
             INC         HL
             LD          (#POINTR),HL
             LD          A,(X#BUFFER)
             LD          DE,#BUFFER
             ADD         A,E
             CP          A,L
             ENDIF
        ;
             IF          #DIR EQ -1
             DEC         HL
             LD          (#POINTR),HL
             LD          A,(X#BUFFER)
             CP          A,L
             ENDIF
        ;
             MEND
```

Well... maybe an example will help. Without getting too carried away, let's see approximately what this macro accomplishes.

In most systems that need for any reason to interact with an operator at a terminal, there is a line buffer (see "key buffer" in Figure 11-4) which accumulates the characters as they are entered from the console. In the simplest implementation, this is maintained by a small routine which uses an address pointer to keep track of the current buffer location, and which typically watches for a termination character such as the RETURN key to signal the validity of the buffer's contents.

That's the basic theme, but there is usually a need for a little more than that. Human beings are rarely infallible, and even the most careful ones occasionally make typographical errors. At the very least, therefore, such systems should provide a backspace or delete key which allows the operator to erase the last character typed from both the screen and the buffer.

As system complexity grows, it can even become worthwhile to maintain more than one buffer. This allows a hierarchical structure without too much data shuffling: the data from the keyboard (or wherever, actually) can be shunted into the appropriate slot as determined by the system's flow of control. Should there be separate subroutine calls with corresponding setup requirements, or could this be handled with a macro? Well, we must have chosen this example for some macro-related reason, so let's look at the code.

The first line, like DISPLA, is the MACRO psuedo-op which announces to the assembler that everything up to a matching MEND statement should be taken as a macro definition. This time, however, there are four formal parameters. They allow specification of any source register (other than H and L), any buffer, any buffer pointer, and either forward or backward direction. Thus, to store register E in the buffer called KEYBUF at location POINT, and then increment the pointer, one would just include this instruction in the source code:

     BUFSAV   E,POINT,KEYBUF,1

See how this looks a little like a higher-level language? The equivalent in-line code would get the pointer from memory, move E to A, move A to memory at the pointer location, increment the pointer, make sure it didn't overflow the buffer space, and finally save the pointer back in memory. Not particularly difficult, but a nuisance. The key here is that macros can do a lot more than replace repetitive little groups of instructions.

There is little need to explore this example exhaustively, but let's look at a couple of high points. First, and trivially, note that the operands passed via the formal parameters are plugged in literally, allowing them to become parts of labels [as in the LD A,(X#BUFFER) instruction]. They can also become parts of instructions: RRCA and RLCA, for example, are types of right and left rotates of the accumulator; the R and L can be replaced by a formal parameter which allows the macro call to express the direction.

But the main point of this example is the use of conditionals. After the first two instructions, there is an IF statement:

     IF   #DIR EQ 1

This means that all code between this statement and the matching ENDIF should be assembled if the value of the #DIR parameter is 1. (The calling specification is DIR = 1 if the pointer is to increment and DIR = −1 if it is to decrement.) If it is not 1, the assembler just skips over that section of code—but then encounters another IF statement which checks for −1. Since the parameter must always be one or the other, either the top or the bottom half of this macro definition will end up being assembled into the object code.

On a couple of occasions, we have made reference to a "matching" pseudo-op, as above, with the ENDIF. This is to highlight the fact that both macros and conditionals can be deeply nested to implement complex logic. If the program just watched for the next MEND or ENDIF after a MACRO or an IF, the utility of this resource would be sorely limited.

One more example, this time from the realm of art.

Back in Section 8.0.1 we made a few brief comments about art versus engineering. The general conclusion was that the tools and disciplines of engineering are powerful resources that can stand unspectacularly on their own or perform miracles in the hands of an artist.

Occasionally, a problem comes along that cries out for elegance. The old reliable pedestrian "turn-the-crank" approach isn't good enough: it's time to have a little fun—not at the expense of future programmers saddled with an update task, of course, but fun nevertheless.

The problem is this: A high-resolution color graphics display system is implemented on the machine, and in addition to producing flow diagrams of an industrial process, it is necessary to display various sizes of text. In order to do this, a *character generator* has to be created somewhere, presumably in the software, which accepts as input a screen position, a character size, and an ASCII byte. Let's not trouble ourselves with anything but the block of stored data which defines the dot patterns that make up the characters.

CRT terminals, of course, incorporate such things quite routinely. Numerous chips are on the market which return in response to an ASCII code and a row number a group of bits which together represent a horizontal slice across the character at the specified level. The venerable 2513 contains 5×7 patterns for the uppercase ASCII subset. The Motorola 6571 contains 7×9 patterns for a full 128-byte character set, with the control characters displaying as Greek symbols. Nice.

Let's steal the internal coding of the 6571 and use it for the graphics display. Each ASCII character can be expressed as a nine-byte sequence, as demonstrated in Figure 8-9 for the uppercase "A." There is also a single bit needed to flag characters that are "descenders," such as g, j, p, q, and the lowercase Greek beta, gamma, mu, rho, and chi—also comma and semicolon.

So how do we represent this in assembler source code such that the data can be made accessible to some character generation logic? It's simple, really:

```
UCA:    DEFB    1CH, 22H, 41H, 41H, 7FH, 41H, 41H, 41H, 41H, 0
UCB:    DEFB    7EH, 21H, 21H, 21H, 3EH, 21H, 21H, 21H, 7EH, 0
UCC:    DEFB    1EH, 21H, 40H, 40H, 40H, 40H, 40H, 21H, 1EH, 0
```

and so on. Each table entry is just a labeled ("UC" means "uppercase") define-byte statement that consists of the nine successive horizontal slices, plus the descender bit.

Let's see ... nine bytes per character, plus a bit. Awkward. Of course, only 7 bits of each byte are actually used, and the descender bit can easily be packed into one of those. Nine bytes. Still awkward—that's 1152 bytes for the full-character set, and the intervals aren't even powers of 2.

Actually, of course, the data requirements come out quite neatly to be 64 bits per character, or eight bytes. The 7×9 character cell makes 63, plus the descender bit. What we need is a clever way to pack the data. Are you ready?

Suppose that we design a macro called PACK which accepts as its input parameters the straightforward horizontal slices as shown above, and then somehow rotates and packs them accordingly to use up the space wasted by the high bit of each one? This would allow the 128-element character set to reside in 1024 bytes of memory, saving 128 bytes and providing us with considerable aesthetic satisfaction in the process. The character definitions would all start on

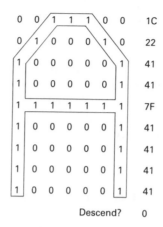

0	0	1	1	1	0	0	1C
0	1	0	0	0	1	0	22
1	0	0	0	0	0	1	41
1	0	0	0	0	0	1	41
1	1	1	1	1	1	1	7F
1	0	0	0	0	0	1	41
1	0	0	0	0	0	1	41
1	0	0	0	0	0	1	41
1	0	0	0	0	0	1	41

Descend?   0

Figure 8-9 The data in the character generator is a 7 × 9 map of the graphic shape corresponding to each ASCII code. Each "1" bit will produce a dot at the corresponding position in the display, and the "descend" bit is used for certain lowercase letters.

power-of-2 boundaries, making the arithmetic of indexing trivial, and any objection to the complexity of the required unpacking process can be dismissed with the observation that the graphics display (a Cromemco SDI) uses a complex byte-pixel mapping that requires some bit manipulation anyway. So we will not be spending our saved memory on complex code that unscrambles the data.
Here's what it takes:

```
Z:        EQU       0FFH
PACK:     MACRO     #P, #Q, #R, #S, #T, #U, #V, #W, #X, #Y
          DB        (Z AND (#P  SHL 1))  OR    ((#Q  SHR 6) AND 1)
          DB        (Z AND (#Q  SHL 2))  OR    ((#R  SHR 5) AND 3)
          DB        (Z AND (#R  SHL 3))  OR    ((#S  SHR 4) AND 7)
          DB        (Z AND (#S  SHL 4))  OR    ((#T  SHR 3) AND 0FH)
          DB        (Z AND (#T  SHL 5))  OR    ((#U  SHR 2) AND 1FH)
          DB        (Z AND (#U  SHL 6))  OR    ((#V  SHR 1) AND 3FH)
          DB        (Z AND (#V  SHL 7))  OR    ((#W      )   )
          DB                (#X SHL 1)   OR    #Y
          MEND
```

Now, if a typical line of source data looks like this:

```
CHI:      PACK      08H,49H,2AH,2AH,2AH,1CH,08H,08H,08H,1
```

then upon its assembly, the macro expansion is evaluated and the 10 parameters are plugged into the formal parameters P through Y. Eight bytes will be produced in the object code, one for each "DB" statement (the same as DEFB) in the macro body.

This is a perfect illustration of the assembler's power to perform Boolean operations above the level of the program code. It treats the parameters as 16-bit values, but we want 8-bit ones, so we precede the macro definition with an equate statement that defines Z as 00FF for masking purposes.

The collective intent of the macro's bit manipulations is the packing of successive data bytes. The first one will be made up of P shifted left one position and Q shifted right six positions: since P and Q are 08H (0000 1000) and 49H (0100 1001), respectively, the result of this first statement will be 11H (0001 0001). In each case, the Z is logically "ANDed" with anything that is shifted left to keep the values within the 8-bit range, the value shifted right is "ANDed" with the number of 1's necessary to get rid of everything but the bits that can be accommodated within that particular byte (in the first case, only one), and then the two values are "ORed" together to produce the result.

The last line of the macro body plugs in the descender bit, and the process is complete.

We recognize that this is just a bit obscure at first glance, but a complete understanding is not crucial to appreciation of this material. We have attempted to demonstrate that the power of some of the assembler's resources makes it somewhat difficult to scoff at it blithely and poke fun at its "level." Somewhere in the creation of a system, it is necessary to work with bits, but macros and conditionals enable elevation above that as soon as the primitives can be coaxed into operation.

Let's turn our attention next to a healthy high-level language and make a few comparisons.

8.3.2 Counting Words in PL/I

Many high-level languages have made their way into the micro world, with varying degrees of success. BASIC, of course, has been around the longest in this context, but its awkwardness in control applications makes it inappropriate for this discussion. FORTRAN—yawn. There are many.

The current fascination with structure that has arisen from some of the more brutal economic and human realities of the business has hoisted into prominence those languages which are best suited to the "go-to-less" style of programming. Among these is PL/I (spelled with the Roman numeral and pronounced "PL-one"). It has spawned some well-known offspring, notably Intel's PL/M and Zilog's PL/Z.

As we have disclaimed a time or two herein, this is not a book about writing programs. We are therefore more interested in communicating the general capabilities and limitations of the three rough classes of languages than in attempting an exhaustive study of any of them. With that in mind, let's skip the preliminaries and look at a program in PL/I which performs precisely the same function as the COUNT routine we discussed in Section 8.3.1. It appears in its entirety in Listing 8-2.

Note first that, even if you have never seen a PL/I program before, it's a little more readable than its assembler equivalent. It is also shorter. The flow of control can be observed by the program's organization on the page, and there are no explicit jumps that have to be followed to their corresponding labels.

The program opens with a named procedure statement. As before, the name of the module is COUNT, and it is herein identified as the point of entry by the OPTIONS(MAIN) attribute. This simply identifies this procedure as the first one that will receive control when program execution begins.

There follows a DECLARE statement, which defines the data types associated with each of the seven names that will be used in the program. TEXT-FILE will be the internal reference for the file-ref with which the program is called, TEXT-STRING is a character string of variable length up to 80, CHR is a single character, LAST is a single bit, and COUNTER, LEN, and BINARY are all binary numbers. As the program progresses, use may be made of these named data

```
COUNT:
        PROCEDURE OPTIONS(MAIN);

        DECLARE
                TEXT-FILE         FILE,
                TEXT-STRING       CHARACTER(80) VARYING,
                CHR               CHARACTER(1),
                LAST              BIT(1),
                (COUNTER,LEN)     BINARY,
                POINT             BINARY(7);

        %REPLACE
                TRUE    BY '1'B,
                FALSE   BY '0'B;

        OPEN FILE(TEXT-FILE) TITLE('$1.$1');

        ON ENDFILE(TEXT-FILE) BEGIN;

                PUT SKIP EDIT('Word Count = ',COUNTER)(A,F(5));
                STOP;
                END;

        COUNTER = 0;

        DO WHILE(TRUE);

                LAST = FALSE;
                GET FILE(TEXT-FILE) EDIT(TEXT-STRING)(A);
                LEN = LENGTH(TEXT-STRING);
                DO POINT = 1 TO LEN WHILE(LEN ^= 0);

                        CHR = SUBSTR(TEXT-STRING,POINT,1);
                        IF (CHR = ' ') ! (CHR = '-') THEN DO;

                                IF LAST THEN COUNTER = COUNTER + 1;
                                LAST = FALSE;
                                END;

                        ELSE LAST = TRUE;
                        END;

                IF LAST THEN COUNTER = COUNTER + 1;
                END;

        END COUNT;
```

Listing 8-2

without concern for their location in memory or their formats (contrasted with DEFB, DEFW, etc., in assembler).

Following the DECLARE, the words TRUE and FALSE are assigned binary values of 1 and 0, allowing them to be used later in testing.

In the assembler version, the process of file initialization required some understanding of the system's file structure. It wasn't too onerous, thanks to the system calls, but it still required the programmer to know something about the structure of the file control block and the proper protocol for opening a file. But in the PL/I equivalent, it takes just one statement:

OPEN FILE(TEXT-FILE) TITLE('$1.$1');

The apparently arcane bit of notation with the dollar signs simply means that the file to be opened is the file-ref included in the calling command from the operating system.

Before operation can actually begin, it is necessary to tell the system what to do when the end of the file arrives. The next four lines accomplish that: on end of file, display the message "Word Count = xxxxx" on the console. The parenthetical expression at the end of the PUT statement simply defines the format of the two components in the message—the string "Word Count = " is ASCII, and the value called COUNTER is to be displayed as a five-digit fixed-point number. Note that it is not necessary to call a routine which translates a 16-bit binary value to an ASCII decimal string, as we did earlier.

Then COUNTER is intialized to zero, and the fun can begin.

DO WHILE(TRUE);—this says essentially that everything between this point and its matching END statement is to loop forever. TRUE is always TRUE. The logic is fairly straightforward: It gets one line of the text file at a time, sets a value called LEN equal to its length, then moves a pointer through it from the first character to the LENth. That is expressed by the DO statement

DO POINT = 1 TO LEN WHILE(LEN ˆ= 0);

The WHILE part of this prevents any attempt to read a line if the length of it is zero.

Note the structure implied by the tabs. Now that we are executing an inner loop, the statements are indented one level further than those immediately preceding. The "CHR = ... " statement just sets CHR equal to the character at the current pointer location within the string, then the next statement checks to see if it is a space or hyphen. If so—indent one more level, and if the delimiter immediately follows a valid word, the counter is incremented. In either case, the value called LAST is set to represent the delimiter condition.

But if CHR had not been a space or hyphen, the innermost statements would not have been executed. Instead, the ELSE condition would be satisfied, and LAST would be set to TRUE, indicating that the character was part of a word.

The final statement before the ENDs takes care of the carriage return, which is also a valid delimiter as long as it is not preceded immediately by another one (null line). Thus

IF LAST THEN COUNTER = COUNTER + 1;

prevents successive blank lines in the input file from confusing the count logic.

The program closes with an END statement which, like the assembler equivalent, reflects back to the starting point.

PL/I notation may seem a bit obscure at first glance, but it offers some spectacular advantages over assembler in many applications. Most notably, the logical architecture of the program can be discerned upon relatively casual examination of the listing. Instead of requiring the design of complex intertwining paths at the machine level, the language allows problem expression in a structure that is representative of the task itself. This can be a real boon to people who aren't intimately familiar with the processor architecture, and it is substantially more convenient even for those who are.

But what of that problem we mentioned earlier concerning gaps in this thick cushion which allow for the inevitable necessity of intimate contact with the machine? In any control task worthy of the name, there is some point at which the standard system-level device assignments become useless and the necessity of bit-level interface arises. PL/I offers no constructs which are particularly useful in this mode, and it is certainly quite impossible to optimize or even predict the execution speed of critical subroutines.

No problem. Like the CDOS assembler, PL/I produces linkable object code. It is a simple matter to define certain procedures EXTERNAL, then create machine-level routines for later linkage with the higher-level code. The net effect of this approach is the best of both worlds: something reasonably close to WHAT for the sake of human sanity and the rest at the convoluted HOW for the sake of machine efficiency.

Before exploring the impact of all this on the industrial microcomputer design business, let's spend a chapter on that magical region between languages like this and those like our own. If computers are so nice, why do we have to go to so much trouble to accomplish such a frivolous bit of trivia as counting the words in a file? Shouldn't we just be able to say, "Uh, BEHEMOTH? How many words are in Chapter 8?"

9

Artificial Intelligence

9.0 INTRODUCTION

"Artificial intelligence." That's one highly charged phrase. Uttering it in random company is likely to trigger responses ranging from the inane ("That's what we've got around here!") to the paranoid ("Oh, that's scary. . . . "). The general public has enough trouble with simpleminded computers—the very notion of "machines who think" is sometimes a little threatening.

Actually, from one point of view it ought to be less threatening than what we have now. How often would a little machine intelligence have softened the impact of "computer errors" on the individual consumer? Wouldn't a modicum of adaptability, a touch of wry humor, and a *soupçon* of grace have a sweeping effect on the popular conception of computers?

AI is not a new idea. Nor is it limited to the science fiction genre. Some of the foundations that underlie today's research were laid in the late 1940s and early 1950s by Alan Turing, whose prophetic comments were considered heretical in his day: "I believe that at the end of the century the use of words and general educated opinion will have altered so much that one will be able to speak of machines thinking without expecting to be contradicted."

Perhaps a discussion of intelligent systems seems to be in sharp contrast to our loosely stated objectives. Sentient solenoid controllers? Why not? There is some level of human interface in just about every system, and the old

WHAT–HOW distinction applies just as forcefully in the industrial control milieu as it does in programming. Consider:

1. Current machines typically involve a seemingly arbitrary set of commands and procedures from the industrial user's point of view. The ability to extract desired parametric and control information unambiguously from an applications-oriented, natural language input would revolutionize operator training and employee relations.

2. Typical control systems are designed to perform the general function suggested by Figure 5-30, in which something called the "process" spends its time modifying the outside world in a fashion calculated to render subsequent observations of it more nearly congruent with a "goal state." But closed-loop behavior notwithstanding, the conceptual space in which said behavior exists is fixed: the system can no more modify the theoretical basis of its own responses than can it step outside the fabric of the software. There is no meta-level of operation. But would it not be a pleasure if the system could determine the appropriateness of its control algorithm and continually adapt it to the task based on an ideal conceptualization of its own performance?

3. It is rare that a control system is programmed so meticulously that there is no abnormal combination of input conditions that can fool it. Even with preset reasonable limits for sensed values, it is difficult to express to the system that there may be combinations of inputs, each of which is in range, that collectively represent a potential failure mode. But an intelligent system with contextual understanding of the process (instead of an instantaneous snapshot of machine conditions) could react in a much more sophisticated fashion to subtle variations in the overall tone of its sensory inputs.

It seems, then, that we are not wandering off into an esoteric domain when we turn our attention to AI. Once a part-time pursuit of a few restless academic visionaries, it has grown into a full-fledged science, replete with robust subspecialties, societies, annual international conferences, and journals. The enchantment of AI is seducing some remarkable minds, and in the process, the computer is being dramatically transformed.

9.1 KNOWLEDGE ENGINEERING

It has been estimated that human knowledge is expanding at the approximate rate of 200,000,000 words per hour. Although it is a little difficult to determine exactly what is meant by "words" in that observation, it does suggest the magnitude of the task that confronts us as we attempt to introduce some sufficiently flexible organizational systems to store the knowledge effectively. On an endless variety of subjects we have accumulated such masses of information that only the narrowest of specialists in any field can realistically claim to be an expert.

Most computer applications in some way represent a distillation of either general or task-specific knowledge. This presents very few problems when the scope of the task is limited to a simple control system, but with the advent of cheaper memory and more powerful resources we are seeing a trend in the direction of complex distributed systems. In Chapter 13, in fact, we deal somewhat with a plant-wide hierarchical network of processors which are deeply intertwined with the controlled processes. It becomes necessary in such a situation to incorporate knowledge into the system beyond the "If A > 100 THEN DECREASE X" level. A production environment is not static; nor are the people who run it. If they are enslaved to a computer network whose behavior is frozen in time at the the moment of conception, then the entire integrated system—corporate bottom lines included—suffers mightily.

But what is knowledge? Can it be summarized as a set of numeric goal states and data structures? Surely there's more to it than that. We can divide knowledge into three distinct types, which differ considerably in their character.

The first, and most obvious, can be called *factual knowledge*. Made up of relatively hard data—textbook-style information and other facts that comprise the bulk of libraries and data-base resources—this type of knowledge is relatively easy to represent in a computer system. In fact, such representation has been going on for decades, in business computers of various sorts which typify the rather pedestrian uses to which the majority of large systems have been put.

The second type of knowledge, however, is a little harder to pin down. Called *heuristic knowledge*, it is the network of intuitions, associations, judgment rules, pet theories, and general inference procedures which, in combination with factual knowledge about a subject, allow a human being to exhibit intelligent behavior. It includes such things as the art of good guessing—not exactly the sort of phenomenon that can be handily embodied in a simple data structure.

The third type, *meta-knowledge*, further muddies the programming waters. It is close in spirit to the heuristic category, being concerned with general problem-solving strategy and nonsymbolic awareness of "how to think," but is unlike the other in not being domain-specific. Where factual and heuristic knowledge are directly involved with a task or application context, meta-knowledge is completely general and abstract.

Even without exploring the nuts and bolts of the AI world, it is obvious that the need for capabilities such as those alluded to above suggests some rather unusual system requirements. In particular, thought processes worthy of the exalted moniker "intelligence" are somewhat difficult to produce in a machine whose information bandwidth is limited by the infamous von Neumann bottleneck. Suddenly, even the cleverest of table-manipulation techniques seems to appear a bit weak.

Human beings, it seems, tend to perform most if not all of their information processing through the simultaneous activation of widely distributed resources. One cannot point at a particular spot in the brain and say "There, that is the CPU." As we discussed earlier in the context of memory, it is much more useful to view the behavior of neural networks in the frequency domain, with individual "thoughts" better visualized as unfocused waves of activity than the firing of particular neurons.

If this is indeed the case, it places a rather heavy burden on computers which seek to emulate their intelligent masters. An analog of that distributed activity has to be funneled through a single byte- or word-wide processing site. It would seem impossible.

Actually, it is a little clumsy, but one thing that computers offer in partial compensation is speed—hardly enough to take the sting out of spatial correlation techniques and other number-crunching attempts to do what we do with nary a thought, but speed nevertheless. With switching times six or seven decimal orders of magnitude faster than ours, computers can occasionally be coaxed into seemingly intelligent behavior.

This general indictment of von Neumann–style sequential machines is not meant to denigrate the work going on in the AI field. Indeed, the tricks and techniques that have been developed to compensate for the fundamental limitation of today's computers are themselves such surpassing examples of applied intellect that their use is almost self-justifying on that basis. They have permitted the development of systems built around carefully assembled knowledge bases that are capable of exhibiting unquestionably intelligent behavior within specific application domains.

Among the more challenging problems facing those who purport to create such automata is the question of knowledge *representation*. In a simple industrial control system, all the necessary data might be stored in the form of a few target values or limits in ROM. But if the system is to be intelligent, then by definition, it needs at least a modicum of heuristic knowledge to guide access to the facts and enable the phenomenon of learning to at least be simulated. This calls to mind something that has been dubbed the "knowledge-search duality" by those of the AI community: the more heuristic knowledge a system has, the less it must search for a specific datum, and the less bogged down it will be when confronted with a highly associative task (such as image recognition) which we, being "parallel" systems, manage to accomplish with a minimum of conscious effort.

Knowledge representation, itself an active subspecialty in the AI world, is simply (!) the formalization and organization of the knowledge in a system. It is not purely a data base, for much of the required heuristics are better recorded implicitly in the system's structure. It becomes, in fact, rather difficult to tell the difference between "program" and "data," for they are each embodied in the other to a significant extent.

There are various approaches to the representation problem. One is called the *production rule*, which is basically an if–then kind of structure (IF the patient appears to have more than a casual obstructive airway disease in combination with significant diffusion defect and higher than normal total lung capacity, THEN there is strong support for a diagnosis of emphysema). Production rules are relatively active entities in a knowledge base.

Another widely used approach, based on more passive symbolic units called *frames* or *schema*, is somewhat more congruent with classic data-base concepts. Here, the representation system consists of various knowledge entities whose interrelationships are defined by lists of properties and linkages. Said properties can include such relationships as "A is a special case of B," "D and F always imply C," and so forth. The individual entities are called *nodes* and the properties are called *links*, resulting in the term *node–link structure* to describe a knowledge base of this type. Forms of this are also known as semantic networks or relational data bases.

Whether through the application of production rules or frames, the selection of a good representation is a central and hotly contested issue in the AI world. One of the primary objectives is a degree of "naturalness" in problem expression which renders comfortable the attempt to deal with the abstract, symbolic constructs which are germane to the knowledge engineering field. Although a highly formal system (predicate calculus) exists, providing a refined set of manipulative tools and a universally accepted means of communication, there is a growing interest in some of the more intuitive approaches.

It is important to note that both types of representation must incorporate "certainty factors" which have a weighting effect on various paths through the network. There is no guarantee, for example, that the two-pack-a-day cigarette habit of the patient was responsible for his pulmonary malaise, but the certainty factor is very high.

Establishment of a knowledge base through either approach does not itself produce machine intelligence. It is necessary to create some symbolic inference procedures which guide the process of drawing conclusions from the knowledge and determine the steps that should be taken to interact with both the data base and the world. Perhaps the chaining of production rules to (hopefully) arrive at a goal is the best approach to a given problem. But sometimes the goal is not known, the input information is incomplete, or there are so many plausible conclusions that exhaustive analysis is counterproductive. All of these situations call for a measure of meta-knowledge which determines the appropriate problem-solving approach, ideally restricting the number of inferential paths actually explored to a minimum by applying constraints and insights gained from experience.

Clearly, causing a computer to exhibit intelligent behavior without the ultra-high-speed exhaustive searches and simulation algorithms that characterize the classic brute-force approach is something of an art. One of the keys seems to be something we alluded to earlier, the incorporation of heuristic knowledge implicitly in the system's structure. The process of learning then modifies the structure rather than the data within it, an effect much more analogous to the brain's activity than the traditional function–data differentiation that pervades the industry.

There are many foci in the AI business, each of which is in some way involved with either spanning that vast gulf between human being and machine or applying the resultant techniques to real-world applications. Rather than catalog the entire field, let's concentrate on a couple of areas which most heavily impact our chosen subject, industrial system design.

9.2 NATURAL LANGUAGE

In Chapter 7 we spoke at some length about the problems of mapping human thought from the WHAT level onto the HOW. These are particularly acute in the task of writing programs, but also restrict the fluidity of human–machine communication at the applications level. There is thus some powerful motivation out there for people to invest computers with a facility for vocal communication and natural language understanding.

These are actually widely divergent kinds of problems. Accomplishing primitive speech recognition is not an overwhelmingly difficult task as long as the desired performance is basically the matching of an input utterance to one of an array of possible words that constitute the system's vocabulary. The problem does begin to appear somewhat formidable as the size of this vocabulary grows arbitrarily large, and performance goes all to pieces when the machine is asked to deal with more than one human speaker, but still, speech recognition is not philosophically remote from the digital signal processing tools of the day.

But once a connected sentence has been assimilated (or simply typed in— let's start small), a vastly different problem confronts the computer. "Uh, now put the other cover plate on top of it, and line it up with the holes."

What is uh? There are three cover plates—which "other" one? On top of what? Line what up? Which holes?

Aha. It seems that the informational intent of that verbal instruction is substantially greater than would be suggested by its explicit linguistic content. If you recall the scenario in Section 7.1, in which human communication was graphically celebrated, you will observe that inextricably linked with the spoken words was a shared context that gave depth to the symbols and antecedents to the pronouns.

This becomes a rather awesome problem for the computer, for if it is to truly understand natural language, its knowledge base must extend far beyond the application domain to include a large body of common sense. When the human being says, "Ah, let's see...," the computer is probably not being directed to fire up its vision system and start frantically performing spatial Fourier transforms on the input array. When the human being says, "Now screw on the cover," the computer must know what screws are, how to make them turn, which way to turn them, how much torque should be applied, and when to quit. It needs to know a lot about screwing—a body of knowledge that is seldom associated with the specific assembly instructions for a carburetor. Everybody knows about that, and if they don't, they shouldn't be trying to assemble automobiles.

Nor is this body of common knowledge a static accumulation of data in a ROM somewhere. Not only does the universe change continuously, but as a conversation progresses the internal states of the participants are modified every moment. The shared context that links them is as fluid and vital as the sounds that fill the air; in fact, more so, since the true communication is an amalgam of expectations, intuitions, visual cues, vocal tonalities, evoked memories, touches, rational deductions, and, oh yes—words. With this larger view of verbal communication, it becomes clear that the information bandwidth of the human–machine link is intrinsically limited, no matter how cleverly we can contrive it. It's going to take a mighty sophisticated processor to note the "body language" of the human converser and subtly modify the derived meaning accordingly.

And what if the human and the computer disagree?

Human: OK, Now let's try to optimize the life of the Blanchard's diamond tooling.

Computer: OK.

Human: Adjust the limits in the downfeed control algorithm such that you get the maximum cutting rate and the minimum tool wear.

Computer: That is a direct trade-off. How shall I weight them?

Human: On the basis of diamond cost and machine throughput.

Computer: As well as downtime for tooling replacement?

Human: I don't think that's significant.

Computer: It certainly is. It takes about six hours for the job.

Human: Sure, but that time can be used for preventive maintenance as well. We wouldn't add any overhead.

Computer: We would if it became too frequent.

Human: It could be scheduled for shift changes.

Computer: Those are not correlated with Blanchard activity.

Human: Well OK, just tip the trade-off to sync diamond replacement with the PM schedule.

Computer: That would force me to ignore grinding time. I prefer to use the downtime figure to adjust the limit values as a function of the machine's work load.

Human: You mean let the diamonds wear faster if there's a lull coming up? That's ridiculous!

Computer: Not if you want true least-cost optimization.

And so forth. Note that the argument is not over facts and figures, but instead over their interpretation. The human and the computer have differing ideas about what is important, and each views the hard facts in that light.

The computer, in order to think at this level, requires sophisticated internal models of literally every process, task, and environmental consideration that affects the factory. Obviously, it becomes exceedingly difficult to view natural language understanding as something distinct from intelligence. Simulation of intelligent behavior with various forms of software trickery (such as the majority of chess-playing algorithms) is a far cry from the creation of truly intelligent systems that can adapt to task domains not originally encoded in their very fabric. Such systems require commonsense understanding of physics, knowledge of human psychology (as evidenced above by the computer's closing *touché*), and readily accessible meta-knowledge that enables the machine to "know" whether or not it knows something without having to exhaustively search a data base.

The natural language problem touches again on the knowledge-search duality that we noted in our discussion of knowledge representation. The fact that communication between person and machine (or person and person, or machine and machine) is even necessary suggests that some level of cooperation is sought that will enable the solution to a problem that a single agent, human or machine, is unable to solve alone. Presumably, said communication takes place during situations in which one agent intends to enlist the aid of the other, and therefore tailors his, her, or its (oboy—I see it coming: you thought sexual bias was a pain?) utterances appropriately. In essence, there is an exchange of knowledge or capability in the task domain for more sophisticated communication skills that potentiate profitable cooperation.

This ties in, of course, with *knowledge acquisition*—that process which takes place once a system is sufficiently imbued with intelligence to begin augmenting its understanding of the world with appropriate queries and investigations. Somewhere during this growth period, there is a subtle shift from programming to teaching, then to mixed-initiative dialogue. The meanings of words are at first inseparable from functions, just as a macro name is equivalent to its behavior, but as the system's intelligence reaches the threshold beyond which learning from the environment can take place, words become more symbolic and less action-oriented. The machine becomes more abstract. It begins to think symbolically, metaphorically—perhaps even appreciating the kinds of context switches that underlie humor, the mingling of levels that gives meaning to art, and the recursive introspection that eventually has to be acknowledged as consciousness. It eschews probabilistic models of intelligence in lieu of multilevel symbolic representation. It takes on a measure of sentience, with language the key: for even with human beings, anything that cannot be symbolized via at least a supra-verbal internal language is of uncertain reality. The thing may be perfectly real, of course, as is the primal "spark" of human awareness, but if untouchable by symbolization it will forever be fuel for theology.

Is this to be the final, subtle distinction that separates biological information processors from those based on silicon? Will it require ever more profound intuition to provide human beings with reassurance of their superiority? Perhaps this, too, will give way, leaving the species feeling cheated once again, just as when an irreverent heretic pointed out that the earth is not the center of the universe and a blasphemer began spouting nonsense about evolution.

We seem to stray. But it is difficult to confine discussion of artificial intelligence within the disciplined realm of engineering. It is comparable to discoursing upon philosophy in the language of neurophysiology: although occasional clever metaphors appear, they are the rare coherent projections of entities existing in at least one more dimension than the domain of the language itself. We can speak of doing a PUSH when we temporarily abandon a line of reasoning to chase down a random thought, but when we sit back and ponder the act of pondering, there is a paucity of engineering analogy.

Language is the key. We have to start thinking of coding depth in an abstract sense, with the information channels between ourselves and our machines as broadband and WHAT-oriented as possible.

Of course, if we look to ourselves for a clue, we discover that the widest information bandwidth in any sensory domain is found in our visual systems. Pictures are truly worth a thousand words: speech, even with a richly textured shared context, has a difficult time competing with vision where description of physical reality is involved. (Can you, unless your features are markedly atypical, describe yourself sufficiently well over the phone that a stranger could recognize you in a crowd?)

Let's consider computer vision for a moment before looking toward the techniques that underlie the quest for machine intelligence.

9.3 VISUAL PERCEPTION

It is a simple matter to hang a CCD imaging array or a standard TV camera on the front end of a computer and scan a scene in such a fashion that an internal representation of the visual field is maintained as an array of pixels. Big deal. That's not information; that's data.

If the computer is to do something with that image other than print it on a T-shirt appliqué, a measure of processing is called for—and a healthy one at that. When the objective is vision, with all that the term implies, the processing requirements become downright formidable. Berthold K. P. Horn of the MIT AI laboratory once commented: "Without the example of human vision, we would have concluded that vision is impossible!"

It's easy to see why. Visual pattern recognition is one of those computational *tours de force* that human beings manage to accomplish all the time with nary a thought. Consider: a friend walks into the room, and you establish his or her (or its?) identity with a casual glance. The accuracy of your decision is not markedly affected by the set of the jaw, the angle of inclination of the head, or the degree of dishevelment of the hair. You simply map an extensively preprocessed visual image in parallel via some feature-extraction hardware onto a gigantic multidimensional associative memory. The answer pops out, linked with an elaborate internal model of your friend. Just like that. Yawn.

A computer, on the other hand, has quite a chore to perform when fitted as above with a TV camera and directed to recognize a face. It must scan the image dot by dot to acquire a numeric representation in memory. Then it must engage in fast and furious number crunching to calculate, with appropriate resolution, the spatial Fourier transform of the face. (Elapsed time at this point might be pushing a minute or more—much more—and the machine still hasn't the foggiest notion of who it is looking at.) Then comes the hard part: one by one, the system must perform two-dimensional correlations between its freshly calculated data and blocks of stored data corresponding to the people it "knows"—in each case, coming up with a number between 0 and 1 which expresses how much alike that pair of images is. The answer, with only marginal certainty, is the one with the largest number.

For if the person in front of the camera parts his hair differently, cocks his head to one side, and takes on a dramatic expression, he might as well have just become someone else.

This problem gives image recognition people fits, and has spawned a number of clever, but so far inadequate, attempts at solution. The catch, of course, is that even with the blinding speed of the computer, a serial approach is necessary, requiring lengthy numeric processes to perform functions that are as natural to frequency-domain biological systems as Boolean logic is to those of the silicon variety. In other words, we are back to the von Neumann bottleneck again: no matter how cleverly we write the program to do this, the machine still fetches one datum at a time from memory and then trundles off with it to turn some primitive logical crank.

But that's not the whole problem, for even the avoidance of this bottleneck with suitable parallel processing techniques would leave us with a conceptual inadequacy. Partially due to the conditioning wrought by years of using computers, we habitually approach a problem with the traditional programming techniques of task partitioning, identification of primitives, and pattern matching. This implies a piecewise conception of the problem even if hardware is contrived that tackles all the pieces at once.

The body of evidence that is accumulating for human vision suggests that this is not at all how we do it. First, instead of pixel-level analysis of a scene, the brain has been shown by evoked-potential studies to react to "gestalts," or global properties of the visual field. By placing an array of electrodes on the scalp of an experimental subject and then repeatedly presenting a stimulus and averaging the response into a buffer, the apparently random noise from the brain can be eliminated to reveal certain characteristic signals that accompany each type of stimulus. This alone is intriguing, but it gets downright provocative when you consider that the evoked response is the same for presented patterns of the same shape but different sizes.

A simple subjective observation will serve to illustrate. Figure 9-1a is a random array of dots formed by going "pfft" a few times with a can of spray paint about 6 feet above a sheet of Mylar film. Part b shows what happens when this is overlaid upon a film positive of itself, then rotated slightly. Your visual system instantly discerns a global pattern, causing you to perceive concentric circles clearly. But local analysis of only part of the image reveals no such grand scheme, but is instead just another array of random dots, as shown in part c.

Computer vision systems of the day would be hard-pressed to extract the

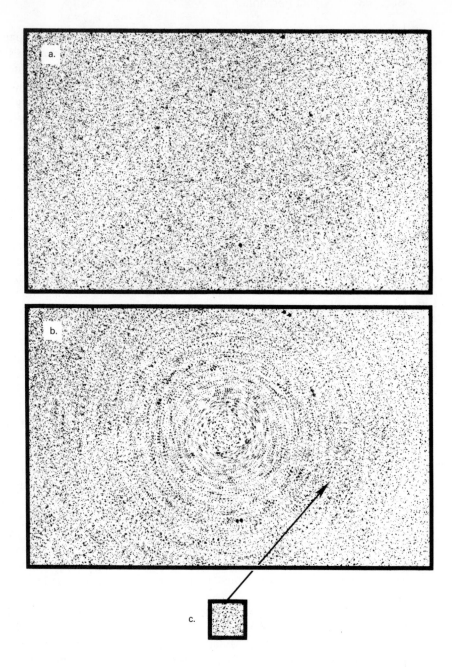

Figure 9-1 The random dot pattern in (a) was superimposed upon a copy of itself and rotated slightly, yielding the circular pattern (b). But the pattern requires holistic analysis to be perceived, something that is easy for humans but difficult for machines. The small patch in (c) shows the sharp diminishing of the effect when the surrounding visual cues are removed.

overall rotation embedded in the random field of part b. An attempt to determine, say, the orientation of close pairs of dots and integrate these angles into a global perspective would likely fail, for only rarely are the dots that give rise to the circular image the closest pairs among their random neighbors. Only a system that sees the entire visual field as a whole could reach agreement with human perception about the nature of the pattern.

The process of hearing, likewise, involves the immediate conversion of sound from the time to the frequency domain at the basilar membrane, which then delivers a continuous parallel wavefront to the auditory cortex. It seems likely that these easily recognized patterns (looking somewhat like real-time FFTs) are comparable to the spatially arrayed data produced by the brain's extensive visual preprocessing activities. It is somewhat enchanting to view this preprocessing as a domain-conversion step, allowing the visual cortex to deal with something loosely analogous to a binocular spatial FFT of the scene.

Some support is lent to this conjecture by a mapping process that apparently takes place between the retina and the cortex. It has long been bemoaned by workers in computer vision that the available repertoire of signal-processing techniques can deal with image rotation only in the most laborious of fashions. Lateral translation is relatively trivial but, as in our opening scenario of human vs. machine facial recognition, the tilting of a head renders the image almost totally alien to the template-matching process. But the human visual system solves that very neatly, with virtually no processing overhead. The cells of the retina are organized in a roughly logarithmic polar grid which is mapped onto a rectangular array in the cortex. As shown in Figure 9-2, this mapping converts rotation of an object in real space into a corresponding vertical translation, and expansion or contraction of the original image into a horizontal translation. Thus, the tilting of the "face" (like its approach or recession) produces no pattern change in the image presented to the brain. (Real-world lateral translation would, but we automatically direct our receptors and visual attention to the subject under analysis, effectively centering it.)

In this logarithmic polar mapping, some of the most classically restrictive problems in image processing are eliminated. As shown in the figure, the circle and line representing the primitive face are invariant upon both rotation and scaling of the original image. As the front end of a computer vision system, an appropriately fabricated fiber-optic array, for example, would have the effect of reducing the number-crunching task of the system to a minimum.

What does that task consist of? Assuming that the intent is pattern recognition rather than abstract scene description, the system's job is the successful correlation of the image with the one that yields the most appropriate information. Until we can emulate our worthy biological paradigm with a holographic system comprised, perhaps, of Josephson junctions in their nonlinear mode, we must, unfortunately, apply relatively conventional array processing techniques (although hopefully, with the aid of some hardware parallelism). If we take the mapped image, transform it into the frequency domain (FFT the rows; FFT the

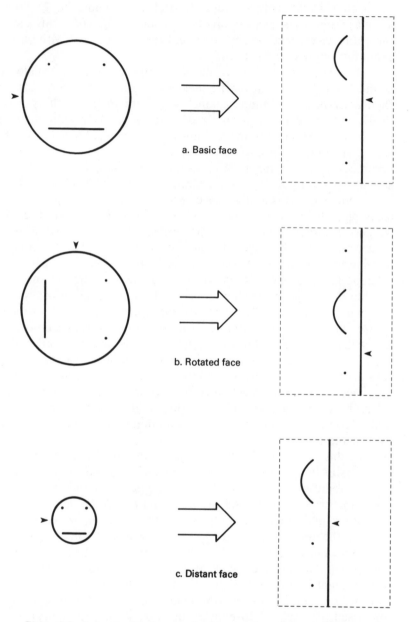

a. Basic face

b. Rotated face

c. Distant face

Figure 9-2 Logarithmic polar mapping between the retina and the visual cortex is partially responsible for a human being's invariant perception of form even upon rotation or scaling of the visual data. This technique can be applied to computer vision.

columns), then apply some heuristic knowledge to restrict the scope of the correlative search (which is simply the relative subtraction of the mapped and transformed images), then we have, *voilà*, a passable image recognizer. (Or, could we perhaps accomplish the mapping in the array-processing step? Hm.)

The catch, of course, is that real-world problems are not simple pattern-recognition tasks. Typically, a system must expect to encounter complex and dynamic scenes that instantly dash our clever algorithms on the rocks. The mapping doesn't single-handedly take care of rotation outside two-dimensional space, nor can it accurately represent objects not centered in its polar frame of reference. Human beings deal with that problem unconsciously, by darting their eyes to and fro to bring the objects of interest into the high-resolution (and centered) macular region of the retina. Present computers must clumsily attempt to follow the scene's highlights with a camera, a task not only plagued by the physical driving requirements but also by the fact that the system, to pull this off, must have enough contextual understanding of the scene to know just what is important and worth focusing upon.

And then there is depth. And color. And texture. And motion. Maybe vision *is* impossible.

9.4 AI PROGRAMMING

Some of this may seem a little esoteric for a book that purports to teach industrial design with microcomputers. It may seem especially so to those conditioned to expect fairly explicit descriptions of engineering tools.

But why? As we saw in our trip through the WHAT–HOW spectrum, tools exist at every level. They may be primitive numerical cranks, or they may be meta-level awarenesses that guide the creative processes that in turn wield the instruments of the trade. It is more satisfying to work at this elevated level: the threat of global action with local perspective is limited when the heuristics are understood.

Artificial intelligence in the factory? You bet. Our entertaining argument between human being and machine was not intended as a cartoon. A factory-wide network with distributed senses and understanding of the "big picture" is of far more value than a dumb DP system that serves as an automatic calculator for management's report-generation amusement.

Natural language in the context of industrial design? You bet. How better to dispense with the clumsiness of coding than to conjure a system that understands the meaning of verbal utterances, and then has the presence of mind to act upon them?

Computer vision? No question. Anything in the general area of robotics needs senses to close the loop of real-world interaction.

Lovely. Now how do we write the programs?

The task of writing a program—be it in assembly level code, BASIC, PASCAL, or whatever—is traditionally characterized by a specific procedure. In essence, the programmer must define some sort of algorithm that will accomplish the objective, then map that algorithm onto the available resources of the computer using a specific set of tools—the programming language. Conversion of an idea from its conceptual form into its coded form involves the donning of an intellectual straitjacket that results in strict formalization of the process.

The generally accepted complexity of this task has spawned some rather elaborate disciplines that reduce the probability of error. Among these is the body of precepts that has come to be known as structured design, born amid a cacaphony of bitter struggles over the relative merits of top-down and bottom-up coding—each of which suggests a controlled approach to designing a program and, hopefully, making it work.

But there are problems with all this, problems that become particularly acute when the programming task at hand involves not the sequential handling of accounts or the polling of a UAR/T, but instead attempts the modeling of human cognitive processes. It is almost prohibitively difficult to write a program to associate data objects via their properties or to perform deductive reasoning when the only languages available require expression of the problem in a fashion that is completely alien to human thought. It can be done, but few have the stubbornness to try. The underlying difference between traditional programming and the type called for by problems of AI is that the latter involves structural interrelationships and abstract symbols, whereas the former can generally be considered in numeric terms. Early attempts to cross this dichotomy resulted in mappings of symbolic problems onto more-or-less equivalent numeric ones, but in the process, most of the naturalness of problem expression was lost.

9.4.1 Enter LISP

The common programming languages can be neatly segregated into the two classifications of "interpreters" and "compilers," depending primarily upon the amount of source text bitten off at any given time by the language translator for conversion into machine code. LISP cannot be placed in either category without being forced uncomfortably into a mold.

Most languages operate on the general principle of inhalation of essentially undifferentiated strings of ASCII, detection of certain delimiters, identification of labels and reserved words, determination of an appropriate object code segment or subroutine call, and either execution (if an interpreter) or addition of the new statement(s) to an output file. LISP cannot be said to behave this way.

The typical programming environment imposes strict syntactical restrictions on the programmer, and presents a limited selection of somewhat primitive (in human terms) tools for use. Although sophisticated tools (macros, subroutines, libraries) can be constructed from these primitives, the language itself is unchanged—almost crystalline in its structural rigidity. APL has been compared to a diamond, which cannot be made into a larger diamond by the addition of a second one. LISP, in this context, is a ball of clay. Its characteristic of *extensibility* permits the language itself to grow without limit, providing the programmer with ever more sophisticated, custom-tailored tools. The language can even be written in itself, personalized totally to the user.

The generation of self-modifying code in almost any programming language is rightly considered bad practice. It is hard to debug, non-reentrant, and generally pathological (although sometimes of dazzling cleverness.) But LISP makes no distinction between functions and data. It is, by its very nature, a dynamic and growing environment—changing its shape and texture to fit the task. Paradoxically, in what seems at first glance to be a hopelessly convoluted mess, LISP outshines 'em all in elegance and simplicity.

(Them's fightin' words in an industry wherein nearly a dozen hotly competing computer languages face off regularly, backed by staunch, almost militant, supporters. Bear with me.)

Just about any language you can name possesses built-in functions for the manipulation of numeric data, with the scope of those functions generally well suited to the application domain of the language in question. But just about any language falls flat on its parser when confronted with the task of efficiently representing symbolic information—not in the sense of "CRLF is a symbol for 0DH," but in the sense of representational thought like: "DOG is a symbol for a carnivorous, four-legged domesticated mammal and MAGGIE is a symbol for one particular DOG."

This has all sorts of implications. One of the most compelling is the image of LISP as a fluid medium for the expression of problems in AI programming. Not only does it allow a process to be written in a relatively abstract way, but it provides, with its lack of function-data differentiation, a means for programs to encode new knowledge by modifying or extending themselves.

It is clear that this language is on a rather different level of abstraction from those of the more concrete variety that abound in the computer world. Sitting down at a LISP machine and building a program is a highly interactive task, wherein the desired system comes about incrementally. The designer shapes tools and data structures, tries things empirically, builds and modifies subassemblies, then gradually integrates everything into a "production system" or finished design. The system supports true debugging, handles details of storage management, adapts well to natural language, and generally acts in a friendly fashion quite unfamiliar to those whose programming lives have been spent banging monolithic aggregates of code into forms that not only compile (or interpret, assemble, or whatever) without errors, but also work—or at least seem to for all the sets of conditions that can be conveniently tested.

A LISP system aids in the implementation of ideas; it doesn't just translate a defined algorithm into machine code.

Before proceeding, however, we should note one small matter. There is not, at present, a vast amount of LISP code being written in the micro world, despite the fact that the language has the dubious distinction of being predated only by FORTRAN. For one thing, LISP is, well, sort of interpretive—meaning that execution of linear tasks is much slower than it would be if written in one of the more pyrotechnic languages that abound these days. Further, it tends to be a memory eater. It also permits (but does not encourage) sloppy structure, at least in a Dijkstrian sense, and isn't widely used as a framework for teaching programming.

This last point is quite a misfortune for the legions of neophyte programmers who are being disgorged in ever greater numbers from ivy-cloaked portals in a futile attempt to keep pace with the requirements of the job market. LISP has something of a reputation for being an academic curiosity, so far removed from the real world that it is seldom even considered in courses outside the AI field. Too bad. Addiction to von Neumann–style languages can forever taint one's understanding of information processing, and also makes LISP harder to learn.

The presence of slick resources like the linker helps us ease a symbolic approach into a typical control environment with only a vestige of self-consciousness. LISP itself may be surpassingly clumsy for fast bit manipulation and machine-level crunching, but you can always reserve it for the more abstract parts of a task and augment it with assembly language routines where necessary.

Let's see why one might be tempted to go to all this trouble.

9.4.2 A Superficial View of LISP Programming

Even in its simplest form, LISP is rather dramatically different from the "standard" languages. Everything takes the form of *symbolic expressions* (s-expressions), which LISP attempts to evaluate. Simple addition, for example, might look like this:

$$(ADD\ 4\ 9)$$

When that s-expression is typed into the system (or is reached during execution), the evaluator notes that the first element of the "list" is ADD, which happens to have an internal meaning—the remaining elements are summed and the resulting value is returned or displayed on the console. ADD, 4, and 9 are called *atoms*, and together they form a *list*.

Lists can just as easily be elements of other lists.

$$(MUL\ (ADD\ 4\ 9)\ (SQRT\ 2))$$

for example, would return the value 18.38 upon evaluation. We could also produce the same result by first doing

$$(SETQ\ THIRTEEN\ (ADD\ 4\ 9))$$

Then

$$(SETQ\ SQUARE\text{-}ROOT\text{-}OF\text{-}TWO\ (SQRT\ 2))$$

at last followed by

(MUL THIRTEEN SQUARE-ROOT-OF-TWO)

Neat. But this is still in the realm of numeric manipulation. Most of LISP's power is in other areas. Since the fundamental data structure is the list, it follows that there are a variety of functions that allow lists to be created and disassembled. CAR, for example, returns the first element, and CDR returns everything else:

(CAR '(THIS IS A LIST))

yields

THIS

Note that the list (THIS IS A LIST) was preceded by a single quote, preventing LISP from attempting to evaluate it. Had that not been there, the evaluator would have attempted to apply the function THIS to the parameters IS A LIST, yielding an error. One could, of course, do the following:

(SETQ THIS ADD)
(SETQ IS 15)
(SETQ A 6)
(SETQ LIST 2.5)

whereupon

(CAR '(THIS IS A LIST))

would still yield

THIS

but

(THIS IS A LIST)

would yield

23.5

Note that there is no difference in LISP's view between a function and a datum.

We could yield to the temptation to plunge ever more deeply into LISP's refreshing waters, but this is a chapter, not a book. It is worth our while, however, to touch upon some of the language's more provocative capabilities.

Just for laughs, consider the following:

(DE COUNT (S)
  (COND ((NULL S) 0)
    (T (ADD 1 (COUNT (CDR S))))))

Remember our twice-presented COUNT program? This, less the file input, accommodation of hyphens, and display of the result, is the same thing. It would presumably be used to count the words on each line of text inhaled from the specified file, yielding values that would be summed into the total.

It is important to note that the list shown above is a DE, which is a function-defining function. Upon evaluation, the new function called COUNT exists in the LISP system with exactly the same privileges as ADD or CAR or SQRT. It has become a part of the environment. Once this has been done, we can say

$$\text{(COUNT '(THIS IS A RANDOM GROUP OF WORDS))}$$

and the system would quite briskly return the value 7.

Let's look at the function, for it contains a couple of aesthetically pleasing concepts.

The first line just establishes the name of the new function and creates a formal parameter, S, which will be used in the function body to reference the input string. The remainder of the function is a COND, or *conditional.*

LISP's conditional structure is elegant and simple. The atom COND is followed by any number of lists, each of which consists of a test and something to return if the test succeeds. In this case the first list will cause a zero to be returned if S is a null string (NULL is a predicate, returning T (true) if its argument is an empty list), and

$$\text{(ADD 1 (COUNT (CDR S)))}$$

to be returned otherwise (as guaranteed by the T).

But what's this? COUNT appears in its own definition! The list that will be evaluated in that last line is simply the addition of 1 to the word count of the same string with its first word removed.

Thus the original order to COUNT cannot be carried out until it creates a copy of itself and hands off the CDR of the string (IS A RANDOM GROUP OF WORDS). Nor can the copy quite evaluate fully until it, too, creates a copy of itself and hands it (A RANDOM GROUP OF WORDS). This continues until the seventh copy counts (WORDS), passes the list () to an eighth copy, and receives a 0 in return as the (NULL S) predicate is finally satisfied.

Now the copies disappear in reverse order as they fulfill their transient destinies and return values to their ancestors equal to one plus the number of words in the strings they handed to their children. At last, 7 is returned as the value of the original list. The execution of the function is shown TRACEd in Listing 9-1.

This is known as *recursion*—the process of solving a problem by simplifying it slightly and then going to work on the new version. What may appear at first as a paradox of self-reference is actually a powerful technique for noniterative problem reduction.

The trivial examples we have shown hardly even penetrate the texture of LISP—beneath the level of primitive functional tools are a variety of powerful techniques for performing association, complex mappings, language parsing, pattern matching, data-driven programming (as in a relational database), frame representation, program generation, and much more. These are the kinds of things that become appropriate as the computer evolves past its presentient bur-

```
> (DE COUNT (S)
      (COND ((NULL S) 0)
             (T (ADD 1 (COUNT (CDR S)))))))
COUNT
> (COUNT '(THIS IS A RANDOM GROUP OF WORDS))
7
> (TRACE COUNT)
(COUNT)
> (COUNT '(THIS IS A RANDOM GROUP OF WORDS))
--> ENTERING COUNT WITH ...
 (THIS IS A RANDOM GROUP OF WORDS)
 --> ENTERING COUNT WITH ...
  (IS A RANDOM GROUP OF WORDS)
  --> ENTERING COUNT WITH ...
   (A RANDOM GROUP OF WORDS)
   --> ENTERING COUNT WITH ...
    (RANDOM GROUP OF WORDS)
    --> ENTERING COUNT WITH ...
     (GROUP OF WORDS)
     --> ENTERING COUNT WITH ...
      (OF WORDS)
      --> ENTERING COUNT WITH ...
       (WORDS)
       --> ENTERING COUNT WITH ...
        NIL
        <-- VALUE OF COUNT IS : 0
       <-- VALUE OF COUNT IS : 1
      <-- VALUE OF COUNT IS : 2
     <-- VALUE OF COUNT IS : 3
    <-- VALUE OF COUNT IS : 4
   <-- VALUE OF COUNT IS : 5
  <-- VALUE OF COUNT IS : 6
 <-- VALUE OF COUNT IS : 7
7
```

Listing 9-1

bling and moves into the kinds of applications that call for something more abstract than an algorithmic approach.

9.5 SOME THOUGHTS ON INDUSTRIAL AI

Before we leave the alluring subject of artificial intelligence to return to some more immediately useful but less exotic subjects, perhaps it would be well to tie this into the industrial world a little more specifically. You can be sure that, like the microprocessor, intelligent machines will become less and less esoteric as the clear-cut economic benefits of their use seduce the corporate decision makers.

The most obvious area (and one of the most colorful) is that of robotics and intelligent manipulators, especially those equipped with vision. There are thousands of production tasks that are too boring, too dangerous, or, let's face it, not cost-effective for human staffing, and the technology is already well past its infancy. But industrial robots have classically suffered from an appalling lack of flexibility, usually requiring special fixtures and precise alignment of workpieces. This inability to accommodate the endless variables of reality is largely attributable to severe myopia in the feedback path: existing vision systems have enough to worry about with single components on static homogenous backgrounds; they are hardly ready to take on the range of environmental complexities that human beings find completely unruffling. Some of the insights now being gained into the wondrous phenomenon of human vision, however, promise spectacular developments in the years ahead.

Vision in a purely restrictive sense can be considered just an input operation, but in any kind of meaningful context is inseparable from intelligence. Just like natural language comprehension, vision is a process that continuously affects the cognitive state of the intelligent being involved. Vision without cognition is a camera; the image, just data.

But the obvious areas of robotics are not the only AI applications in the industrial engineering world. Computer-aided design of logic, networks, and printed circuit layouts is becoming commonplace, and will become more sophisticated as intelligence begins to grace the processors involved. Problem-solving systems will simplify the selection of control algorithms. Expert systems with appropriate knowledge bases will serve as intelligent assistants to design engineers.

This last point deserves a bit of elaboration. Practitioners of most of the world's professions have been using computers for years, of course, but the scope of their use has been generally limited to that relatively small subset of their problem-solving requirements that can be handled in a mathematical or routine data-processing fashion. Although this has certainly been a big help in the reduction of tedium, there are many larger and more pressing problems that could be attractively approached with computers if only they could reason symbolically and draw upon a comprehensive base of professional knowledge.

Why use a computer simply to host a text editor and language translator when it could instead accept a verbal description of the problem and interact with you as necessary to produce a set of ROMs that plug into an SBC?

Or why build a factory-wide network of dumb subordinates to a single big system when the entire hierarchy can possess a collective intelligence that renders it self-maintaining?

None of this should be misread as an assertion that machines can do all tasks better than human beings, freeing us to turn our minds to loftier pursuits. We seek not to be Tralfalmadorian biological superfluities, notified by the machines which we have built to replace ourselves that we have, indeed, no higher purpose. There are some touchy sociological issues here, which shall receive only scant public attention until a profusion of intelligent machines is a *fait accompli.* (It should be interesting, seeing the neo-Luddite factions, the "silicon lib" crusaders, the high-IQ pocket associates. . . .)

AI is coming out of academia; it is spilling over, crumb by precious crumb, into the hungrily waiting industrial marketplace. It will be a wide-open field for years.

But now, let us shift our focus. In three chapters we have spanned, perforce too quickly, the whole of the WHAT–HOW spectrum and some of the tools that enable us to move within it. With all that behind us, we can turn to some of the considerations that affect the development of real systems, of real software. We have reached a peak of abstraction, and now begin the gradual descent into development of some tangible industrial machines.

10

Approaching a Design Project

10.0 INTRODUCTION

We have amassed a respectable armamentarium of tools, yet there is still something missing. Despite shelves full of SBCs nestled snugly in their conductive bags, a development system patiently spinning in a keyboard wait loop, and a fresh batch of wirewrap wire gleaming amid the clutter of the workbench, we are yet ill-equipped for an all-out assault on a real project.

Somewhere between the initial specification of a system and the coding of its software, there is a large and unquantifiable region comprised of magic, intuition, creativity, memories of past successes/failures, and guesswork. In a rigorous philosophical sense, perhaps these could all be classed as tools along with everything else in the WHAT–HOW spectrum, but it is somewhat more gratifying to think that it is our minds that provide the spark. Those of a Skinnerian bent might disagree.

Hard-core engineers might disagree as well, for there exist a number of disciplines which purport, in theory, to strip a design task of the vagueness suggested by intuition et al. These disciplines, like all good tools, can in some respects serve in a compensatory fashion when inspiration is lacking. But is that what we want?

It is tempting, in this technological age, to take one's exotic tools too seriously and let them define the shape of one's works. It is especially so when said implements are internal and can thus seductively suggest conceptualization of the problem on their own terms. But it is our position in this text that they are not stand-alone solutions. Can we not take advantage of sophisticated and well-refined disciplines within the much larger framework of art? Or is engineering just a mechanical, crank-turning process?

We have talked of artificial intelligence, and can dream of the day when an automatic programming system can create, on the basis of an unstructured chat, a finished piece of code.

But that's relatively remote, and still isn't quite the answer. A design is far more than a program.

10.1 THE FIRST VAGUE STEPS

When all the wheeling and dealing is over, when the customer or manager walks out the door, you are left with the task of converting a vision into a working system. Maybe many working systems; our discussion is by no means limited to one-of-a-kind custom controllers.

Now what?

Often the first step, before the application of any formal methods, is simply to discalceate and cerebrate. Scrutinize your fingernails. Inflict small, tooth-shaped indentations upon your Pilot Razor-point. Have more coffee. Clean off the desk. Call the time service, change the batteries in your calculator, and take a long lunch.

Once all that is out of the way, it becomes possible to make meaningful marks on paper. Some of these might relate to the constraints of the job: if a system is intended for production, the maximum acceptable unit cost is probably specified. There is doubtless a deadline, certainly unreasonable. There may be size or weight limits, and the presence around your shop of leftover hardware from other jobs might suggest specific components. Unquestionably, you have some indication of performance requirements.

But before anything of a highly analytical nature can be done with this agglomeration of random notes, it is necessary to apply a measure of intuition and common sense. This is one of those areas in which attempts to formalize the design process take on the general character of "n equations in $n + 1$ unknowns." The job begins, slowly, to acquire some substance—its hazy conceptualization drifting now and again into focus as you reject unreasonable possibilities and absurd notions. The Blanchard controller will not require an Amdahl 5860. The deadline may suggest the use of an off-the-shelf I/O module for the prototype, after which an in-house design costing a third as much can be substituted. You make rough cabinet sketches and try to imagine the front panel layout.

This is admittedly very vague, but creating a system out of thin air does not begin to lend itself to definitive heuristics until the grosser characteristics have been roughed-in by whatever combination of unquantifiable phenomena is invoked by the nature of the task.

This seems, unfortunately, to be leading us nowhere. Are we saying that you cannot know how to do this until you have done it?

Not exactly. But we *are* saying that more than a few of the heuristics that are involved with this cogitation process are based on the wealth of intangible psychological assets that accumulate over years of work in the field (and of simply being alive and sentient). Consider, for example, the logarithmic polar mapping scheme we discussed in Section 9.3.

The body of literature on the subject of digital image processing, at least until about 1980, made little or no mention of the fact that such a coordinate transformation could solve the classic problems of rotation and scaling in a spatial correlator. But suppose that some engineer is scratching his head over the fact that he is supposed to build a system, based on a CCD imager, which examines products coming down a conveyor for proper alignment of labels. If his design process were totally derivative—that is, based on the surrounding tenets of the image processing discipline—the project might be scrapped after due consideration as being too complex for the budget.

But suppose this head-scratching engineer had once worked as a technical support person in eye research, and suppose further that he had read a lot about the brain and fiddled around in his basement with fiber-optic image conduit— why, then one of his natural and immediate responses to the problem might lead directly to the mapping scheme.

Anyway, the point of all this is simply that we cannot exhaustively document magic. That's up to you.

But we can make a few suggestions for dealing with some of those vague first steps in the design process, such as the determination of the hardware–software balance.

10.1.1 The Big Trade-off

As we have intimated at various points herein, particularly with regard to the Blanchard controller and the insertion/withdrawal force tester, there is a key trade-off of hardware versus software in the design of many (most?) systems. Often this manifests itself as a reluctance to write the code that would be necessary to do the job of an off-the-shelf module, or weak-kneed submission to the particularly alluring qualities of a new device.

But the key factor is cost. In a system designed for production beyond a few units, the use of, say, a 9511 floating-point processor might be an unacceptable alternative to a few dozen lines of code, however awkward the code might be to write and debug. But one-of-a kind custom jobs are much more likely to justify the extra hardware cost, especially in the light of the unamortizable alternative of human time.

The industry is finally seeing a departure from the obsession with "doing it all in software" which accompanied the introduction of microprocessors. Originally, people were so enamored of the micro's capabilities that they would bend over backward to implement all of a system's logic functions in code. After all, they might have reasoned, that's what the devices are for. But it has become clear that the advantages of some measure of distributed intelligence can frequently outweigh even the economic bias in favor of code.

In many situations, there is another variable in this equation. We mentioned hardware as a replacement for complex segments of code, but probably more common is the situation in which the system spends the bulk of its time performing some relatively trivial operation, such as polling a keyboard. This can be almost totally eliminated with the addition of hardware that singlehandedly tests for a depressed key, then interrupts the processor with a key code whenever appropriate. This frees the computer to do other things (in the music keyboard interface of Figure 5-8, for example, the dedicated scanner made possible real-time Fourier synthesis and envelope generation).

What we have here, then, is not a simple trade-off of the old gain–bandwidth variety. There are many conflicting requirements that must all be resolved and synthesized before a system configuration can be settled upon. Matters are worsened by the fact that typically, this decision is made before any of the software is designed. You know where that leads: patches and kluges scar a once-elegant design. Somehow all the requirements of the job must be thrown simultaneously into the air, there to congeal in some kind of rough organization and drop back into your waiting hands.

10.1.2 Processor Selection

Along with all that, it is usually desirable somewhere near the beginning of a project to establish the type of processor that is to be used. This problem receives considerable attention in the press, much of it from the manufacturers of the chips involved. They each, evidently, feel that their product is the best.

This is not the place for the old, reliable "bigger is better" philosophy of design. With increasing complexity comes increasing overhead, and the use of a robust SBC with a real-time multitasking executive for a simple control system might end up requiring 64K of RAM where 4K would otherwise suffice quite neatly.

Further, there is always a lot of talk about speed, with ever faster superzoomers appearing monthly in the pages of our trade journals. If you are attacking some monstrous task, fine: even though the introduction of parallelism into the system is a better way to increase information bandwidth, it may well be necessary to go for broke in all aspects of the machine. But the speed–power trade-off hasn't been obsoleted by cool new technologies, and pushing the accelerator to the floor also worsens noise problems. Perhaps it's time to acknowledge the inevitability of an ultimate processor speed limit (enforced by the time it takes electrons to get from one place to another) and begin, as an industry, to approach time-sensitive tasks with distributed resources. Our brains, after all, manage some rather impressive performance with logic elements having propagation delays of many milliseconds.

The raw CPU cycle time of a system, by the way, is of little use as an index of the actual throughput that can be expected. As has been amply proven by the comparative execution of "benchmark" programs on different systems, the net performance is much more a function of processor architecture and the style of the program itself. A register-oriented machine, for example, might appear readily to run circles around a memory-to-memory design, but rewriting the code to efficiently make use of index registers might precisely reverse the relationship. Choosing a micro on the basis of CPU cycle time is analogous to choosing a stereo system on the basis of its peak output power.

And then there is the bus width question. Four bits? Eight bits? Sixteen? Thirty-two? Something oddball made out of bit slices? This can be a tough one, for like speed, there's something almost pheromonically seductive about whatever new technology is currently making headlines. Inappropriate conversion from an established board family to a more sophisticated one can involve dangerously long learning curves.

Of course, there is much to be said for wider buses—the bandwidth is higher and the von Neumann bottleneck slightly less severe, precise number representation is simplified, and address manipulation is much easier. But it is not uncommon for people to shift to 16-bit systems just to get more out of an overloaded bus, when simply housing all the overhead functions of an 8-bit system on one card (see Section 4.2.2) can nearly eliminate bus activity that is unconnected with the specialized hardware of the application.

At the chip level, processor selection is probably most reliably based on the experience of the person doing the programming—at least if no high-level language is being used. You tend to develop favorites, and since there is really very little functional difference between devices of the same broad families, the resulting preferences are often the best criteria.

At the board level, there are more variables. The amount of on-board RAM and ROM, the availability of local I/O and peripheral controllers, the type of bus interface, the power requirements, the physical size, and the cost all enter into the decision. As a general rule, every data transaction that crosses card boundaries has a penalty associated with it (in system speed, cost, and reliability), and the more self-sufficient the CPU board is, the better.

10.1.3 Random Considerations

Even after the decisions described above have been at least tentatively made, there are a host of others that must precede the actual logic design and coding. Let's touch on a few just to indicate their general flavor.

What logic family? If power is critical, perhaps CMOS is the way to go. Is the thermal environment an issue? If so, it might even be necessary to go to MIL-spec parts.

What level of field repair is going to be provided? Should the chips be socketed to facilitate component-level service, or will card swapping be more appropriate? This train of thought will also yield a few clues about the mechanical assembly of the parts, something that dramatically impacts later attempts to dig out a suspect bit of hardware.

What kind of cooling will be required? Is the environment clean enough for filtered forced air, or will something more esoteric (and expensive) be required?

Any notable vibration? It might be necessary to shock-mount a few things, pin down boards with retaining clamps, and build strain relief into wiring harnesses.

Is future expansion of the system likely? If so, a somewhat more robust power supply than would otherwise be indicated will be highly appreciated in the future. Some leftover space on a wirewrap panel, a few spare card slots, some uncommitted wires in the cables—similarly.

Is vandalism likely? Contamination? Theft? Serious power spiking? Danger to human beings? The possibility of "logical runaway"? Consider the comments in Chapter 6 and remember that nothing is ideal.

Will the customer be encouraged to isolate problems? A few key LEDs on the front panel cannot only facilitate this but are an excellent selling point: people love blinking lights for reasons beyond the practical. An assembly machine gaily blinking away can become a showpiece on factory tours instead of just another boring old dirty gray box streaked with coffee stains.

Much of this is automatically suggested by the originally stated requirements of the job, and will be intuitively obvious. But all these things and more must be considered together and used to arrive at a fairly comprehensive image of the system-to-be, even before any real logic design is undertaken.

10.2 THE BIG PICTURE

Ah, but we have left out an essential element—another one of those key factors that has to be simultaneously considered with everything else. We speak of selecting enclosures, picking a processor, and all those other early design decisions, but what's the thing supposed to do? Inevitably, that will have some effect on the hardware.

To illustrate the process that leads up to the actual software design, let's use a specific example, starting with the initial contact with the customer.

10.2.1 The Patrol Tour Recording System

We receive a call from the plant engineering manager at a large distillery. He has heard of our custom engineering firm and wonders whether we might be able to help solve a problem that he has been having with security. We raise our eyebrows.

It seems, he explains, that they are paying a substantial monthly charge for a predominantly mechanical system that logs calls from guards as they make their rounds and check in via telephone. But the system is inflexible, breaks frequently, and produces only the sketchiest of printed records. What they would like is a machine that randomly assigns patrol tours and then logs any deviations therefrom. The tours should be chosen such that no area is left unchecked for too long, and the printout corresponding to each checkpoint visitation should show the date, time, station number, telephone extension, and the name of the area.

In a classic patrol tour environment, guards are given a sequence of checkpoints that are to be visited at prescribed intervals. But there is little guarantee that time and sequence requirements are being met without some form of supervisory system, and there is always the chilling possibility that sinister forces can observe the guard's activity over a period of time and thus know with some certainty when key areas will be left unprotected—or when the guard will be in a particularly vulnerable position.

We chat on, arriving at some sort of reasonable understanding of the job's requirements. He wants a quote, but we cannot do that until we settle with some certainty upon the hardware and software requirements.

So let's look at the general information-flow requirements. Even without defining the internal logic, it is obvious that some method must be provided to interconnect the distillery's PABX with our machine, so that the selected telephones (upon the dialing of a certain code) will be detected and identified. Fortunately, the original crude security system provided this function, allowing us to build a little scanner to monitor a set of status signals in the PABX. Unfortunately, this equipment is about a half-mile away from the intended location of the patrol tour recording system (PTR).

In addition to input from the phones, the box will need a small keyboard which will allow the time and date to be set, as well as yet-to-be-defined special functions to be invoked. There's no real need for a display, since a printer is called for. The tentative overall information flow is indicated by Figure 10-1. We don't know anything about the logic yet, but we can see where it fits in the grand scheme of things.

So far we have made few decisions, for the grosser aspects of the hardware configuration are pretty well determined by the requirements of the job. But the guy wants a couple of numbers: how much, and how long.

So we enter into the pondering process. It seems reasonably clear that some communications logic will be necessary to shuttle data between the PABX scanner and the PTR itself, and for that reason as well as the general desirability of using a hardware keyboard scanner, we manage to postulate an interface card. There is also the CPU, of course, and although we yet have no idea what kind it will be, we will certainly use an off-the-shelf SBC. The printer becomes something of an unknown at this point, so we dig around in the literature, call a few vendors, and finally settle on a small panel-mount matrix job that comes with a microprocessor-based controller (using a 6502).

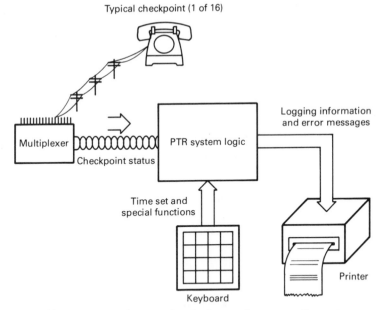

Figure 10-1   Overall information flow in the patrol tour recording system: a remote multiplexer continually provides the system itself with the status of 16 phone lines representing checkpoints used by guards.

A visit to the distillery helps establish a few more key factors. We are shown the intended location of the machine, giving us some maximum values for height and depth. We meet the guards and gain some unofficial insight into their feelings about the system—something well worth considering in the design. We are shown the PABX room, and discover through inquiry that there are some unused telephone lines connecting it with the guard shack. The power and thermal environment seem reasonable, and there are no unusual concentrations of airborne contaminants.

Now we can pick a box and think about the arrangement of components. Also, about this time we start to accumulate an intuitive understanding of the software complexity, and *guess* that it will fit within about 4K of ROM. The I/O requirements are fairly straightforward, and there is certainly no need for a disk, extensive RAM, or external access to the bus.

How did we guess that the code would fit in 4K? That smells suspiciously of black magic.

Well, let's sit back and ponder. It will obviously need something called INITIALIZATION, as all programs do at one point or another. Even though we have no firm idea of the program's operation, we can assume that some space will be given over to an executive loop together with interrupt handlers and other closely associated support logic. We will need some real-time clock software, a module that handles the data from the multiplexer in the PABX, a module that prints activity lines, some logic involved with keyboard handling (including clock set and a self-test routine), a block of code that assigns tours, and a few miscellaneous support routines. In addition to the code, there are ROM data requirements (all the area names, the stored tour sequences to be randomly selected, printer messages, etc.).

If we give some thought to the operations that need to be performed, we can "ballpark" (based on past experience) the ROM requirements in the 2 to 3K range, so we'll plan on 4.

Similarly analyzing RAM requirements, we find that there is very little need for data storage outside flags, time values, key buffers, and so on. Without writing code, it is well nigh impossible to pin this down, but 256 bytes is probably more than adequate.

Remember the Intel SBC 80/04 we talked about in Section 4.2.1? It looks like it will do the job just fine—we care not about providing an external bus interface, masses of memory, or master-slave operation. Why get carried away with high technology when a $200 card should be more than adequate?

Now we can start to think about a price. The starting figure is the hardware cost:

Intel SBC 80/04	$ 200.00
Enclosure	100.00
Power supply	75.00
Printer	700.00
Interconnects	50.00
Interface logic	200.00
Multiplexer	250.00
Fan	20.00
Keyboard	20.00
Total	$1615.00

This, of course, represents only our cost on the parts, or at least an approximation thereof (as of 1978–1979 estimates only). The amount quoted to the customer should be somewhat higher. We should probably at least double the hardware cost, then add a value representing the estimated time to be spent on the software. Two hundred hours, perhaps, at $35 per? (Remember overhead and expenses—this is, alas, not a salary figure.) That's $7000—probably a bit much for the code alone, but there will be a good bit of hardware design as well and doubtless some frustrating problems with installation. The total comes to $10,230, which is not a bad figure but sounds a little too close to a very round number. Gotta make the customer think the price was derived analytically ... let's see, how about $11,475? (See, it ends up being rectumological anyway, but it saves a lot of later pain if the guess has some reasonable basis to it.) We get brave and promise delivery six months after receipt of order (ARO).

The customer casts his eyes heavenward upon receipt of the proposal, utters a futile complaint about highway robbery, is reminded of the payoff, shakes his head, sighs in resignation, looks to his secretary for moral support, finds none, and finally says with a look of thinly veiled anger, "I'll get you a PO."

10.2.2 The PTR System Gets Serious

Now, suddenly, our design efforts are no longer speculative. We are about to project ideas upon matter, in the exalted fashion of Figure 7-2. The PO arrives. Parts trickle in.

What's the deal here? Do we have to build and debug the hardware before it becomes possible to test any software? That could be disastrous. This is one of the central problems of the microcomputer system development world: How can we maximize the efficiency of our time? If we simply take a linear approach, wherein you park on your *nates* while I build the hardware, then I do likewise while you build the software, then we each waste a monstrous amount of time. Even if we had enough other projects to keep us busy, the job of interest—the PTR system—would take much longer to get off the ground.

But what are the alternatives? We could begin simultaneously on the code and the hardware, then hammer them together and debug the resulting abomination. We could interact somewhat, you wishing that you could test at least a few basic routines, I wishing to test the printer interface. Or we could somehow simulate the hardware in the development system, then build the real thing while the code development proceeds at a leisurely and well-supported pace.

These differing approaches are shown graphically in Figure 10-2, with elapsed time in all cases indicated by horizontal extension to the right.

In the first case, we see what happens when a project is attacked by a single person: no processes can take place concurrently. The resultant switching back and forth between hardware and software debugging can be incredibly time-consuming, but is amenable to some relief with techniques that we will discuss shortly.

Figure 10-2b shows what happens when a second person is added. As soon as the initial rough code is written, the programmer sits around waiting for a machine to test it on. When it is finally available, both hardware and software need debugging, so again confusion reigns.

But let's add a new concept: *simulation.* If the development system (such as BEHEMOTH) can be equipped with hardware which in some fashion allows the code that is being written for the target system to be tested in something even remotely approaching an environment akin to the one at which it is aimed, then some meaningful measure of concurrency can be achieved. Working with a simulator frees the programmer from the need to await the completion of the target hardware, dramatically reducing the time necessary to integrate and debug the complete package. This approach is suggested by Figure 10-2c.

Now if we really want to get fancy, we can marry the development system and the target system, then simultaneously develop hardware and software without having to make a separate step out of creating a simulator. As shown in

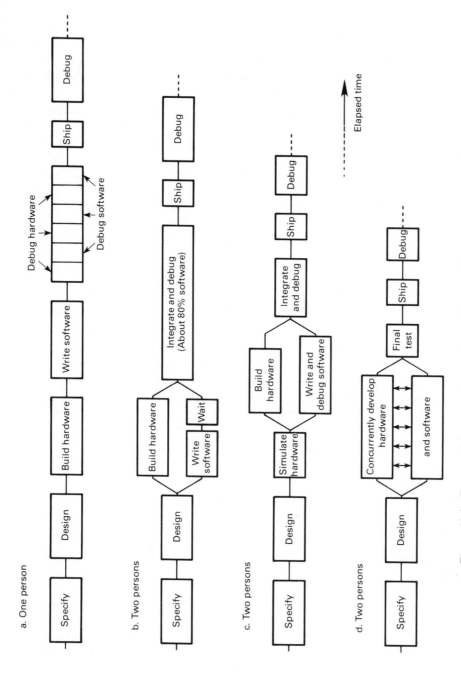

Figure 10-2 Four possible approaches to a hardware and software development project, with elapsed time in all cases represented by horizontal extension from the starting point (specification).

254

part d, there is constant interaction between the logic designer and the programmer, with each completed module becoming available to both parties along with quickie test routines and scope loops. We shall have occasion to describe the hardware and software tricks that can be used to implement such a design scenario in Chapter 12.

One of the nightmares that this last approach helps avoid, by the way, is the sudden discovery at integration time that the hardware and software don't fit. This is not as absurd as it sounds. The original specifications and design can usually be interpreted a number of ways, some of which are mutually exclusive. It is not at all uncommon to find significant timing differences, reversed logic polarities, incompatible handshaking schemes, tests for levels where edges are the issue, undefined states, and so on. The general effect of this problem is graphically depicted in Figure 10-3, and should be recognized as the classic situation that it is.

Let's assume for the purposes of patrol tour recording system development that we have at our disposal a system that is interconnected with the SBC in a fashion that allows the approach in Figure 10-2d. There is a mysterious cable coming out the back of BEHEMOTH that terminates in a 24-pin DIP header. It is plugged into one of the SBC 80/04's ROM sockets. The table next to the computer begins to become littered with wirewrap wire, tools, and half-finished sheet metal constructions. The contents of the wastebasket begin to consist more and more of continuous form paper astrewn with assembler error messages and handwritten patches.

What exactly is going on here? We know what we want the machine to do: keep watching the data from the multiplexer (which is a continuously repeating map of the 16 input lines, each represented by its 4-bit binary address and a single bit indicating its current status). When one of them becomes active, indicating a phone call from a checkpoint, the system should check the currently assigned tour to ascertain the call's correctness, then print an "activity line" consisting of the pieces of information mentioned previously. Meanwhile, whether or not calls are

Figure 10-3   There is a distinct danger in the attempt to develop hardware and software independently of each other. The deadline is two days away.

coming in, the system must keep an eye on a clock interrupt which serves as the time base for its real-time clock software, updating it accordingly. On each update, it must check for special times, which indicate the need to issue new tour assignments, or for excessive elapsed time, indicating that the guard may need some assistance. It also needs to watch for keyboard interrupts, in case someone is trying to reset the clock or invoke a special function of some sort (paper feed, self-test, etc.).

Clearly, then, the system has three things to watch for at all times: a character from the mux, a depressed key, or a clock pulse. This, perhaps, suggests the top level of our program. The processor can just spin continuously in a tight loop testing for those three events, taking appropriate action in each case. But instead of tearing blindly into the code, let's step back for a moment and think about it.

We have arrived, finally, at a place wherein some disciplined structure is a desirable alternative to head-scratching and intuitive response to the nature of the task. It is time to do some software design. Should we just sit down and code whatever part seems most obvious? Should we decide what the most primitive modules are and get them working first, moving to progressively higher levels as we amass some larger-scale components? Should we ignore the low-level trivia and start at the top? These questions have inspired volumes. Let's take a shot at it.

10.3 SOFTWARE ENGINEERING

In our various discussions of art versus engineering, it might have seemed that we were inveighing against the disciplines of the industry and proselytizing for the magic of intuition. Not at all. We might as well suggest the abandonment of stress analysis and finite-element modeling in the design of bridges: hey, man, just conceive the overall aesthetic magnificence and build it!

You can, of course, approach software that way if you want, but it is almost doomed. It gets tricky as the project grows beyond the level that can be conceived at one time in its entirety.

The author recalls with some embarrassment his first, uh, custom engineering job. It was also his first program of over several hundred lines—it was, in fact, well into the thousands. It was a disaster.

Oh, it worked all right (still does, in fact), but the "integrate and debug" cycle was so lengthy and painful that there was some doubt, especially on the customer's part, whether it would ever be completed. The deadline was meaningless.

No concept of software structure, however primitive, was applied to the task. There were no separable modules in a listing well over an inch thick. Data values were interspersed with the code, routines modified other routines and then called them, and any attempt to follow the flow of control led quickly to confusion. One little section of the code, a subroutine package, became a classic example of dangerous practice. It was extensively self-modifying, dangerous but dramatic. But the customer's requirements changed, and the system had to suddenly become interrupt-driven. Now it was possible for partially executed routines to be interrupted by tasks that needed to use the very same routines! It was therefore necessary to save the entire status of the mangled code, effectively pushing chunks of RAM onto one of the three simultaneously maintained stacks, in order to achieve performance that looked as if it might be acceptable.

Old war stories like this evoke a measure of satisfaction, but it is too easy to forget the endless 16-hour days of slaving away at incomprehensible complexes of subtle interacting bugs, while perspiring under the disapproving gaze of the customer, already $30,000 poorer. Any pleasure in the seemingly substantial contract amount evaporated long before project completion, with the final flourish lost amid legal grumbling and swallowed pride.

Ah, those were the days, though. Let me tell you about the pattern-sensitive memory.

The code was perfect. It had to be. But there was a persistent problem that appeared to be associated with system initialization: every time the machine was reset and told to GO from the beginning, a short block of code somewhere in a communications routine was wiped out. Zap. The garbage that replaced it was the same every time, but bore no resemblance to 8080 instructions, or, for that matter, to any of the data types that were being manipulated by the program.

On a hunch, it having been concluded that the problem was not software, the memory boards were swapped around (this was 1975, and the boards were those newfangled high-density 8K jobs, available in kit form from two or three forward-looking manufacturers). Fascinating. The problem went away. The cards were restored to their original configuration and the problem returned. This swapping, it should be noted, was just the reassignment of the 8K address ranges occupied by the four boards.

Unfortunately for the desperate system designer, this hardly constituted a reliable fix. He might have gotten away with it, though, but for the fact that a new problem was introduced: error reporting no longer worked. A new arrangement of the boards was tried. The system crashed heavily. Yet another arrangement was attempted, and this time the real-time clock showed the time as "%gK;=\."

It was patently obvious that there was a bad memory board in there. A spare was brought in, and used to replace the suspects, one at a time. No combination produced correct system behavior, although for a couple of relaxed and idyllic days we thought we had it. But the phone rang.

All of the boards were replaced, at the customer's expense. He was disgruntled, but finally shared the designer's conviction that some obscure environmental problem, lightning maybe, had caused widespread pattern sensitivity in his system.

Same thing. Different problems. The customer no longer believed that it was hardware. Over a coffee-sodden period of two days, interrupted only by fitful sleep and hastily gulped burgers, a notebook was filled with a detailed trace of system operation, accomplished by single-stepping through the entire mess.

Something became a little suspicious. A sort routine, invoked upon startup to place the elements of a data file into a sequence amenable to binary search techniques, seemed to be swapping one extra data field at the end. It finished the real sort, then picked up an address, wandered off into a totally meaningless memory location and undertook a 12-byte replacement of the data therein found with similar data located elsewhere—such as in the middle of the program. Why? The sort was told to terminate when one pointer was greater than another, instead of when they were equal. The fix was swift and sure. It worked. The fact that 2102 RAMs have the habit of powering up with the same ''random'' data every time explains the apparent pattern sensitivity—the garbage in an uninitialized area was used to decide what addresses would be involved in that final superfluous swap.

We engage thus in ''true confessions'' to illustrate a crucial point. The software was so intricately intertwined that it was almost impossible to isolate individual modules for testing. The sort routine should have been exercised on dummy data before being integrated with the rest of the system but it depended so heavily on its surrounding environment that such testing was too much trouble. A single instruction, a conditional jump, was the cause of all the anguish.

There are some well-established ways around this sort of thing.

10.3.1 Structured Design

Some of the tenets to be expounded here may well appear contradictory to earlier rhapsodies on the general theme of art. But they aren't. Like an assembler, a milling machine, or the technique of surface grinding, structured design is a tool. It can be used by laborer, craftsman, and artist alike, with corresponding degrees of inspiration.

The study of software structure arose from the plethora of problems in the programming world: There is a vast and ever-expanding body of knowledge which few can use efficiently. Too many existing techniques depend for their successful implementation upon wizardry, and wizards are in short supply. Individual applications have become too large and complex for a single programmer to handle.

It has become clear that an engineering approach might buy some measure of organization, hopefully yielding the twin benefits of fewer programming errors and greater productivity (with the latter arising almost automatically from the former).

Errors. Where do they come from? Typos, obviously, but they are easily located. The kinds of errors that are disturbing are those arising from improper problem specification or inaccurate (or incomplete) comprehension on the part of the programmer. There are also a number of problems that can result from simple carelessness or sloppy structure, but they are relatively minor. The real lulus seem to appear when the scope of the task is so large that it is impossible to maintain it all in active consideration at any given time.

Of course, one well-established ''school'' of thought suggests that the best approach is to just get something to assemble (or compile—whatever), and then hammer it into shape as necessary. Once it can be tested, after all, the application of good debugging technique should allow us to home in on the majority of the problems without too much difficulty.

The problem here is that the attempt to fully test a complex piece of code that is riddled with multiple errors tends toward the impossible. In a system such as the one that introduced this section, the "percent correctness" probably asymptotes at about 98%. Avoiding this rather unacceptable condition involves something other than all-out attack on a great monolithic aggregate of code—we must somehow attempt to make it error-free in the first place. Writing a correct program is a more noble objective than a long but successful debugging marathon.

One of the most obvious ways to accomplish this is to decompose the problem into a number of smaller, semiindependent *modules*, each of which encompasses one central idea. Before any actual coding takes place, the imposition of modularity upon the original complex task can free the designer from the need to somehow arrange all the details of the problem into a meaningful internal model. It also makes it much easier for more than one person to share the programming task. Modules can be fully defined and assigned to staff programmers without the need to communicate the complete system specification beyond the "big picture" level.

Within the modules themselves, there are techniques of structured design which further improve correctness and programmer efficiency. These are based on a small repertoire of basic flow-of-control structures, shown in Figure 10-4. Note that each has but one entry point and one exit point. This suggests that the individual structures can be used as components (replacing boxes) in other structures, as suggested by part e of the figure.

If an entire program is built from these basic elements, an interesting phenomenon occurs. The flow of control becomes obvious. Unlike the highly convoluted paths through a flowchart such as the one in Figure 10-5, the logic of a structured system can be viewed at any level of magnification without dissolution into amorphousness. This has a number of fringe benefits, for if the logic can be readily understood and isolated, then testing and modification become relatively trivial. *Divide et impera.*

Nonstructured code, such as the Figure 10-5 flowchart, often represents a failure on the part of the programmer to consider all logical cases before beginning work: he started coding too soon. It also suggests numerous patches, attempts to make the code "efficient" by reusing segments (usually associated with flags representing the present meaning of a given group of instructions), and general unclear thinking. It is not particularly difficult to imagine confusion upon later attempts to debug a complex unstructured system—and there will be later attempts, for such code is almost impossible to test thoroughly.

But a question arises. How, precisely, does one tear into a massive and complex task (or even a PTR system) and separate it into modules to which structured techniques can be applied?

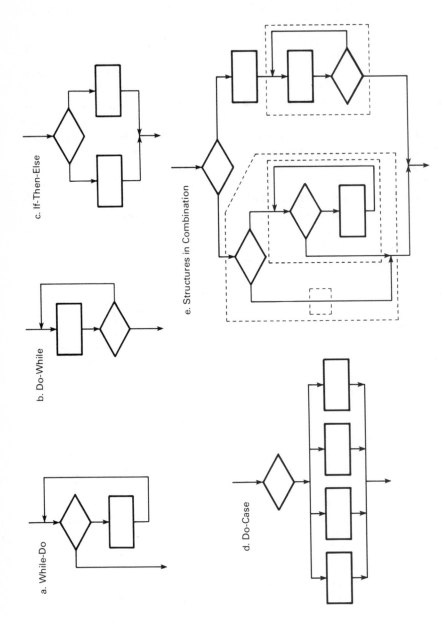

Figure 10-4 A key tenet of structured programming is the use of specific flow-of-control structures which each have only one entry point and one exit point. Each can be treated as a single "process" block within the body of another structure, rendering the entire system modular and free of random cross-connections.

a. While-Do

b. Do-While

c. If-Then-Else

d. Do-Case

e. Structures in Combination

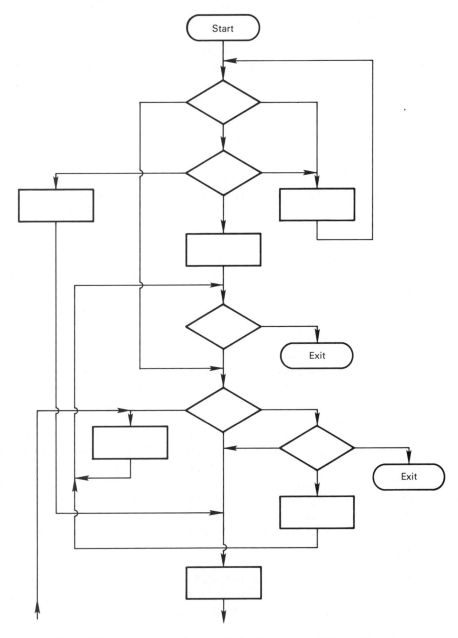

Figure 10-5 An example of the confusion that can result from a blatantly unstructured approach. The design here might well be brilliant, but it is very difficult
to understand (and thus to debug).

10.3.2 Top-Down Design

It is not sufficient to just dive in, isolating obvious functions and writing routines to perform them. Some thought at the earliest stages of the design on the overall structure can result in clean divisions between modules. This is important, for the advantages to which we have paid homage depend in large part upon the ability to separate modules easily for testing.

How can we approach the problem in a fashion that allows this? Let's look at a few approaches.

First, there's the old reliable (?) method known as amorphous design, beautifully exemplified by that lamented system that constituted a not insignificant part of the author's growing pains. We cannot really recommend this.

Next, perhaps, might be "bottom-up" design. This is actually a fairly reasonable approach, wherein the most basic functions are designed first and then incorporated into ever-larger assemblies until the final one represents the entire system. Although we didn't explicitly explain this in Chapters 8 and 9, LISP (and FORTH) encourage this kind of approach, even though the original conceptualization might come about differently.

Another technique could be called "middle-out" design. This receives very little serious attention in the trade press, for good reason. It is nearly as haphazard as the first type: "Ah, let's see ... we'll need a random number generator, so we'll write that ... then we'll need something to handle the keyboard. ... "

The one we have been leading up to, of course, is called "top-down" design. Here, we start with the whole problem, then break it into major logical components. Each of those is further divided and subdivided until the obvious "component-level" structures are arrived at.

The really neat thing about this is that testing can begin as soon as the top level module is written. Since each module's design incorporates specifications defining calling parameters and returned data, it is a trivial matter to simulate them with *stubs* to facilitate testing. A stub is simply a dummy subroutine that either returns known test values or just returns control. Using this technique, the control structure can be exercised repeatedly, increasing confidence in its behavior by the time the complex of subordinate details is woven into the overall texture of the system.

Stubs are ideally embodied in the form of skeletal modules containing a complete definition of the logic intended to take their place. Expansion into lower levels then takes the form of replacing these dummies with real code, minimizing the need to simultaneously change a number of conditions for each test/debug cycle.

There are naturally situations, especially in assembly-level coding, wherein it is difficult or impossible to break the task into an obvious hierarchy of modules. There may be a central data structure with which almost everything else interacts, a multitasking executive, and so on. In situations like these, some perversion of the classic top-down approach is justifiable.

This whole scheme has advantages outside error reduction and increased programmer efficiency. It is a software management dream, allowing such unaccustomed luxuries as meaningful monitoring of progress, early demonstrations of the rough behavior of the system, increased interchangeability of programmers, and vastly improved documentation, perhaps even to the point of the often-fantasized "self-documenting code." Clearly, the disciplines represented by this and other aspects of the software engineering field can wreak havoc with some of the classic manifestations of programmer autonomy: job security embodied in obscure code, apparent wizardry likewise, and the satisfaction of cute—albeit absurd—design. Rather than being cause for alarm, this is encouraging, for it is perhaps worth trading that feeling of surpassing cleverness for a respite from the agonies of extended, seemingly hopeless debugging.

Another major benefit of this type of design philosophy is the ability to involve more than one person on a job without increasing the time it will take to complete. In the Old Days, programming projects could be shared only in the grossest of fashions, with no easily discernible task boundaries to lift the burden of complete system responsibility from the shoulders of one bleary-eyed denizen of the computer room. But now, programming teams can be assembled in a fashion that reflects the project's architecture.

So how does all this affect our patrol tour recording system? Let's take a quick look.

10.3.3 PTR Reprise

Armed with a notion of structure, it is a fairly simple matter to modularize this system, even if we don't fully understand the logic. Figure 10-6 shows a fairly rough division of functions: after initialization, an executive will continuously poll the conditions we mentioned earlier. One of three types of operations will take place, depending on whether a perceived event is a clock pulse, a received character from the PABX, or a key depression. Note in the figure that each of these has as subordinates one or more support routines, themselves sometimes drawing on even-lower-level routines for assistance. Some of the very low level ones might be called utilities, and can be shared among many others (such as the PRINT LOGIC).

This drawing by no means represents the ultimately detailed modularization of the project, having been kept reasonably uncluttered to demonstrate the implied functional hierarchy achieved in this fashion. It also totally ignores the data flow between modules—a significant part of the system's operation. The TOUR OK? routine, for example, must draw upon the stored sequence specified by the TOUR SELECT LOGIC, then pass data upward to the PABX DATA handler so that it may appropriately transfer values to the ACTIVITY LINE PRINT module.

The process of designing this system continues, once all the decisions discussed earlier in this chapter have been made, with the successive refinement of this structure chart to include data flow and full specification of the modules.

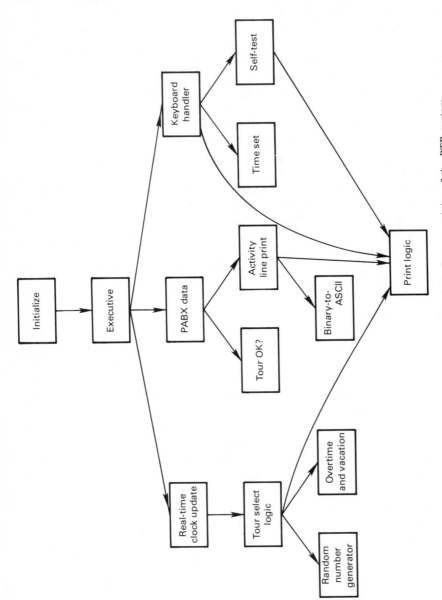

Figure 10-6 The general relationship between the software modules of the PTR system can be visualized with the aid of this skeletal structure diagram (not to be confused with a flowchart). Each block shown here can be written as a module and "plugged in" to those adjoining it.

When this is complete, there should be a comprehensive understanding of the system's operation.

Then it is coding time at last. The initialization process will take care of such trivia as defining the ports, clearing the RAM, setting up the time-base generator for a pulse interrupt every 10 ms, resetting the keyboard handshake logic, and presetting certain key parameters. It then passes control to the executive.

Here, we can stub out the three branches, write the basic exec, and test it. Perhaps the stubs can return 1, 2, or 3, depending on their identity, whereupon

```
CLOCK SET
READY?
00
59
20
16
12
SAT
?ENTER  W*MM*DD*HH*MM*SS*


TOUR COMPLETE
STATION #13    DAY/TIME = SAT 12/16 16:40:05    EXT #268    BOTTLING FLOOR 1.5
STATION #12    DAY/TIME = SAT 12/16 16:37:22    EXT #282    STAMP ROOM
STATION #11    DAY/TIME = SAT 12/16 16:22:08    EXT #283    GOVT OFFICE - BOTT
STATION #10    DAY/TIME = SAT 12/16 16:18:42    EXT #216    4TH FLOOR BOTTLING
STATION #09    DAY/TIME = SAT 12/16 16:15:36    EXT #277    RECEIVING OFFICE
STATION #07    DAY/TIME = SAT 12/16 15:59:10    EXT #264    WATER TREATMENT
STATION #06    DAY/TIME = SAT 12/16 15:56:50    EXT #293    RESEARCH BLDG
STATION #04    DAY/TIME = SAT 12/16 15:46:13    EXT #262    MAINTENANCE SHOP




50% COMPLETE
STATION #05    DAY/TIME = SAT 12/16 15:43:52    EXT #256    DRYER HOUSE
STATION #03    DAY/TIME = SAT 12/16 15:41:21    EXT #258    BACTERIOLOGY LAB
STATION #02    DAY/TIME = SAT 12/16 15:33:47    EXT #242    TUNNEL - NORTH END
STATION #01    DAY/TIME = SAT 12/16 15:29:04    EXT #233    CENTRAL OFFICES
STATION #08    DAY/TIME = SAT 12/16 15:24:57    EXT #257    WINE ROOM
STATION #14    DAY/TIME = SAT 12/16 15:20:31    EXT #267    CONVERSION AREA
STATION #15    DAY/TIME = SAT 12/16 15:16:33    EXT #245    BLENDING TANK ROOM
STATION #16    DAY/TIME = SAT 12/16 15:09:56    EXT #278    FINISHED GOODS




BUTTON-UP TOUR
```

Figure 10-7 The PTR system works—this printout from a typical guard tour and sample clock set procedure demonstrate the general behavior of the machine.

the EXEC can appropriately notify the development system, which then displays the status on the console. Simpleminded, perhaps, but a test of the executive can then be undertaken without the obfuscating effects of all the other code to confuse things.

Presumably, while all this is going on, someone is building the hardware. As system development progresses, it becomes desirable to begin the integration and make certain that hardware and software can indeed work together in harmony.

We shall go no further into the body of the PTR code, leaving it at this relatively superficial level so the thunder of our later case study will not be rudely stolen. Suffice it to say that continued effort at last yields something that works—something that can, in fact, be delivered to the distillery and put to use. A few problems mar the unrealistic smoothness of that misleading statement, of course, such as specification changes ("Hey, can you guys fix this thing so that it will handle overtime?") and problems with local power-line glitches—but in essence, it works. The output from the printer for a typical tour and the clock set procedure are shown in Figure 10-7: the listing reads from bottom to top.

The net effect is a successful industrial application for a microcomputer. The machine works reliably, is reasonably compatible with the guards, and provides the customer with detailed records of security activity which can be of significant value if a loss ever leads to an insurance claim.

So let us press on, now that we have reeled through the strange mixture of magic and engineering which constitutes the initial approach to a system design. We look into a grab bag of software techniques, then spend some time tracing the evolution of a piece of code through a development system and ultimately into a ROM on an SBC.

11

Software Tactics

11.0 INTRODUCTION

All the foregoing notwithstanding, we still need something other than a healthy notion of structure before actually converting a modular conceptualization of a system into code. In this chapter we look at some of the major issues involved in the creation of a program—not the bit-level techniques, for they would tie us too much to a specific implementation—but the architectural considerations instead.

There are certain inevitable bridges to be crossed whenever we sit down and begin converting one level of description to another. It is easy enough to say, "Input the command from the console and do what it says," but precisely how are we to convert a text string into an action? Similarly, there is no problem with the concept of an "internal model of the world" in a control system (remember Figure 5-30?), but the method of actually implementing it may not be at all obvious. Much of the value of programming experience lies not in facility with a particular assembler or compiler but in an understanding of what works: what collection of relatively philosophical techniques is likely to yield a smoothly operating and elegant system.

To aid in this regard, let's spend some time considering, in no particular order, a potpourri of programming concepts.

11.1 OVERALL STRUCTURE

"Structure" quickly becomes an overused word in this business, for it has been applied to bus architecture, logic design, the approach to a project, and the predominant style of programming. Now we encumber it further, by associating it with the interrelationships between modules and subroutines in a program. Could all these uses for this essentially general term suggest a notion that indeed permeates all aspects of this business?

Perhaps. Consider: the process of spanning the WHAT–HOW spectrum in order to convert a concept into a reality results, not in a random mass of logic at the HOW level, but in an overall system design that is a silicon-and-software analog of your ideas. You envision, for example, the patrol tour system as a box that inhales information from the PABX and prints reports on guard activity. That is the highest level of the design, and is reflected in the overall placement of units in the plant—a gadget in the telephone equipment room connected by a current loop to the machine in the security shack.

At the next level, you see some logic in the processor checking for events and reacting to them appropriately. That is the executive.

At a lower level, you imagine code that assigns tours, keeps track of the time, and watches for inappropriate phone calls. At a still lower level, you see something that prints lines. Lower still are some little I/O routines that actually talk to the UAR/T, the keyboard, and the printer. And way down at the bottom, there are the chips and wires and edge connectors—the guts of the box which allow your exquisitely contrived structure to constitute a working system.

So we really aren't stretching the meaning of the word, are we? Everything that we have been talking about to date fits into a harmonious whole.

In programming practice, "structure" suggests the creation of a hierarchical relationship between different blocks of code, such that the higher levels are spared the trivial details that tend to be obfuscating upon human perusal. The main body of your program, for example, may include a few lines that basically say, "Find the billing address for the customer." The routine responsible for this may determine that the customer is Putridata, Incorporated, whereupon it hands off the task to a utility that interprets the whole problem as, "Fetch bytes 45 to 98 of record 19 in the file CUST.DAT."

Another "structural" question that arises occasionally is not so much concerned with the relationships between procedures as with the trade-offs between procedures and data. Almost any task can be approached in two general types of ways: with the parameters embodied either in the program itself or in a data structure of some type. A believe-it-or-not example shall serve to illustrate.

Once upon a time, in a "DP shop" somewhere in Bluegrass Country, there was a neophyte programmer. She was charged with the task of writing a program (in COBOL) to convert all of her company's current five-digit customer codes to a new seven-digit system. No predictable mathematical relationship existed between an old number and its corresponding new one. There were about 3000 numbers to change, and they were located in a customer master file.

As the project progressed, people began noticing that she was carrying around some mighty hefty listings, and the rumors began. "What on earth is she working on?" She prided herself on the fact that her program was the biggest applications job that had ever been run on the company's computer. The operations manager even had to assign more core to her job, since the system began producing "out of memory" errors.

Eventually, someone became a bit suspicious, and asked to see the program, by now a massive listing over an inch thick. It was beautiful. Customer by customer, the program would read the master file, extract the old number, and charge into a series of over 3000 IF–THEN statements:

> IF IDENT = "35729" MOVE "7446201" TO IDENT.
> IF IDENT = "19408" MOVE "7859915" TO IDENT.
> IF IDENT = "60043" MOVE "8628219" TO IDENT.

Page after page. After page. Whenever it found one, it would simply make a new value for IDENT, then write IDENT to the output file after finishing the whole series of now-superfluous IF statements. Understandably, it took a rather long time to run.

A much saner approach would have been the simple creation of a table, through which a relatively trivial routine would search in order to find the old customer number and the corresponding new one. The program would then be completely general, capable of performing a similar task on any file if given the appropriate table—and it certainly would have been more efficient.

Yes, that really happened.

There are also situations in which other factors affect this code–data trade-off. Suppose that you are writing a little routine to control the temperature of a waterbed.

The problem is simply stated: if the temperature goes below 90 degrees, turn on the heater. If it goes above 91 degrees, turn off the heater. So you make a little program that looks like this:

> DO   FOREVER:
>   INPUT TEMP
>   IF TEMP < 90 THEN HEAT = ON
>   IF TEMP > 91 THEN HEAT = OFF
>   END

Great—as long as nobody ever wants to change the temperature setpoints. If this were compiled, stored in ROM, and marketed as a waterbed controller, it wouldn't sell (even though it could be billed as a product of space-age microprocessor technology!).

This is an absurdly obvious case, since any waterbed controller worth its salt would provide the user with some means for establishing control parameters, which would then be compared with the observed temperature.

But in a realistic software environment, there are dozens of parameters that are not obviously variables. They include such things as the length of a keyboard buffer, the number of bytes comprising each entry in a data table, the address of the top of RAM, and so on. Should these things be built into the code? What if

the length of a keyboard buffer is defined as 20 bytes by some instruction buried
in the software, and somebody wants to change it? It shouldn't be too much trouble to shuffle through the code until the instruction is found … but what if that
"20" appears in two places? What if one of those places—the obvious one—
simply causes a beep and prevents entries over a certain length, but the other one
is used far away in some other part of the code which clears the buffer prior to performing some activity? Then, reassembling the program with the first occurrence
changed to 15 would cause the system to blow up as soon as that other instruction
is executed, wiping out the five following bytes (which may be flags, data, or anything else).

   This all boils down to an observation. In most cases, it makes much more
sense to impose a strict separation between program and data. If that had been
done in the keyboard buffer case, the "20" would have appeared in a statement
like this:

$$\text{BUFLEN: EQU} \qquad 20$$

Changing it would involve a change of only that one statement, and with that
would have come a guarantee that all occurrences of the value in the program
would also be changed—for the label BUFLEN now means "buffer length"
throughout the entire system.

   If we accept the general desirability of maintaining data apart from code,
then let's proceed to look at some techniques that can make such a structure op-
timally efficient.

11.2 DEALING WITH DATA

We have to consider two categories of data in any given system: ROM data, consisting of permanent values, messages, parameters, and so on, and RAM data,
comprised of variables and other manifestations of actual system operation.

   ROM data are often associated with the concept of "table-driven software,"
wherein essentially general routines and macros derive their operating parameters
from bodies of data assembled with the program. This offers the operational advantages alluded to above in the keyboard buffer example: modification of the
software package for a slightly different environment (a different customer?) involves not a change of code and subsequent documentation and standardization
nightmares, but instead just the inclusion of a different set of tables.

   This proves to be a remarkably useful concept when a software product is
aimed at a more general market than would normally be suggested by the word
"custom." Suppose, for example, that you were sufficiently moved by this book
to wander off and begin manufacturing Blanchard control systems (go to it!). You
locate a model 42HD-84 somewhere nearby and base the design on it, then begin
actively selling the systems to other owners of the same machine. But one day,
the owner of 60HD-120 calls, having heard about this dazzling technological in-
novation, and orders one.

Now this machine is identical in principle, but the chuck (turntable) is 36 inches wider, it can handle 18,000 pounds more at a time, and the grinding head is 18 inches larger. Should it be necessary to go back and rewrite code just to accommodate differences in parameters? Of course not—you simply grab a "60HD-120" ROM off the shelf, plug it in, and gleefully ship the unit (doubtless charging more than for the "smaller" version). The difference between that ROM and the others is in the tables—the lists of machine parameters that are referenced by the control algorithm to determine the limits of its operation.

RAM data are another matter entirely. In a typical system, there will be dozens of flags, counters, and chunks of data reflecting the instantaneous status of the process as well as our much-touted internal model of the world. In the patrol tour recording system, there are 34 single-byte values maintained in RAM (ranging from obvious stuff such as HRS, MINS, and SECS to rather obscure items such as JUST1, which directs the tour selection logic to print only one tour after an odd number of overtime hours). In addition to the byte data, there are five 16-bit values—all address pointers. Then there are three bytes for the binary–ASCII conversion algorithm, three bytes for a keyboard buffer, four bytes for the random-number generator, and 32 bytes for something called the "anti-repeat logic table."

Managing all this is no real problem in a small program, especially in a case like this where the agglomeration of data hardly constitutes a "structure." But as system complexity grows, it becomes more and more troublesome.

Suppose, for example, that the code in the patrol tour system were a little more sophisticated, and retained a map of the guard's activity during an entire shift instead of just printing a line whenever he or she happens to call in. This internal record would probably consist of a list of movements, identifying the time, the area in the plant, any irregularities noted, and so on. The program would simply update this data structure as the guard moves about, yielding a complete image of the tour and perhaps enabling some higher-level security functions as well.

OK. Now we get a call from the distillery: "Hey, what would it take to add another 16 checkpoints to this thing? And some way for the guard to report on unlocked doors, trash, and other irregularities?"

"No problem," we reply, not thinking.

The purchase order arrives, and we tear into the code. Changes are made to the data structure, but we find ourselves spending a long time digging around in the original code, making patches, each a potential (actually, guaranteed) debugging problem. It's a bit depressing, especially since we remember reading that using a data structure can save a lot of trouble.

The problem can be visualized with the aid of Figure 11-1. In part a, the data structure is not sacrosanct—any routine that needs to examine or modify a piece of data is free to do so. This results in a complex marriage between the code and the data, which is more or less OK until someone changes one or the other.

Then all hell breaks loose. Have we found all the places that access the data structure? By adding more bytes to each record, have we created potential addressing problems that will not become evident until the thing has been in the field for six months? ("Uh, hello? We just noticed that if it is a Friday and there have been more than two hours of overtime, then the machine prints 'Happy New Year' when the guard checks in from the north end of the tunnel, but only if the tour before that one happened to include the bacteriology lab.")

A better way is illustrated in Figure 11-1b. Here, access to the data structure is strictly limited to a small group of subroutines, each of which is very rigorously defined. In the PTR system, the group of routines might include ADDREC,

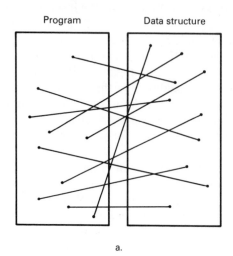

a.

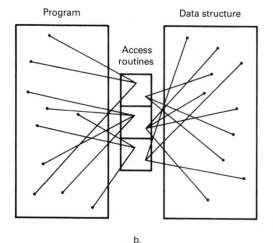

b.

Figure 11-1   The use of a limited number of "access routines" between a program and its data structure greatly simplifies later change of either by formalizing the interconnections and restricting them to a small number of logical paths.

which adds a new record to the structure, GETREC, which retrieves one and passes it to the calling program, TOUROK, which returns a byte indicating the correctness of a check-in, and so on. Now when we change the data structure, all we have to do is change this small group of access routines, and we can be assured that the rest of the system will not be disturbed. Obviously, there will be code changes to accommodate the new features that we might add, but they might well be restricted in scope to a single module.

If a policy is adopted which strictly forbids access to the data structure by any code other than these special routines, much madness can be avoided.

Before we leave the rather broad subject of "data," it would be well to touch briefly on some of the techniques that are of use in the attempt to locate one datum among many.

We spoke back in Chapter 9 of the "knowledge-search duality," that catchy moniker for the general idea that the more you know about your data, the less you have to search through it to find what you need. That is as applicable in a list of names as it is to a relational data base: if the names are in alphabetical order or are otherwise arranged in some predictable fashion, finding a specific one can be accomplished much more handily than with an exhaustive search.

It thus makes a certain amount of sense to exploit whatever relationships exist between data items, because exhaustive search—the brute-force alternative—is slow and clumsy.

One rather neat way to take care of this is to order the elements of the data file (list) so that a much more efficient search method can be used. The ordering can be made an intrinsic feature of the list, or it can be dynamically sorted—it all depends on what else the system is doing and how stringent the access-time requirements are.

Probably the best known of the search methods is the *binary* type—so named for its characteristic of successively discarding half of the list until it is left with the entry it was seeking. This can be readily understood with the aid of Figure 11-2.

At the start of the search algorithm, two pointers are established: BEGIN, which is set to address the first location on in the list (0), and END, which is set to one greater than the last. Then, a third pointer is created which references the midpoint, by simply taking the integer value of (BEGIN+END)/2. A quick test is made to see if POINT and BEGIN have crashed into each other, a condition that would indicate a lack of success in finding the entry of interest. In this case, FFLAG (the "Found Flag") is cleared.

But if this is not the case, the input datum is compared with the one located at POINT. If they are equal, the job is done—it has been found and FFLAG is set. Otherwise, either BEGIN or END is set to the value of POINT, effectively discarding the half of the list that does not contain the entry sought.

The beauty of this is its speed. If a list of 1000 entries were scanned sequentially to locate one item, the average number of comparisons necessary would be 500—the maximum, 1000. But if the binary search is used, the worst-case

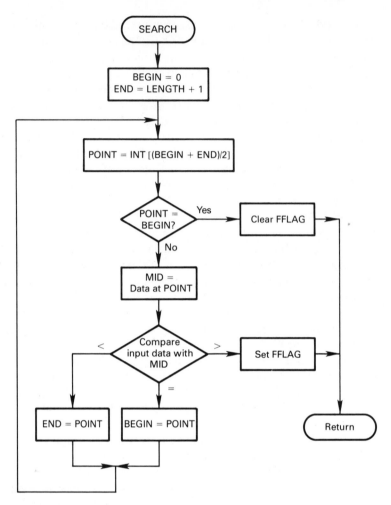

Figure 11-2 Flowchart of a binary search routine, which successively discards halves of the search space until it is left with the datum sought.

number of comparisons is only 10. The difference could spell the success or failure of a real-time system.

There is a much easier alternative, of course—provided that you already know where within the list the particular element of interest resides. Called *indexing*, it simply involves the calculation of the element's address via the general form: $ADDR = BASE + N*L$, where N is the number of the entry and L is the length of each one. If $BASE$ is the start of the table, the performance of the calculation will yield a pointer to the desired datum.

11.3 EXECUTIVES

During our discussion of the PTR system, we noted that the highest level of the code constitutes an "executive"—a loop that examines a battery of conditions for something that justifies further action. This could be as trivial as a keyboard wait loop ("Did anyone hit a key yet? No? Did anyone hit a key yet? No? Did anyone ...") or as complex as the top level of a multitasking production batching system.

The simple loop approach that we used begins to become somewhat inadequate as the complexity of the system grows. A lot of time can be wasted on the overhead of the executive itself, and it certainly lacks the flexibility that would be desirable in an environment characterized by changing user requirements. For these reasons, other methods have evolved.

One of the most obvious is probably an *interrupt* structure, but it can have some mighty convoluted implications. Essentially, hardware is provided which forces CPU response (in the form of a CALL to a predefined location) on the basis of external events. A keystroke, for example, might generate an interrupt that causes the system to suspend its current activity and go perform the code beginning at location 20.

The first requirement in an interrupt scheme is *prioritization*, which is normally provided by the hardware itself. It would be unfortunate if a keystroke could suspend the operation of a real-time process that must be completed in 200 μs— the keyboard activity should have the grace to wait until a more expedient time. But if the processor is in the midst of the lower-priority job, it should have no qualms about taking a moment to deal with more urgent matters. This prevents anarchy.

It's easy enough to solve the problem of inappropriately timed interrupts by simply disabling them whenever a critical section of code is entered. This seems to eliminate the objection noted above, but all sorts of complications arise when there is more than one critical section of code that must be so accommodated. It is impossible to predict the system's response time to a given peripheral, and it certainly becomes messy when modification becomes necessary.

No, the kinds of applications within the range of modern microprocessors (especially 16- and 32-bit units that were designed with much thought on this subject) suggest the need for a more reliable and flexible method of dealing with the real world's annoying asynchronism. Although the scheme outlined above can be further modified to disable only lower-priority interrupts within a service routine, it is still cumbersome: patches become even more complex and the data structures involved in the resulting "context switching" become large and unwieldy. A different approach is called for.

One method that has gained favor is known as *time slicing*, and makes use of a *kernel* whose task is the dynamic creation and destruction of processes as necessary to handle the system's activities. As shown in Figure 11-3a, each process (or

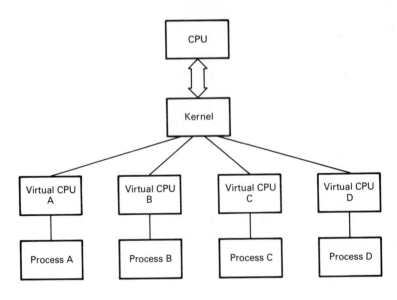

a.

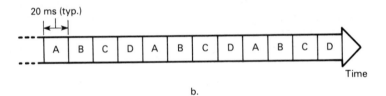

b.

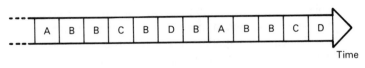

c.

Figure 11-3 Methods of time sharing a processor among multiple tasks, giving each its own virtual CPU (a). In (b), each is given time slices in turn, but this can allocate too much time to a relatively lazy task and too little to an urgent one. A modification is suggested by (c), in which prioritization is used to permute the normal sequential time-slice allocation.

relatively independent activity) is given the illusion that it has a whole computer to itself—virtual CPUs exist for their sole benefit.

Time slicing in its essential form is based on a clock interrupt, which decrees every few milliseconds that the time has come for the process in progress to be mothballed and replaced by the next one down the line. If the processes are users pounding on CRT terminals, this is quite acceptable, and is, in fact, a very common method of implementing a time-sharing computer system.

The problem that arises in typical industrial systems is that there is usually some real-world activity that is time-critical. One could simply speed up the clock interrupt, perhaps from 50-ms slices to 10, but there is a brutal trade-off associated with this: every process change involves saving current registers and status, determining the next process to execute, inhaling the new registers and status, and possibly checking "mailboxes" for interprocess communication. There obviously comes a point whereon a decrease in interslice interval so burdens the system with overhead that nothing else gets done.

But it is well known to sampling theorists that one must sample an event at a rate greater than twice the highest frequency of interest contained therein. It may be that simple time slicing cannot accommodate this, in which case it becomes desirable to marry the two techniques and add interrupt prioritization to the scheme we have just described. This is illustrated in Figure 11-3c, with process B the most demanding of the four.

It is important to note that this time-sharing activity, however it is carried out, does not affect the internal details of the processes executing under its auspices. Each takes up precisely where it left off during its previous slice. It is also worth noting that an architecture such as this offers a considerable bonus over the more convoluted methods of the past, for there is no conceptual difference between a system with a single CPU and one with many. A well-conceived real-time operating system built on this foundation can benefit without any software change by the simple addition of another processor card or two.

Now that we have spent a little time on a brief look at the two somewhat grandiose subjects of data structures and executives, let's look at a few techniques that are more limited in scale, but nonetheless extremely useful.

11.4 A SOFTWARE SAMPLER

Although the number of programs that could conceivably be written is infinite, there are a few standard problems that seem to crop up again and again. How many console command decoders should one design from scratch before generalizing the function and stashing it in a library as a relocatable module? One of the most pleasant aspects of the computer world at this stage of its dynamic evolution is that techniques are well established for linking modules together into a program, even if they were originally written in different languages. A rich and growing resource of library macros, subroutines, and modules is one of the most powerful assets available to a system designer. Maybe now and again the age-old fantasy will become reality, and a program can be created entirely from items already in stock.

We certainly make no attempt herein to catalog the possibilities exhaustively, but with reasonable comfort we can at least spotlight a few perennial favorites, hopefully reducing somewhat the number of occasions on which someone, after lengthy consideration, settles on "round" as the optimum shape for a wheel.

11.4.1 The Quintessential JUMP Table

If a microcomputer is being ensconced in a NEMA enclosure and consigned to a (hopefully long) life of dedicated industrial control, there is probably little need for an elaborate operator interface. It may be enough to accept START and STOP commands.from a pair of battered pushbuttons.

But it is not at all uncommon for system specs to call for some measure of user interaction. This probably involves a repertoire of commands, some of which expect operands (you don't just SET LIMIT, you SET LIMIT 500), others of which initiate relatively lengthy dialogues. There may be just two or three, or there may be 10 times that: the problem that arises is how the input commands can be parsed, decoded, and implemented along with any associated parameters.

One could always approach the problem in the manner of the programmer in Section 11.1 with the 3000 IF statements, but that's a bit crude for the late twentieth century. Somewhat cleaner is the idea of extracting from the command string the substring that represents the command itself (usually the first word) and then scanning a table for a match. This is illustrated in Figure 11-4.

Here, someone has typed "PRINT AQL SHIFT" into the system console, and the string has landed in a key buffer. Upon depression of the carriage return key, a command parser is invoked, which scans through the buffer using spaces to separate words.

The first word, the command itself, is successively compared to the entries in a table until a match is found. Associated therewith is the address of the subroutine that handles that particular function—in the example of the figure, the printing of reports. This address is typically used as the target of a JUMP.

The target routine can quite handily dig through the keyboard buffer and extract any operands of interest, but this contributes to inefficiency and sloppy structure. As we noted earlier in our discussion of access routines, it is much cleaner to isolate a data structure from the main body of the code. We therefore include in the parser some logic which grabs any remaining words (or numbers) in the buffer and places them in a parameter list. This is then available to the target routine, which in this case would discover that the report to be printed is the quality control data for the current shift.

The last command in the table shown is HELP, a useful adjunct to the system which simply lists on the screen the 12 command options. An operator, especially a new one, can easily forget the syntax or spelling of some of the available choices.

That problem can be more readily solved, however, by using a *menu* approach (Figure 11-5). The options (in something other than cryptic mnemonic form) are presented simultaneously on the screen, and the operator need only enter the number of the one desired or, alternatively, move a cursor or light pen

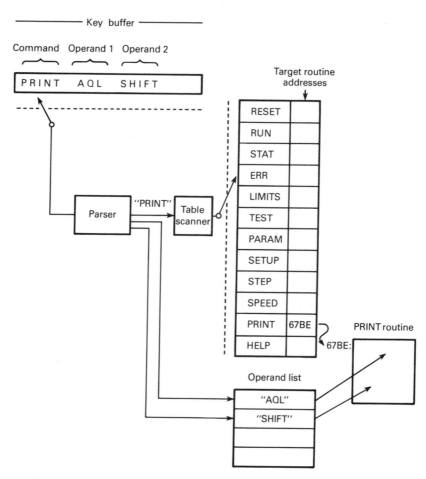

Figure 11-4 The use of a "jump table," together with a command line parser and a method of passing operands, can yield a very efficient operator interface. Here, the PRINT command is located via a scan through the table, yielding a target routine address to which control is passed.

adjacent to it and somehow indicate that that is the one desired. The commands that execute directly (such as the status display) simply go ahead and do so; the others present menus of their own to lead the human by the hand through the process. A cleverly designed system can be fully self-documenting, requiring only a few minutes of familiarization before a new user can effectively operate it.

This highlights something we have talked about quite a bit since we turned away from hardware at the beginning of Chapter 7: friendliness and user interface. It is unreasonable to ask a semiskilled machine operator to understand complex command formats and option switches—for him the WHAT end of the WHAT–HOW spectrum is probably as far from "computerese" as one can get. "RUN/G: SYS0,OPT21-5" probably doesn't make much sense, and it would be

```
 1 RESET
 2 RUN MACHINE
 3 STATUS DISPLAY
 4 ERROR REPORT
 5 SET OPERATING LIMITS
 6 SELF-TEST
 7 CHANGE MACHINE PARAMETERS
 8 PERFORM SETUP PROCEDURE
 9 SINGLE-STEP OPERATION
10 SET MACHINE SPEED
11 PRINT REPORTS


ENTER FUNCTION NUMBER __
```

Figure 11-5 A "friendly" user interface often calls for a menu of options which eliminates the need for the operator to remember the full repertoire of commands.

much kinder (and far less conducive to errors) if the system were contrived such that all necessary functions and parameters could be expressed in something approaching the natural language of the anticipated users.

11.4.2 *Handshaking and I/O Logic*

Interface with I/O devices raises the specter of all sorts of problems. Once signals start crossing the secure boundaries of the system's enclosure, they become far more susceptible to noise, overload, mechanical abuse, and inappropriate interconnection.

Much of this, of course, can be handled with the techniques described in Chapters 5 and 6, but sooner or later it becomes necessary to extend the interface into the software itself. It can be as simple as a single input or output instruction, or it can involve extensive timing and conditioning logic. Let's look again at some of the circuits of Chapter 5 and consider them from the programmer's viewpoint.

First, we should discuss one of the most common of interface software tasks—the implementation of a UAR/T for serial data communication with a console, modem, or other system. The device takes almost all of the pain out of the job, especially in the light of old-fashioned "bit-banger" schemes which handled serialization and timing within the code itself.

Section 5.1.1 detailed the behavior of the UAR/T, but at the time we ignored the software end of things. To illustrate how it might work, we shall create a program called CHAT which allows conversational exchange between one system and another.

As depicted in Figure 11-6, the problem is fairly trivial. We can exploit the fact that human beings are abysmally slow creatures: even with the serial link operating at 300 baud (30 characters per second), it is highly unlikely that the user of the terminal emulator will type too fast. The UAR/T provides a handshaking flag for those occasions when the data flow must be held in check to avoid overrunning the serial line, but in this simple application we will not need it (the flag is called TBE or TBMT—Transmit Buffer Empty).

Examining the flowchart, we see that the first step is one of initialization. This could be done more simply in hardware, but let's assume a flexible system

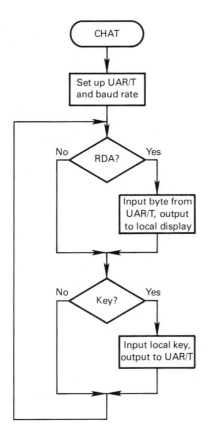

Figure 11-6 Flowchart of CHAT routine illustrates the use of status lines to synchronize the processor with external events. The program continually tests for UAR/T "Received Data Available" or a local keystroke, performing actual data movement only when one or the other has occurred.

with software-selectable baud rates and UAR/T options. The latter consist of the number of data bits in each word, whether or not parity is generated and tested, and if so, whether it is odd or even. This initialization is probably a pair of output instructions that load external hardware latches with data appropriate for the intended operation.

Then begins the loop. The program needs to find out if a character has been received by the UAR/T, so it issues an input instruction of the typical form,

IN A,STATUS

At the hardware level, this generates a "Status Word Enable" signal that places the device's five status bits, instead of its output data, on the bus. If the wiring in Figure 5-2 were actually used, the data available bit could be handily interrogated by rotating the accumulator to the right after the completion of the input instruction and checking the Carry flag.

If a character is available, the program obtains it by issuing another input instruction—this one enabling the data rather than the flags onto the bus. Presumably, the byte is immediately shuttled off to a local display, which may well be a CRT terminal interfaced by another UAR/T. For simplicity's sake, we leave out these details.

After this (or immediately, if no character had been received), the program checks the local keyboard for a character and undergoes the reverse process if one is found. Transmitting via the UAR/T is a simple output operation, generating at the hardware level a strobe commonly called TDS (Transmit Data Strobe). Somewhere in here a test could be inserted which aborts the entire program if a certain key (such as ESCAPE) is depressed, returning to the loop point otherwise.

There is nothing particularly complicated about this, a fact that has taken much of the pain out of I/O programming since the introduction of the UAR/T in the early 1970s. In Chapter 13, we'll have a field day with these devices, using 16 of them on a single bus structure to shuttle data between a host computer and a bevy of satellite processors scattered about a factory.

Most of the circuitry presented in Chapter 5 stands alone well enough that any discussion of the software involved is fairly trivial. In the keyboard scanner of Figure 5-7, for example, a simple handshaking scheme is used: When the logic has detected a key depression, it issues to the processor a "Data Available" signal, which may be the input to the "KEY?" test in the CHAT program we just discussed. The hardware suspends its scan while the processor discovers the flag and subsequently issues an ACKNOWLEDGE pulse in the act of picking up the key code. There's just not much else we can say about that.

But sometimes, as in the IDAC terminals, it is not advantageous to build keyboard scanning logic. If a machine is intended for volume production, or if there are unusual interface requirements (or if the designer just happens to lean toward a software solution), tasks like this may be better approached with the resources of the microprocessor.

Take a look at Figure 5-10. The circuit shown is much simpler than the scanner, consisting of nothing more than a matrix of switches and diodes connected to the CPU input data bus via a group of buffers. The matrix is arranged in three rows, which are connected to "select" signals generated by the control decoder.

In operation, the processor can determine which switches, if any, are actuated by successively taking the three control lines low and examining the data on the input bus. This is fairly straightforward, but as we mentioned much earlier in this book, certain realities associated with switches call for some additional processing effort. In particular, the physical contacts must be debounced to prevent the illusion of rapidly repeating depressions, some scheme for rejecting simultaneous depressions must be implemented, and keys not currently lit should be ignored. Clearly, this gets a bit hairy—in the IDAC software it requires 356 bytes of code plus some overhead at the executive level.

I/O programming covers a wide range of complexity, and its solidity is critically important in an industrial setting. An example will serve to illustrate.

Once upon a time, a microprocessor-based control system company began experiencing some very frustrating problems with units in the field. Essentially, it developed that the system was "hanging up" in a loop that read the A-D converter used to monitor certain plant conditions. The loop existed as an error-reduction trick, taking two successive snapshots of the A-D's value and comparing them—assuming that any difference suggested a transitional state. If no difference was detected, it would return the observed value; otherwise, it would go around again.

The problem was finally traced to a change in chip vendors for the 8216 buffers used to interface with the analog device. The original Intel parts had a fairly high leakage current and floated high when they were in the disabled state, whereas the new National parts had no such tendency.

This would not have created any difficulty but for the fact that the converter was a 12-bit device and thus the software had four extra bits to deal with on its 16-bit input operation. The comparison loop tested all 16 bits. Not until a device change eliminated the happenstance of fairly dependable input floating did the programmer's oversight create problems, but when it finally did, systems manifested the problem as extreme sensitivity to heat and noise.

11.4.3 Self-Test Facilities

One of the classic problems with writing code is that the ability to state with certainty that it is completely correct decreases dramatically with project complexity. This, perhaps, is the prime motivation for the software engineering disciplines that have materialized in recent years. A 64K address space chock-full of deeply intertwined chunks of logic does not lend itself well to fast and comprehensive testing.

Of course, the development of self-test programs does not purport to solve this problem; it still makes more sense to modularize the system and subject each easily defined entity to isolated analysis. But a self-test module linked with the rest can force some system-level consideration of performance criteria and also become a lifesaver when the system has been relocated from the bench to the field.

That's the real motive behind this. Few things are more frustrating than a sealed, ROM-filled box that makes mistakes. The unit's behavior is so thoroughly rooted in the software that meaningfully poking about with a scope is nigh impossible, and attempting to find a bad buffer on a tri-state bus is worse than chasing down noise in a dc-coupled amplifier. It's nice to be able to push a button and put the machine through its paces—one thing at a time.

The data collection terminals that shall occupy us in Chapter 14 are representative of both the problem and the solution. Buried within the confines of the padlocked box, mounted to a piece of aluminum angle supporting the interface card, is a three-position toggle switch, momentary one way. Throw it to the momentary position, and all the lamps in the 20 lighted pushbuttons become illuminated, while the top text display says "LAMP TEST." There's a twist here: even though the switch position is read and acted upon by the processor, the illumination of the lamps is purely a hardware function (see Figure 14-8). The primary value of this, then, is finding bad bulbs, but it also allows one to determine whether a problem in the lamp circuitry is in the drivers or the preceding bus interface logic.

Throwing the switch in the other direction invokes a much more comprehensive self-test. Depressing individual pushbuttons lights their associated lamps, inserting a badge displays the data read on the lower display, and a communications loopback test is automatically performed on the serial I/O channel that represents the link between the terminal and the host computer. Armed with these functions, a field engineer can quite handily isolate a problem to a relatively small part of the system—often to a single component.

There are a few obvious problems associated with the attempt to produce a thorough self-test facility. First, the von Neumann bottleneck taints all that touches it, and there is little hope of pinning down the location of a short on the bus through this means alone. Also, a dead CPU will not even begin to execute a self-test, leaving considerable uncertainty if the problem is central.

It is wise in some systems to incorporate an exerciser that is designed to produce some dependable test patterns for a signature analyzer. This is a piece of test equipment first introduced by Hewlett-Packard, enabling identification of even the most subtle deviations from the norm on a given circuit node. As long as a correctly functioning system has been used to establish the signatures for key test points, the technique can be reliably used to pinpoint faults. Since it requires the unit under test to be self-stimulating, however, it is sometimes considered along with self-test program design.

11.4.4 Miscellaneous Utilities

Somewhere down in the low levels of a structured system can be found any number of utilities. They may be math routines, general I/O handlers, random-number generators, or good old GEORGE, which rotates the accumulator three times to the right and then increments it.

It is worthwhile to sprinkle these about liberally, for it is far more pleasing both to the programmer's aesthetic sense and to any persons who may someday find themselves doing a patch if the higher levels of the code do not focus on bit-level specifics. A program should not include a "multiply" routine in the midst of mainstream logic: it should call some menial service routine to perform the support function.

In systems written in assembler, a question often arises regarding the relative desirability of macros and subroutines where low-level functions such as this are involved. There is a trade-off: although macros appear in the form of expanded code wherever they are invoked, thus eating memory space, they offer substantially more flexibility in parameter passing than do subroutines. It might well be unwise to use a macro for a frequently invoked math operation because of this code expansion, but in cases such as those presented in Chapter 8, it can be highly advantageous.

This is a good point to mention again the idea of creating a library of utilities, each well documented and generalized to the point whereon its use requires no modification. It is the kind of thing that can otherwise be a never-ending pain: "Where's that old listing of the distillery job? Wasn't there some anti-repeat logic in there?" The key to the success of a software library is documentation, with each module accompanied by a textual explanation, a list of calling and return parameters, the stack depth used, the registers modified, and so on. Without this, use of the sketchily identified blocks of code can be exceedingly dangerous.

We must accept the near futility of attempting even a quasi-comprehensive discussion of software techniques within the relatively limited space of one chapter, and move on regardless. Only programming can make a good programmer, and it is somewhat counterproductive to plunge too far below the philosophical levels unless we are willing to totally redefine our stated objectives.

But there is a lot more we need to talk about. Once we pick a processor configuration, somehow contrive an interface, conceive a surpassingly brilliant software structure, and hammer it all together—how do we get it from our head to a ROM? How do we debug the code *in situ*?

Let's take a look.

12

System Integration

12.0 INTRODUCTION

Somewhere during the development of a system, it becomes necessary to convert logic diagrams into circuitry and listings into ROM contents. There is a temptation to consider this phase to be relatively trivial (after all, the design is done, right?), but it is riddled with booby traps that can have dismaying effects upon the unwary designer.

Most of these are in some way connected with debugging, for only with the utmost rarity does something work the first time. There are actually people out there who design and build hardware, write a program, burn it into a ROM, then plug it in and try it. "Hmmm. Doesn't work. Maybe if I try this...."

It's an agonizing procedure for two reasons. First, the bandwidth of the feedback path (see Figure 7-2) is minimal when so many unknowns are being tested together that the meaning of individual indications is vague. Second, and closely related, is the fact that the difficulty of a debugging task tends to increase exponentially with the number of bugs that exist. As we have pointed out, methods exist that will dramatically reduce the number of errors in a piece of code—but be assured, there will still be many if nothing is tested until everything is "done."

We open thus with an emphasis on debugging because that is, indeed, the predominant activity associated with the system integration phase. But it is also a time to establish more firmly than ever the relationships between hardware and software, the agreement between actual and predicted performance, and the overall congruence between the neonatal system and the tattered, oft-bemoaned specs. Let's look at some of the techniques that can smooth this critical process.

12.1 A SYSTEM INTEGRATION DISASTER

Perhaps we should begin with a portrayal of the way it should not be done. This would be an occasion for some humor were it not for the fact that this is the way it frequently happens: without some precognition of the agonies in store, an inexperienced system designer is easily tempted to develop the system in the manner we are about to describe. The sole consolation is that only rarely does a person go through this a second time.

Remember our scenario of Chapter 4, beginning with the placemat specs? We shall use that as the paradigm, but look a little more closely at some of the methods used to arrive at such an unhappy state.

It's a big project, occupying over 50 pages of handwritten source code. The designer (we shall assume only one) alternates between hardware and software development as whim and exhaustion dictate, one day wirewrapping logic and sawing bits of aluminum angle, the next writing code and cluttering the blackboard with ever more obscure notes. It is actually making a fair amount of sense, with the exception of a strange routine or two that can be left until later.

Somewhere well along the line, the hardware assembly is completed, and power is applied. There is no way to check for functionality, of course, since all of the intelligence is embodied in software, but primitive scoping eliminates at least the grossest errors and offers some assurance that things are OK. Meanwhile, the software is coming along tolerably well, and the designer cranks up the text editor on the development system and begins keying in the massive assembler-level program. The source file occupies 72K bytes on disk.

At last, with a flourish, he directs the system to assemble the program. Page after page, the listing pours from the printer—punctuated frequently by error messages: undefined symbols, bad expressions, duplicate references, division by zero, missing labels, out-of-range relative jumps, and syntax errors. There are over a hundred of them.

This isn't particularly dismaying, however, since the assembler is kind enough to show precisely where the errors occur. So it doesn't take long to clean them all up and obtain a listing that looks deceptively correct. With shaking hands, the designer invokes the debugger (which contains a facility for programming EPROMs) and begins casting the fresh object code in concrete. The hardware awaits.

He plugs in the chips. He applies power, and notes with a surge of excitement that nothing smokes.

Nor does anything work. A scope probe on an address line or two reveals that the processor is indeed "working," but there are no external indications to that effect. After considerable head scratching and poring over listings, our fearless design engineer concocts a theory, tosses the ROMs into the ultraviolet eraser, and wanders off to edit and reassemble the program.

But what hope is there for success? There is no certainty that the hardware design is correct, for checking the circuit board simply involved verification that the wiring matched the schematic. The software surely contains numerous interacting and nested errors, and until there is something other than a "dead" indication, only guesswork is available as a guide. Even after the stack crashes and I/O problems are solved, subtle combinations of bugs can masquerade with impressive realism as totally different problems. Our hapless designer is going to be at it for a while—so let's leave him mumbling and look at some more effective ways to convert a design from a stack of papers into a shippable unit.

12.2 SOME BETTER APPROACHES

First, before we even consider alternative approaches to the system integration process itself, we must note some of the other sins committed in the foregoing scenario. It was never explicitly stated, but it seems quite likely that the system design itself was quite amorphous, with no overriding structure to minimize the obfuscating effects of giant aggregates of code. Had it been developed differently, it would have been possible to begin testing as soon as the highest levels were written, with debugging proceeding at a modular level as the project progressed. This would obviously have had the effect of reducing the burden that the designer now faces.

Another bit of poor practice is the creation of one massive source file, which we identified as being 72K long. Not only is it simply clumsy to manipulate, but even the most minor correction involves a long assembly process. Further, later addition of afterthoughts and updates will require a fairly solid understanding of the entire system. This is hardly fair to other programmers, who might suddenly find themselves saddled with an incomprehensible and convoluted jumble of unfamiliar code. Even if the design is well structured, it causes confusion if it is all crammed together into one intimidating file.

And what about the hardware? Doesn't it seem a bit absurd to vastly compound the number of unknowns by introducing untested code into an untested circuit? There may not even be any assurance that the memory is being addressed properly, that the I/O bits are correctly connected to the interface logic, or that device outputs are not fighting each other furiously. The meaning of the already obscure debugging indications is further eroded when the program's environment is uncertain.

We are arriving at a suggestion that there should be some closer connection between the software and hardware development processes—and between the hardware and software themselves. The general scenario we have described is suggested by Figure 12-1, obviously a rather awkward way of doing things.

12.2.1 Use of Target Hardware

One of the most obvious solutions to the problem, if you have the time and money to implement it, is the addition of *target hardware* to the development system.

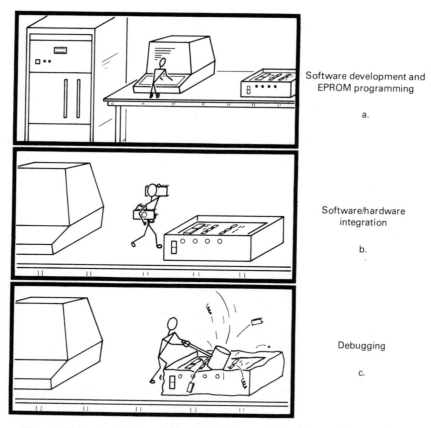

Software development and
EPROM programming

a.

Software/hardware
integration

b.

Debugging

c.

Figure 12-1 A three-step development process that treats hardware and software as totally independent entities generally involves a complex and frustrating debugging effort.

This, in the case of the PTR system, might consist of the little matrix printer, a keyboard with its interface, and the communications logic that inhales data from the PABX. The danger in this lies in the added overhead time associated with the customization of the development system, not to mention the risk of destroying something that is necessary for programming and ongoing business activity. In a small operation, the same system is probably the word processor and the bookkeeper as well as the software development station, so such intimately wedded test hardware is a bit risky.

But there are numerous situations in which such a setup, illustrated in Figure 12-2, can be quite convenient. In particular, an engineering firm that specializes in the creation of paper mill control systems might well discover that the general requirements of most jobs are similar enough that this approach is profitable: the system is outfitted with a flexible and well-debugged collection of I/O devices typical of the final products. Since it doesn't have to be designed each time, it

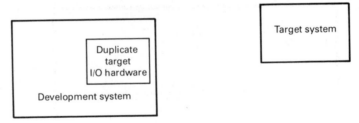

Figure 12-2 Software development using a full duplicate of the target system hardware built into the development system.

presents little danger, and the convenience of having the plant interface hardware at hand during software development is significant. A routine that watches an optical sensor and closes a valve when light is sensed can be tested *in vivo*, so to speak, then transported to the actual target system with reasonable assurance that it works.

12.2.2 Simulation of Target Hardware

A somewhat orthogonal approach is represented in Figure 12-3. Instead of fitting the software into the hardware configuration, this consists of just the opposite: the target hardware is simulated and used as the development environment for the actual code. The simulation takes place entirely within the development system.

This is rather handy, because it can allow the elimination of interdependence between the hardware and the software development teams. It is also dangerous, because there is no guarantee that the simulation is precisely representative of the real world. It also allows timing problems to go unnoticed.

In essence, simulation is performed by defining the various I/O ports and interrupts that characterize the target system and, depending on the sophistication of the simulator, providing some logical interconnection between them. In the simplest case, every input instruction simply creates a "prompt" on the development system's console, allowing the designer to specify the data that are to be plugged into the program at that point.

A clever simulator will also aid debugging by trapping excessive stack depth, memory addresses outside a reasonable range, and so on. Like any good de-

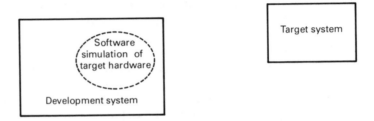

Figure 12-3 Software development using a simulation of target hardware.

bugger, it will support the introduction of *breakpoints*, which allow the program under development to be "halted" at certain specified points for examination of variable conditions. ("Is the Carry flag really set here? What is in the E register just before this subroutine is called?") Similarly, the simulator might allow a *trace* mode, in which the exact logical path through the program is printed, together with the values of all pertinent variables at each step of the way. The instruction opcodes can be shown in mnemonic form—perhaps even with labels—to simplify the human task of keeping it all straight.

These techniques are both very useful, but wouldn't it be better if we could actually interconnect the development and target systems to absolutely minimize the number of surprises that await us when we try to make the real thing work?

12.2.3 In-Circuit Emulation

A technique pioneered by Intel in the early days of the 8080 dramatically improved the lot of designers faced with that problem. The CPU chip in the target system is removed and replaced by a DIP cable originating in a block of logic embedded in the development system (Figure 12-4). This cable carries all control commands and bus signals normally generated by the CPU chip, and enables the designer to plunge into the environment of the system under development with the full debugging resources of his machine to aid him: breakpoints, symbolic disassembly, trace, and so on.

This provides considerable power. A typical robust development system already hosts facilities that allow the human being in charge to deal with code in some manner that is friendlier than binary, and a good emulator package will include even more. Some systems even allow the symbolic debugging of real-time systems and those based on more than one processor. As project complexity

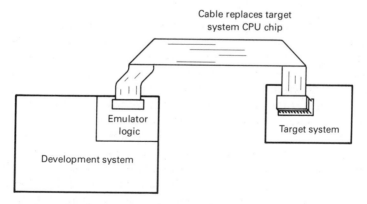

Figure 12-4 Software development using an emulator—a replacement of the target system's CPU chip with a cable connected to emulator logic within the development system itself. This provides considerably more power and flexibility than either of the first two approaches, but is expensive.

grows, the ability to work within the application context becomes more and more desirable until it is an actual necessity.

It must be remembered that the objective here is the reduction of the number of surprises that lurk in the process of merging hardware and software. The scenario of Cartoon 10-3 is hardly atypical; you can always find someone busily whittling and hammering to install the proverbial square peg in the round hole. The whole system suffers when this happens, for any modularity that once existed is sacrificed in the frantic "kluging" that characterizes the attempt.

The emulation technique allows almost complete elimination of this problem. It does not require the complete target system to exist; in fact, the entire system under development can be emulated, with real hardware plugged into the growing system as it is developed. Figure 10-2d illustrates the effect of this.

Before leaving the subject of emulators, we should note another area in which they are of immense value. Field repair of a system is subject to many of the same headaches that plague design debugging, with the possible exception of uncertainty about the viability of the underlying theory. Most servicing schemes depend in some way upon the system's ability to be self-stimulating—from the use of a scope or logic analyzer to observe the bus to that of a signature analyzer to check the congruence between actual logic streams and those that once were. In any of these approaches, things get a little slow if the whole machine is dead.

But an emulator (especially one that is married to a signature analyzer) quite handily eliminates this difficulty. Given the most primitive functionality of the CPU board, an emulating system can selectively stimulate individual parts of the system under test with an appropriate repertoire of debugging routines and "scope loops." Organized analytical techniques thus gain some precedence over the more traditional brute force, part-swapping approach.

12.2.4 The RAMport

There is another technique of interest here—one that grew out of one of the author's development projects back in 1976. Emulators were too expensive, starting at about $20,000, including the system in which they were ensconced—but BEHEMOTH was already in-house, costing less than half that amount and certainly no slouch in the development support department. Could it be used to efficiently support the creation of a system based on an SBC, whose RAM and register contents were invisible outside the card boundaries?

Well, maybe. A bit of thought spawned an approach that subsequently proved to be remarkably useful not only for the original problem of system development but for many interesting interface tasks as well. Represented in its basic form in Figure 12-5, the S-100 card dubbed the 'RAMport' is the key: simply a field of RAM (2K × 8) which, under the control of suitable logic, can be made to appear either within the development system's address space or on the end of a 24-pin DIP cable. It is no coincidence that the arrangement of pins on the 24-pin header is identical to that of a 2716 EPROM.

In Chapter 5 (in the general vicinity of Figure 5-16), we spoke of DMA, and that drawing happens to be very similar in spirit to the design of the RAMport. It consists of sixteen 2102s (it's been awhile...), an array of data path switches (which, for purposes of package count, turned out to be more efficiently handled with multiplexers than with tri-states), and some arbitration logic. There's very little arbitration to be done, actually, for the "mode" of the RAM is determined solely by a single bit that makes up 12.5% of an output port. Why bother with simultaneous access?

The value of all this is the implied ability to plug the card into the development system, plug the cable into the ROM socket of a single-board computer, and switch the RAM back and forth to pass progressively more correct versions of some piece of code in a manner more comfortable than endless, tiresome invocations of the EPROM programmer and the UV light. In typical operation, an appropriately scaled application program, module, or debugging routine is assembled, moved into the RAMport's address space, then handed off to the target system. Debugging can then proceed with enhanced facility for making changes.

But if this were all it could do, it would not be flexible enough for the kinds of tasks to which such a tool must be put. One of the most annoying parts of the SBC world—especially those minimal jobs without bus interfaces—is the fact that the CPU registers and the on-board RAM are quite invisible to the designer without the creation of a monitor to host a primitive set of I/O facilities. That gets in the way. So the RAMport idea was extended to allow the SBC to write into it, via the "program" pin of the pseudo-2716. With that little fillip, it is a fairly trivial matter to interrupt the SBC's execution of some application program, inhale and store same, pass over a RAM and register copy utility, cause it to execute (moving the data into "ROM"), inhale and store THAT, restore the original program, and allow it to continue. This only takes a fraction of a second once it has been packaged as a debugging command, and, with the addition of some software breakpoints and other niceties, pretty well gets into the same corners so effectively penetrated by the emulator—but at a small fraction of the cost.

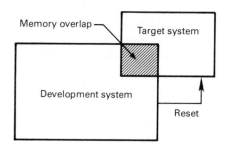

Memory overlap

Target system

Development system

Reset

Figure 12-5   Software development using a "RAMport"—a shared region of memory between the development system and the target hardware.

We highlight this not as a celebration of the author's design but as a demonstration of the tricks available which can be used to replace some of the traditionally expensive facilities. There is never anything sacred in a system configuration (unless some vendor is sensitive about tinkering and threatens withdrawal of a maintenance contract), and every development environment presents unique problems. It is possible for some mighty small investments of time and money to yield dramatic returns in usefulness over time. (The RAMport, for example, has also served as the basis of an envelope generator for music synthesis, a high-speed data link for burst-mode communication between processors, and a graphic display refresh buffer.)

12.3 MISCELLANEOUS INTEGRATION NOTES

12.3.1 Debugging Routines

We have mentioned more than once the idea of using debugging routines as the initial step in hardware–software integration. This turns out to be incredibly useful, for simple little test loops present far fewer unknowns than even a skeletal version of an application program, and they can be made general enough to support the testing of hardware facilities that are not even being used (yet).

The insertion-withdrawal tester (IWT; see Figure 5-17), for example, was characterized by a relatively complex combination of code and hardware logic. There were miscellaneous control lines, an analog input channel, a CRT, machine sync detection logic, a solenoid driver, and a 9511 floating-point processor, not to mention some real-time interrupt-driven software. This is obviously not the sort of thing that is best approached all at once. But even a top-down approach required too many simultaneous conditions to be dealt with, so a quick and dirty debug package was written.

This initially consisted of six test loops, which were assembled and made resident in the RAMport. The hardware engineer was free to work with them while other software development was taking place in the main system.

The six tests were as follows:

1. Continuous rotation of a single bit through the data out port (shown in Figure 5-17 as the output bus). By syncing the oscilloscope on bit 0, the hardware designer was able to trace the path of each bit by noting its time delay.

2. Similar rotation of a bit through something called "OUTLATCH," which is the source of the control lines to various parts of the machine.

3. A binary count of the control port (which is decoded to produce all the internal system timing commands) to allow determination of correct bus interface activity.

4. A continuous copy of the status buffer (input) to the OUTLATCH, enabling a quick check of both I/O function and the logic signals on the input.

5. Continuously repeating A/D conversion, allowing the entire analog front end to be checked with a voltage reference and a scope.

6. An echo of UAR/T characters, which simply accepts characters one at a time from the CRT port and then turns around and sends them.

By the time all of these had been put to use and appropriate circuit modifications made, the hardware of the IWT was adjudged to be correct. The entire debugging package occupied 135 bytes, and substantially simplified the development process—each isolated feature of the logic circuitry could be exercised alone without the obfuscating effects of closely related events (with uncertain cause–effect relationships).

12.3.2 Debugging Technique

This is one of those areas in which everyone is different, and we will not attempt to foist our favorite methods upon you. There is one fairly universal requirement, however, and we shall foist that shamelessly: an organized approach.

It is occasionally tempting to charge headlong into a nonfunctional system and seek one's fortune in buried glitches. There's something noble about the attempt to do it all at once, but it is not only surpassingly difficult but less likely to yield truly correct results.

Part of the reason for this is that interacting bugs can masquerade quite effectively as something they are not, leading one to implement a patch that solves the phantom problem. If the testing process extends only to the boundaries of the application program's functionality, it is easy for such misunderstandings to occur. One good way to prevent it is the use of simple, dedicated debugging routines such as those described above; another, best implemented in concert therewith, is a carefully organized approach.

This consists of testing first the most central parts of the system: the I/O channels, the data paths, the timing logic, and so on. It helps immeasurably to have a few known-good pieces of hardware and software to use as leverage while pushing ever deeper into the obscurities of the unknown system.

Another critical part of this organized approach is the simple (and seemingly self-evident) use of pencil and paper to note each test, each modification, and each observed effect (no matter how trivial). Patterns can emerge from the notes that might well have gone unnoticed in human short-term memory, and it can be helpful to flip back a few pages and say, "Hmmm, that worked yesterday—until I inverted the status flag of the A/D. ... " There is something about this kind of technique that tends to take the desperation out of a monstrous debugging project and replace it with a measure of structure—not unlike that which presumably underlies the system's design.

12.3.3 Testing

At some indistinct point, the implementation task changes subtly from debugging to testing. Once "all" the errors are found, and the people involved start getting jovial again, a difficult phase must be undergone.

You have to try to break it.

This is not a trivial matter. Unless you are a faceless crank-turner in a large and impersonal engineering organization, you are going to have an emotional investment in that contraption. That's just part of the territory that goes with creativity.

The temptation, of course, is to be very forgiving of the system's minor faults, and to handle it gently and lovingly every step of the way. As earlier chapters suggested, this is not at all representative of the care it is likely to receive in a factory setting, receiving brutal abuse day and night. Before it goes out the door, you might as well see if it can take it.

Mash the switches hard. Subject it to shock and vibration. Run SCR-controlled motors on the same power line. Build a Jacob's ladder out of an old oil-burner transformer and set it alongside. Cover up the filters and place the machine in the sun (under power, of course). Turn off the power at random intervals while it is performing its function—sometimes with a few glitches thrown in. Short the communications line. Generate static discharge to all exposed points.

If it still works, leave it running for a week or so, preferably under continuous functional exercise. This increases the probability of catching such subtle problems as "stack creeps," in which an imbalance between PUSH and POP instructions allows the stack to wander slowly downward until it begins obliterating data.

This raises a question that is unrelated to the classic interpretation of the term "reliability." Who says that only hardware fails with time? It is counterintuitive to suggest that a piece of code that once worked might someday fail, but it happens: conditions change. How can it be tested exhaustively?

If you think about it, it becomes obvious that it cannot—unless by some magic you manage to derive mathematically, and then produce, every possible combination of logical and temporal influences that impinge upon the routine under test in any way. This is actually approachable with small, isolated modules, but if you try it with a large and complex program, you may be wasting your time. We are talking about an exhaustive test, with the same astronomical numbers of possible states that we discovered in our discussion of memory testing.

Where does all this lead, then? Right back to an old subject: top-down, structured programming. A comprehensive test is feasible for a small routine, and the relationships between those routines, if embodied in a high-level structure, can be tested independently as well.

12.3.4 Exposure to the Naive Operator

If you thought that abusing your beloved creation was bad, try this: turn someone loose on it who has no idea how beautiful it is. You may have been subconsciously avoiding command sequences that obviously make no sense, or those that could confuse the logic. But the end user will place no such constraints upon his random keystrokes, especially in the learning phase.

One such situation arose during the development of that environmental control and energy management system—the one with the "pattern-sensitive" memory and the giant amorphous program. A call from the customer one day revealed the familiar news that the system was down, so our fearless designer hopped in the company car and drove to the site. Examination revealed about 20 cleared memory locations immediately below the keyboard data buffer.

"Did you hit backspace a bunch of times just before it died?" he asked the operator.

"Maybe."

"You're not supposed to do that."

The operator just shrugged. One key was like another—it wasn't his problem if the stupid machine couldn't take it.

The cause, obviously, was the lack of a test for "beginning of buffer" before the backspace key was allowed to write a 0 and decrement the pointer. No excuse, sir.

The naive operator test can help you locate things like this—ridiculous things that you might never try yourself. It hurts to watch, but there is little choice.

12.3.5 Out the Door

Eventually, confidence in the solidity of the system grows to the point whereon you can finally give in to the threats of the customer and ship it. It has been around the shop for so long that permanent indentations from the mounting ears of the enclosure now scar the bench.

The machine is burned in, well tested, and absolutely reliable. Missing screws and PC board retainers are located and installed, the box receives a thorough dusting, the little pile of mostly metallic detritus in the corner is carefully removed, a couple of extra cable ties are added, and the obligatory photos are taken. Then—then it's off to its final phase of testing.

Huh?

Of course. It is highly unlikely that, no matter how sophisticated your development and simulation facilities, the system's environment to date has had much in common with the one "out there." All of the phenomena discussed in Chapters 2 and 6 suddenly manifest themselves in concert when it arrives at the customer's location, with the real opto sense lines from the process probably somewhat sloppier than the nice debounced switches you used on the bench for debugging. Marginal input thresholds?

Might the inertial load on a stepper motor cause slippage that never occurred during development with an unencumbered sample? Could the age and wear of sensing contacts trick your logic with bounces and intermittencies? Will the baud rate of the customer's communications line to a host processor be just a little different from the one that you used for testing—just different enough to cause occasional transmission errors but not so much that it's obvious? Could, despite thorough testing, the combination of real-world conditions somehow reveal heretofore undetected software errors?

It is, indeed, the final phase of testing.

But don't tell the customer that.

Part   III

Part III

13

A Distributed Industrial System

13.0 INTRODUCTION

According to the COUNT utility, we have so far spent 113,046 words talking about microcomputers. They have been distributed amongst 149 chapter and section headings, and occupy at present the better part of 13 diskettes, including their backup copies.

All this just barely to penetrate the surface of a subject that didn't even exist a decade before this book was written!

We spoke back at the very beginning about the importance of application contexts in the learning process, and even discussed some of the mechanics of that in Chapter 7. Hopefully, such examples as the Blanchard controller, the insertion/withdrawal force tester, and the patrol tour recording system have helped you create associative links between your own reality and the various concepts associated with microprocessor system design that we have presented herein. At this point, assuming that you came aboard without too much prior knowledge of the field, there should be a balanced but somewhat vaporous understanding of everything from hardware design and interfacing to software strategies and system integration.

So we can now begin the wrap-up—that reinforcement of the material brought about by its application. Throughout the remainder of the book, we shall devote ourselves to the solution of real problems with our newly acquired tools and techniques.

We will focus on a system that contains other systems, a distributed, factory-wide network of processors that vastly extends the resources of the plant computer system and contributes dramatically to the quest for that holy grail: 100% uptime.

13.1 THE DATA COMMUNICATION CRISIS

Long before computers came on the scene, those concerned with the efficiency of a manufacturing operation knew that the flow of information between management and the plant was critical. In those days it was normal to find VPs in their shirtsleeves standing amid the din and the grime, shouting orders and examining samples randomly drawn from the production flow. This willingness to get hands dirty was only rarely a hollow employee–relations gesture: it was necessary for that all-important closing of the loop that prevented the production of garbage.

Along came DP. It didn't require too much imagination on the part of the people in the three-piece suits to concoct a means of decreasing the frequency of those visits to the sweatshops. The computer was not spending every available second on payroll, inventory, and accounting—so why not let it analyze data from the plant floor and produce reports detailing quality, product flow, material requirements, and employee effectiveness? Quite reasonable, actually.

But computers are not, and never have been, omniscient. To pull this off, some means had to be arranged to transfer appropriate information from the plant to the machine. Primitive computers, like the clichés they engendered, gobbled punched cards by the thousands, so a reasonable approach was to somehow encode production reporting data onto 80-column cards and then feed them into the great thinking monster.

But look what happens in practice: Joe Smith, operator of a ceramic furnace, notices that the dishes coming out are slightly brittle and off-color, probably indicating trouble up the line with a mix formula. He makes a note on his shift report, circles it for emphasis, and has the "cycle man" run it up to data processing.

It sits on a keypunch machine together with numerous other smudged reports from the plant, until one of the operators finds it and converts the information to punched-card form (minus the emphasis). A couple of hours later, the cards are digested by a batch program that produces a printed report.

This is carried to the plant manager's desk by a DP clerk and deposited in his IN basket. During a lull in the day's activities, he glances through it.

"Omigod!" He checks his watch, notes that at least four hours have elapsed since the defect was noted, dons his hard hat, and races out to the plant. The furnace, an immersion-arc type that burns energy at the rate of $206 per minute, has been producing scrap all afternoon at a total cost of at least $49,440—to which must be added the costs of lost materials, human time, deviation from production schedules, and basic plant overhead.

This scenario is obviously pitiful, but it has happened in at least one instance known to the author. But aside from such obvious "red alert" conditions, how can feedback from the plant's processes be generated in a timely and accurate manner? Simple implementation of a data-processing system is not enough:

somehow it must obtain the data to process. This has classically been left to human beings, who, after the novelty wears off, may well consider production reporting to be a boring and time-consuming routine. So even though a computer (gee whiz!) is being used to produce sophisticated reports, the information in them may be ancient and peppered with errors.

This was the original problem that spawned the development of the IDAC/15, the industrial data collection system to which we have made fleeting reference at various points herein. Data collection was the objective, but as the system developed, it became obvious that a variety of related problems existed which had at their roots the same basic requirement: communication.

It is evident, for example, that the key to almost any process is closed-loop control, but only rarely is this widely accepted philosophy recognized on a factory-wide scale. A local process, such as the Blanchard downfeed problem, is quite readily accommodated with a micro and a clever program. A widget assembly machine can be made to reject its own bad parts. But we have just spoken of production reporting—the feedback loop to management. Could this not encompass the moment-to-moment operation of the plant? In the scenario described above, the time constant of the loop is ridiculously large—depending on human footsteps, coincidence with lunch and coffee breaks, and whim. Even though it is slow and potentially haphazard, it fits into the general form suggested by Figure 13-1, an obvious throwback to the drawing of a generalized closed-loop process presented in Figure 2-2.

Let's speed it up. Further. Further still. Let's make the loop delay a few seconds rather than a few hours. Suddenly it is no longer quite appropriate to call it "production reporting," for it has become a closed-loop factory control system.

What, exactly, have we accomplished by doing this? In a philosophical sense, nothing has changed: the factory operation still receives commands based on evaluation of finished products and other indications of ongoing performance. But if it is making subtle instantaneous adjustments of the process as a function of widely distributed operations and, more important, carrying out those adjustments by means of equally distributed controllers, then we have the beginnings of an intelligent hierarchical system.

One possible outgrowth of this type of design is illustrated in the figure. Here, a hypothetical manufacturing operation consisting of numerous independent control systems has been integrated into one large automated loop. The high-level commands entering from the left consist of such things as customer orders and management-introduced changes to the normal operation of the system. The feedback returning to the plant computer is made up of information from data collection terminals, high-level reports from process controllers, and QC data on the outgoing products. Based on these, the algorithms in the computer system produce an ongoing set of commands and operator messages which together trim the operation of the factory in a manner consistent with the objectives: maximum efficiency, appropriate product flow, and optimization of the myriad trade-offs associated with such a multifaceted operation.

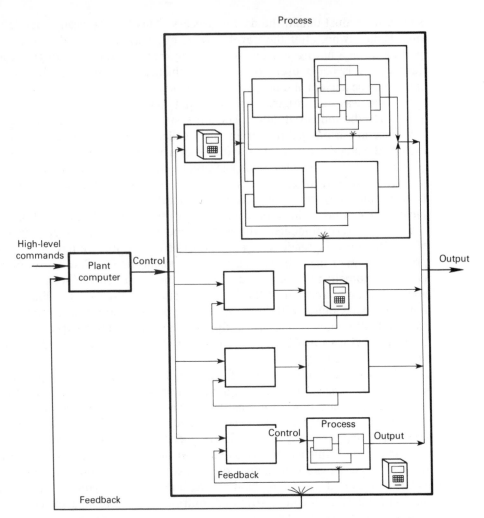

Figure 13-1   A complex factory control system can be viewed as a number of local processes integrated into ever larger ones which eventually encompass the entire plant. At the outer level, the input commands consist of management decisions and customer orders.

Given this structure as an objective, we can readily see that the most potentially limiting factor is communication—both to and from the main computer system. To some extent, the requirements can be relaxed with an increase in the amount of task distribution, wherein greater system responsibility is assigned to lower-level processors. In any case, some high-level central processor is assumed, very loosely analogous to the human cortex, and we can think of it in a fashion perfectly equivalent to the implied role of the top block in a good structured design. The fact that it is a computer rather than a section of code is immaterial—

one of the hallmarks of a structured system is the relatively equivalent modularity of hardware and software.

Indeed, early approaches to this idea embodied all control algorithms in the host computer, requiring very wide bandwidth communication to and from the plant. It is because of microcomputers that we can talk of distributed processing, along with the resultant profound relaxation (and increased abstraction) of the "feedback" and "control" channels. This is by no means a new idea; it's just that pre-micro technologies rendered the entire notion of computers actually located in the plant absurd—they naturally belonged in air-conditioned computer rooms. The resulting high-speed communication requirements led to all sorts of absurdities, such as parallel TTL-level lines running through conduit along with high-impedance analog sense signals. But by moving most of the active "real-time" control to the lower, distributed levels, we can trade this bandwidth requirement for the increased coding complexity of high-level information.

13.2 THE COMMUNICATION NETWORK

Let us begin by considering the requirements of the *network*—the communications "environment" that hosts the various functions we would like to integrate into the system. For illustrative purposes, we shall consider it in the context of a large foundry, whose objective is the manufacture of precision refractory blocks.

The "heart" of the system, if one can be said to exist, is the plant computer, a more or less traditional "big system" along with its associated peripherals. There are disk drives, printers, terminals, modems linking the facility with other company divisions, and miscellaneous other resources. Perhaps the system is a DEC Industrial 11/40, functioning under the RSX-11 real-time, multitasking operating system. Perhaps the resident language of choice is (gag) FORTRAN.

Enlargement of the system's sphere of activity to include the plant is hardly a novel concept. It is a simple matter to hang various serial ports onto the mainframe which interconnect it with remote terminals and printers. For sensing and control application, DEC provides something called a "UDC," or Universal Digital Controller, which provides up to 4032 individual digital I/O points, including relay outputs, interrupt lines, D/A converters, and so on. This Unibus-resident subsystem, in concert with the standard system facilities, comprises a highly capable real-time industrial processor.

The only problem is that continued burdening of the machine with the details of terminal I/O, satellite processes, and communications overhead has the very noticeable effect of slowing it down. It also represents the creation of a classic Achilles' heel: if the entire factory is dependent upon one system, no matter how nice, those inevitable crashes and maintenance periods stop the whole show. This is not only expensive, but dangerous as well—especially when the downtime is unscheduled.

So let us use some of our newly acquired microprocessor expertise to minimize the risk and overhead of such expansion.

13.2.1 Basic Network Requirements

As we set about to create a factory communications network, we can assume that we are subject to some mandate concerning cost. After all, most of the problems such a system purports to solve affect the "bottom line" in one way or another, and we can be certain that someone in the company is busily doing calculations concerning the payoff period.

Naturally, this places us once again under the tyranny of the trade-off. How much do we want to spend on ruggedness and reliability? How much of the communications overhead burden do we want to leave on the shoulders of the customer's application programmers? How much do we want to spend on wire?

On wire? Isn't that pretty well defined by the placement of terminals? Not really. Consider Figure 13-2.

Here we show three ways of accomplishing the same basic function: communication between something called the "host" and eight remote terminals. In part a of the figure, and *loop* scheme is represented, perfectly analogous to the old series Christmas tree lights. This is one of the obvious approaches to the minimization of the amount of wire (and probably conduit) required, but is an invitation to serious trouble.

The author once had the misfortune of spending a summer maintaining such a system built by a now mercifully defunct manufacturer. With perhaps 20 "job information" and "time and attendance" terminals scattered around a mile-long shipyard, the system required a high-voltage current loop almost 3 miles long. A terminal could die—anywhere—and if its failure mode resulted in irregularities on the comm line, then the entire system would fail. Then the fun would begin. "Which terminal is causing the problem? Or is it the mainframe, perhaps? Maybe water in a cable somewhere?" Thus would ensue a full afternoon's frantic dash about the yard, unplugging and testing terminals (whose keyboards, incidentally, were made by a well-known toy manufacturer). It was endlessly challenging work, not only because of the fascinating puzzles posed by intermittent failures, but also because the system always had to be operational during shift changes. The problem here was the fact that the "time clocks" were terminals on the loop, and at such times as 4:00, great angry queues of powerful steelworkers in need of beer would place considerable pressure on any hapless field engineer preventing their departure.

So let's forget about the loop configuration. Bad memories.

Figure 13-2b depicts another type, called *multidrop*. Like the loop, this allows minimization of the amount of wire needed, but also like the loop, certain failure modes of individual terminals can have the unfortunate effect of crashing the whole system. It is worth noting that this vulnerability extends to the communication line itself: it is not overly uncommon for a zealous lift-truck driver to accidentally sever a piece of conduit that appears to be somehow involved with "them computers," and it would be better if the extent of the effect were local.

Incidentally, we should note another disadvantage of both loop and mul-

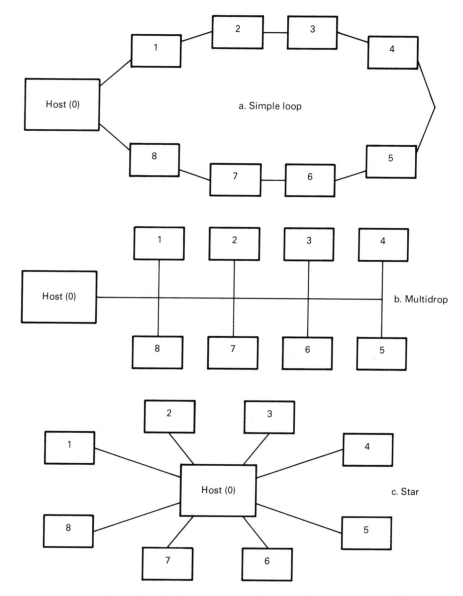

Figure 13-2   Three ways of interconnecting elements of a communications network: (a) is a series loop that can fail completely if one terminal dies, (b) is a "multidrop" scheme in which all units share the same line, and (c), used in the IDAC system, dedicates a line and channel to each terminal.

tidrop configurations. Since all terminals share the same communication line, there must be some rigid protocol that prevents everybody from talking at once. This is not difficult to accomplish, but it requires that terminals know their own address, requiring some code to be set when it is necessary to replace a broken unit. The personnel requirements are thus somewhat more demanding than would be the case if a unit could simply be replaced without concern for any address selection.

It should be admitted that these disadvantages can be outweighed in some environments by the wire cost associated with the third alternative in the figure. It is not uncommon for conduit installation to run as high as $5.00 per foot, and some environments simply cannot justify a dedicated line to every terminal in the plant. But consider the advantages.

Shown in Figure 13-2c is the *star* configuration. Each terminal has its own communication line, obviating the need for address decoding and preventing widespread failure in the event of even a catastrophic failure in any individual unit. Terminals can be swapped by unskilled employees, for it involves merely the unplugging of power and communication cables from the offender, a physical exchange, and subsequent reconnection. For our system, then, let's use this approach.

Now, what kind of communications lines are they? We can obviously reject serial TTL because of noise problems, and even RS-232 has been known to become flaky over large distances. To keep noise at a minimum, let's use current loops, as illustrated back in Figure 5-4. Each "line" shall consist of two loops, one each for send and receive—a bit of a luxury, perhaps, but something that potentiates roughly twice the speed in situations characterized by the need for extensive handshaking. Our basic IDAC terminals will make only lazy use of this feature, but other devices, like the oven controller, might use it to good advantage. It's a matter of discretion.

Given the rough characteristics of the interconnection, let's consider the behavior of the system as a whole.

13.2.2 System Configuration

One of our objectives is to avoid saddling the host computer with all the overhead trivia associated with adding a number of remote terminals. This consists of polling the channels (or watching for interrupts), handshaking to determine correct reception, and so on. It would be ideal if it could send and receive messages with no consideration of the details involved in their communication.

To do this, we shall create a multiplexer, connected to the host by a single 9600-baud serial line and connected to the remotes by 1200-baud lines. The general layout of the entire system is shown in Figure 13-3. The multiplexer is shown inside the plant boundary, where it can be mounted in some relatively out-of-the-way location (it will need no operator action once the system is fired up).

Let's take a moment to look the system over before considering the details.

Note first that there are 15 terminals (including subsystems more easily thought of as controllers). This is a convenient number, for with the addition of the communication line to the host computer system, 16 serial channels are supported by the multiplexer (hereinafter called mux). The corresponding devices are:

0	Plant computer (9600-baud line)
1-8	Data collection terminals, each providing 64 characters of text display, 20 lighted pushbuttons, and a badge reader. Some units also offer a selection of digital and analog I/O channels for interconnection with production machinery.
9-10	Time and attendance terminals, consisting of a clock display, IN/OUT pushbuttons, and a badge reader.
11-12	Blanchard control systems, as described in Chapter 2.
13	Furnace control system, with a local support processor. This is a self-contained unit that governs immersion-arc furnace operation.
14-15	Modems equipped with UHF business band FM tranceivers and associated transmit/receive control logic. These correspond to terminals located on a Clark lift truck and on a cab crane used for material handling.

The key to the information flow into and out of the plant is obviously the mux. It has the thankless task of continually scanning the status lines of its 16 UAR/Ts, responding to received characters, empty transmit buffers, and errors as appropriate. The messages themselves, to keep things simple for the FORTRAN applications programmers, are ASCII strings of the general form

\ - Terminal - Device - ...Message... - CR - Checksum

Assuming for the moment that this is a message from the host computer, the opening backslash alerts the mux that it should assign some buffer space (32 bytes) to the forthcoming message, then the first two locations in the buffer are filled with the terminal ID# (the channel number, such as 13 for the furnace controller) and the device ID# (the address of a particular logical or physical device in the terminal). Following this preamble, the message itself is sent, a variable-length ASCII string of up to 30 bytes, followed by the code for carriage return and a checksum.

This last byte deserves a moment's explanation. In a noisy and generally troublesome environment like a factory, it is difficult to fully trust the data arriving on a communications line, even if the line is a well-isolated current loop. It is thus worthwhile to embed some means of error detection in the data itself. In this system we use two methods: *parity* and *checksum*. The first is accomplished by the UAR/T hardware, and simply consists of a bit associated with each character reflecting whether or not the number of "1" bits in the rest of the character is odd or even. The receiving UAR/T makes its own determination of this and then checks the parity bit to determine whether or not it is in agreement with the

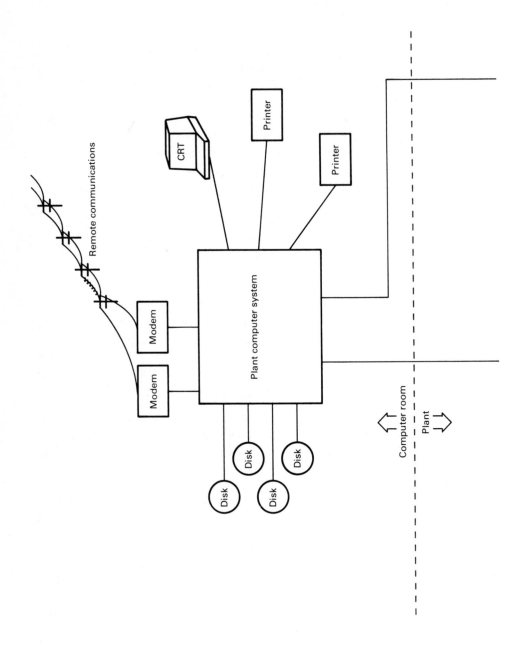

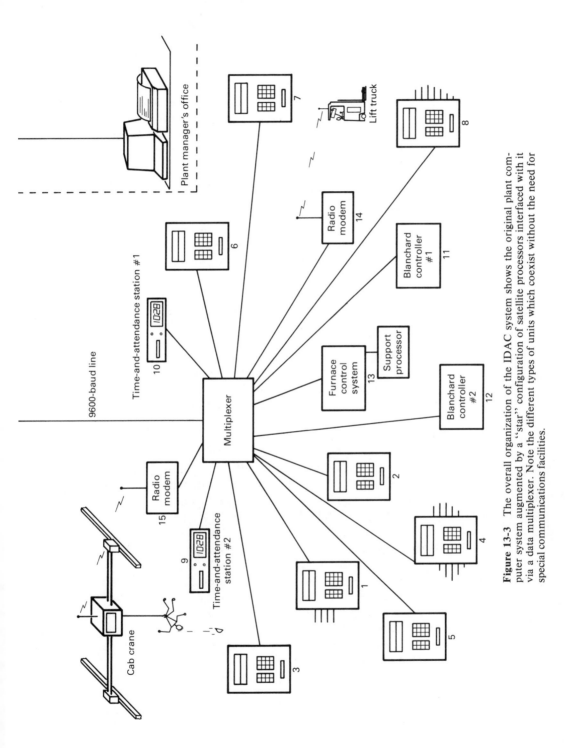

Figure 13-3 The overall organization of the IDAC system shows the original plant computer system augmented by a "star" configuration of satellite processors interfaced with it via a data multiplexer. Note the different types of units which coexist without the need for special communications facilities.

311

sender. The probability of detecting a single-bit error is 100% with this scheme, but that of catching a two-bit error is 0. Parity, then, is a reasonable glitch-catcher, but we need something more.

A rather easy method of checking the integrity of an entire character stream is the use of checksums, wherein both the sending and receiving system maintain a running arithmetic total of the bytes as they are passed through the communication link. Only the eight least significant bits of this total are actually saved, and when the message is complete, the total accrued by the sending system is sent. The receiver knows from the carriage return that the message is over, so when it inhales the very next byte, it compares it with the checksum that it has been accumulating, and if there is any difference, it is safe to assume that there has been a communication error. Its reaction to this condition is the return of the ASCII "NAK" character (for Negative AcKnowledgement) instead of the ACK which would have signaled correct reception. This scheme is quite reliable, with a substantially higher probability of catching any error within the message stream to which it is applied.

Referring again to the general form of a message, we see that both a terminal and a device ID are transmitted by the host when it chooses to initiate a message. The mux's job, after this has been received and adjudged correct, is to retransmit the same message to the specified terminal, less the terminal ID number. This occurs at a slower baud rate, but precisely the same sequence of events occurs.

If the message is initiated by one of the terminals, as when an employee inserts a badge, the message to the mux is stored, then transmitted to the host with the terminal ID added. The whole scheme is actually very simple.

13.2.3 Self-Diagnostics

We have not yet looked into the functions of the various terminals, the types of messages that might flow through the system, or the internal operation of the mux. It is desirable that we first dispense with our examination of those aspects of the system's design that pervade the entire network and are difficult to functionally localize within a given processor.

We spoke of error detection, but there is another level of this which becomes evident when we initiate communication to a terminal via the mux and abruptly kill power to the terminal while the communication is in progress. Because RAM space in the mux is limited, we allow the 16 channels to share six 32-byte data buffers, forcing a pending message to wait in a queue in the rare but quite possible event of all buffers being busy. A rather convoluted status flag scheme is used to determine the current availability of buffer space and the activity of each channel.

None of this is intrinsically objectionable—until communication in progress is interrupted in a fashion that renders impossible its successful termination. This can result from the loss of terminal power, as we noted, or from the abrupt failure of one of the processors. It calls for an almost supervisory level of task management in the design of the mux, for ignoring the problem could leave precious message buffer space tied up in the truncated attempt to communicate—a condition which, if repeated sixfold, would cause the entire system to "lock up." This is clearly unacceptable.

The solution can be handily provided by initializing a timeout counter whenever a message has been sent, but before the ACK or NAK has been received. At prescribed intervals, the counter is incremented, and if no response has been received from the terminal after about 100 ms, the mux knows that something is amiss. It also has a buffer that is tied up with a message in progress.

Now what?

The mux can accurately assume at this point that it has a sick terminal on its hands, and we can add considerable power to the system if we cause it to take some appropriate action. How about this: the mux generates a message to the host computer which we can call the "Request for CONFIG request."

When the host computer sees this, it knows that something is amiss. Depending upon the types of activity in progress in the plant, it begins an orderly shutdown of the communications system, first completing any other message transactions that are in progress, and then notifying any particularly sensitive remote units that they are on their own for a while. Once activity has effectively ceased, it issues a CONFIG request to the mux.

Now the microprocessor in the multiplexer begins testing each of the 15 terminal channels, in every case issuing a test message that initiates a local self-test in the processor at the other end of a given line. At the conclusion of each local test, a message is returned to the mux, indicating the health of the terminal. This message (or the lack of response from dead or missing terminals) is used to add bytes one by one to a map which, upon completion of the 15 tests, is transmitted to the host as a report on the overall condition of the system. It is also maintained internally for future reference.

At this point the host can compare the information in the CONFIG map to a stored pattern representing the way things should be, and automatically generate a work order for the maintenance department should anything be broken. In some cases, the simple process of performing the tests will clear up any problems, for the terminals undergo a "soft" reset when this occurs. In other cases, a printer somewhere in the plant will spring to life, directing a plant electrician—or somebody—to cart a new terminal to some specified location and swap it for the one that is ailing.

It is easy to see how this kind of operation, only possible because of the distributed intelligence of the various elements in the system, can contribute to a highly reliable and flexible network. Depending on the extent to which we want to carry the idea of self-diagnosis, we can even remotely test the lamps on the terminals' front panels (by sourcing all lamp current from a monitored point, then scanning through all 20 of them one at a time while checking for appropriate values).

With that, let's turn our attention to the internal design of the mux.

13.3 THE CENTRAL-SITE MULTIPLEXER

The job of the mux is simply the expansion of one serial I/O channel into 16. As we noted in the description of the message format, each transmission from the host contains an embedded terminal ID code, telling the mux on which channel the message should be retransmitted. Aside from the CONFIG function we just described, the mux logic performs no functions that are not directly related to the shuttling of ASCII messages back and forth.

This lends itself to an elegant and simple architecture, for there is no need for the random logic that clutters so many other types of systems. The design is shown in Figure 13-4.

On the left is the processor card—an Intel SBC 80/04. Actually, it turns out that the software design is seriously complicated by the shortage of scratch-pad RAM space (only 256 bytes) on this card, demonstrating the advisability of doubling or even quadrupling one's expectations of memory requirements. The absolute RAM space restriction does provide us with the aesthetic satisfaction of using every last bit, but that's hardly the objective.

That limitation aside, however, the 80/04 is fine for the application. It provides three ports—two of 8 and one of 6 bits—allowing the creation of the "fake bus" scheme as shown in the drawing. Each UAR/T is as shown in Figure 5-2b, with "Transmit data in" coming from the processor's data out bus and "Receive data and status out" impinging upon the input bus. Everything remains more or less passive, however, until one of the three decoders connected to the control port sees fit to issue a command to one of the UAR/Ts.

As long as no communication is actively in progress, the processor spends the bulk of its time scanning the status words corresponding to the 16 channels. This is accomplished by setting the high 2 bits of the 6-bit control port to "01" and counting the low 4 bits. The high two, as shown in Figure 13-5, have the effect of enabling the "SWE decoder," and the others determine which of the decoder's 16 output lines will change from its resting "1" state to a "0." The scan, to save time, is permuted to that subset of the 16 possible channels that has been determined to be active by the CONFIG process, which takes place not only on terminal failure but also on power-up or routine host request.

Anyway, when the Status Word Enable command of one of the UAR/Ts becomes active, it responds by placing the five bits representing its current status on the bus (they are: Receive Data Available, Transmit Buffer Empty, Receiver Overrun, Parity Error, and Framing Error). The processor can then read these by simply doing an input on port 2, whereupon it can check to see if any data have arrived or, if a transmission is in progress on the channel, if a character has completed. In either case, appropriate action is taken: the establishment, continuation, or completion of either an incoming or outgoing message.

In the event, for example, that the processor detects a status bit on channel 7, indicating that a character has been received, it responds by checking the flag

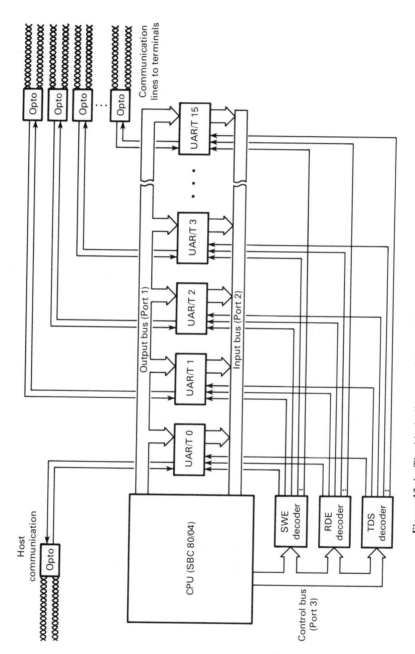

Figure 13-4 The block diagram of the multiplexer reveals its simple structured architecture. Sixteen UAR/Ts are connected to the data buses, and controlled by the software via the three decoders.

315

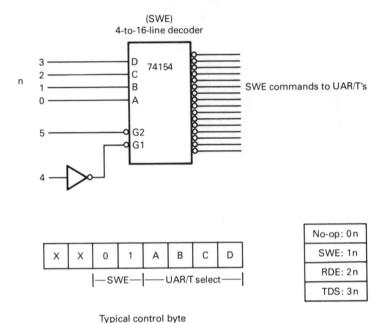

Figure 13-5 The action of the decoders in the mux is illustrated with this detailed view of the one that produces "Status Word Enable" commands to the UAR/Ts.

byte (in RAM) corresponding to that channel. This byte contains the following information:

bit 0   NAK or ACK just sent
bit 1   Last byte sent
bit 2   Character in progress
bit 3   Awaiting checksum
bit 4   Message fully received
bit 5   Awaiting acknowledgement
bit 6   Transmit in progress (TIP)
bit 7   Receive in progress (RIP)

If, upon detection of a received character, the processor discovers that there is no indication of a receive in progress, nor any expectation of acknowledgment of a recently transmitted message, it activates the RDE decoder to begin the process of obtaining the character that was received. Once it has been inhaled, it is compared to the backslash (\) character that must always precede a message (to prevent the system from taking garbage seriously), and if it is indeed the same, the machine can assume that the channel has just become active. It sets the RIP flag, assigns a buffer to the incoming message, and initializes a checksum. As subsequent characters are received, they are appended to the buffer, until the carriage return is detected, indicating that the next byte will be the checksum (flag bit 3 is set). When the checksum arrives, it is compared to the internal value, and an ACK or NAK is returned as appropriate.

At this point, assuming things went smoothly, the mux is holding a message from some device that is intended for shipment to another. It checks the "terminal ID" character, and if that channel is not already busy, begins the retransmission via a sequence of TDS commands and appropriate manipulation of the flag byte.

In the unlikely event that all communications buffers were busy when channel 7 began pumping data, the mux would simply return a NAK immediately after its reception of the opening backslash. It would then be up to the transmitting device to wait a discreet amount of time and then try again. An alternative scheme, not implemented here, might be the setting of a flag indicating that the channel desires a buffer. When one becomes available, the mux can notify the device that it may begin transmission. To be done properly, however, this scheme requires some form of queueing to prevent high-numbered terminals from being "shut out" for long periods during heavy channel traffic—and that would involve more RAM than we allowed ourselves with the use of the nonexpandable board.

13.3.1 Mux Packaging

Since we have postulated that the multiplexer will be living within the boundaries of the plant (primarily to keep wire costs down—this is totally optional), we have to apply some of the principles of industrial equipment packaging to its design. This raises the cost. The world is full of trade-offs.

With the exception of the decoders and their few associated gates, the logic is nicely modular. Each channel consists of a UAR/T (a single-supply IM6402 CMOS part from Intersil), a Monsanto MCT-6 dual opto-coupler, and a few resistors involved with current limiting and collector load in the loop interface circuitry. We can thus design the mux circuitry with the devices resident on identical four-channel cards, allowing convenient field replacement and relatively simple expansion beyond the basic configuration (the host plus three terminals). The decoders can be on a "control logic" card, rendering everything else completely modular.

A question always arises in situations like this concerning the relative desirability of housing all the circuitry on one large card, thus eliminating the cost and potential noise of a backplane, or splitting it up as we have described. There are various overlapping and conflicting considerations. The large card would be much more expensive to fabricate and stock, with replacement units costly and difficult to carry because of their size. The small cards, however, require a motherboard. But they also make it possible to sell someone a minimal version of the mux and then follow up later with optional, field-installable enhancements. With the large board, this would involve carrying in stock numerous versions differing primarily in the number of chips installed. As if all this were not enough to juggle, physical packaging restrictions have to be considered—in this case weighting the decision in favor of the roughly cubical card cage with a maximum of five boards.

If we are clever about this, we can make them physically match the CPU board, thus preventing the need for a somewhat clumsy cable between an SBC hanging on an inner wall of the enclosure and the card cage. Perhaps use of the smaller card sizes found in the STD bus standard would make this a little more attractive.

Heat is a problem, of course, although not quite as acutely as in the IDAC terminals we will be discussing in Chapter 14. The use of CMOS UAR/Ts and very little "hot" logic allows us to get away with a fully sealed enclosure, as long as we are careful about the thermal coupling between the open-frame power supply and the walls of the cabinet (which, incidentally, is a NEMA-4 padlocked steel unit from Hoffman).

Interconnection with the outside world is something of an issue, because in addition to the ac power, we need 16 four-wire current loops. It is tempting to simply make everything available to the customer on a board full of barrier strips with screw terminals, but that raises the specter of people poking about with screwdrivers during the installation of cabling, people drilling holes in the cabinet without adequately protecting the boards against flying metallic debris, and people leaving the cabinet open. Better to provide solid connectors of the classic Cannon variety (yup, military style)—one for each channel—and keep curious fingers away from the circuitry.

A coat of OSHA-orange paint tops it off. We have a multiplexer.

13.4 THE SYSTEM IN ACTION

Let us return to Figure 13-3, now that we know something of the multiplexer's behavior and the nature of the communication lines, and discuss the system within the context of the factory.

In Section 13.2.2 we briefly identified the types of terminals found in this particular installation. Let's look at them a little more closely.

13.4.1 The Host Computer

It seems a little silly to call the PDP-11/40 a "terminal," but from the standpoint of the mux it is not all that special. It does possess the high-speed line, and enjoys favored status in buffer assignment, but it is, more or less, about like the others. In the original design of the system, it seemed logical to make all messages from terminals automatically go to the host, and vice versa, but there are occasions when an appropriately clever satellite processor might have reason to communicate through the network without bothering the big system. So the mux software is general enough to handle traffic from any destination, even though this is a rarely used feature.

The details of the host computer are of little interest to us, but it is worth noting that the characteristics of its use had much to do with the design of the system. In particular, the amount of wizardry required of company programmers must be kept to a reasonable level.

It is not at all uncommon for a large factory to purchase a gee-whiz custom system which can be clearly shown to save money. All too often, however, that which appeared up front as a permanent solution to the company's business problem turns out to require the extended services of a $75/hour consultant for every minor change in system requirements. This is hardly consistent with the intent of industrial microcomputers, which, as I believe we have stated, is to cut costs and improve efficiency.

In a system like the one pictured in Figure 13-3, we can assume that the original objective could be characterized as some set of task-specific procedures: optimize Blanchard use, handle mold inventory, track orders as they pass through the plant, and so on. It is tempting to skip all the generality and just build machines that do those things. Even if the host computer is to be involved in all activities, it is a fairly simple matter to write a custom application program that drives a group of specialized terminals and satellite processors.

But nothing ever stays the same for long, even in the staid world of low-tech industry. If the company's customer requirements don't change, the materials do; if those don't change, the people do; if those somehow remain constant, something will break and need to be replaced by a newer machine since the manufacturer of the old one is out of business. No matter what, the information-processing requirements of the firm are guaranteed to be dynamic, something that is probably well understood by management in the first place.

The problem is that a communication system like this, being contrived with microcomputers, tends to scare people who have spent their lives speaking FORTRAN to large and faceless time-sharing systems. If we embody application-specific software in our network, we have trapped the customer into dependence on outside help—something that appears at first glance, perhaps, to be a bit of gravy for our side. The truth is that it leads almost inevitably to endless handholding and bitterness, sometimes culminating in the abandonment of the system as the company's management becomes acutely aware of their vulnerability to economic conditions that might spell the demise of the brash young entrepreneurial firm that created the custom machine.

So. Our treatment of the entire system reflects this awareness. The mux is a totally general device; the terminals are completely undedicated, leaving only the local controllers to possess task-specific knowledge (at the HOW level, of course). Further, we make things easy for the customer's programming staff by allowing all data transfers to take place with simple ASCII character strings over a system port that is precisely like that of any standard CRT terminal. If a programmer wishes to display the message "WHAT IS THE ORDER NUMBER?" on the upper of the two 32-character text displays on terminal 5, he need only send this string to the multiplexer port:

\EAWHAT IS THE ORDER NUMBER?

terminated by a carriage return. The device handler in the PDP-11 takes care of the checksum, and the mux and terminal take care of everything else (incidentally, note that the terminal ID in that message is shown as "E"—this is because the actual ID codes are mapped onto the first alpha column of the ASCII character set).

Because no weird binary addressing schemes and convoluted handshaking techniques are involved at the application programmer's level, the whole network can be operated in FORTRAN, C, assembler, or even BASIC without creating any obfuscating problems. This allows the customer to feel that he has control over the whole system, something which is politically, if not emotionally, expedient.

13.4.2 Data Collection Terminals

Most of the devices occupying the multiplexer's channels are terminals designed primarily for human interaction—things that we have herein called the IDACs. In Chapter 14, we discuss their innards, but let's see what they're good for.

Much of the information that flows between management and the plant is in some way involved with people, or is at least best accommodated via same. It follows that much of the "feedback" and "control" involved with industrial processes is actually a matter of communication with plant employees.

Many schemes have been attempted over the years to perform this efficiently, but unfortunately not all employees recognize that they have a vested interest in the smooth operation of the plant. It would sometimes seem that they are at cross purposes. If you don't believe that, put an office-grade CRT terminal in a steel mill and see how long it lasts. It acts like a powerful source of "gravity waves," somehow attracting all sorts of foreign matter ranging from sticky drinks to fork-lift trucks.

Our very first criterion, then, is ruggedness—the type of ruggedness that can withstand wanton brutality and hostile aggression. Every depression of a pushbutton might call for a fair amount of energy absorption, hopefully in the hand and arm of the person making a miniature attack on management by using a haymaker where a light touch would do. Unfortunately, most plant operations need to be restricted to certain personnel, so a badge reader is necessary to determine the identity of the operator. Since this represents a rather vulnerable point of interface with the outside world, we can expect all manner of strange things to be inserted. Some of them are organic.

But that's not all. Not only must the terminal be tough, but it must be friendly as well—both in the context of the JOKE program we mentioned in

Chapter 5 and in the sense of easy and smooth operation. It should be fast but not too fast (that's threatening), it should have easily read displays, and it should be positioned for comfortable operation.

A typical exchange between the host and a terminal might go something like this. Joe walks up to the terminal and crams in his employee badge, which identifies him through Hollerith-style punches as No. 1673. The terminal sends a message to the host indicating that a badge has been inserted. A quick exchange acknowledges the message and requests the badge data, which the terminal agreeably sends. In response to the system's message,

\EGR

[terminal 5, device 7 (badge control), and "R" for "Read the badge"], the terminal replies,

\@JXXXXXXXXXXXXXXXXX001673

[= device 0, J is device 10 (badge data), followed by the full 22-byte data ensemble of the badge reader, of which only the last six are used by the company].

A supervisory program in the PDP-11 consults a lookup table to determine the "type" of employee, any special privileges he may have, and whether or not he is allowed access to that particular terminal. This information is used to "attach" an appropriate FORTRAN job.

After a moment's delay, the system begins executing the job, which, let's say, is for shift supervisors in the mold fabrication area. It sends the message,

\EASELECT REPORT

immediately followed by

\EENNNFFFFN

The first places the prompt on the top display, and the second delivers a "lamp map" to the group of eight lighted pushbuttons on the left half of the terminal's front panel (collectively device E, or 5). For each lamp, N means on and F means off, so the message will light the first three and the last of the eight buttons.

At this point, it is the human operator's move. The buttons may be labeled in a dedicated fashion or, in keeping with our intended generality, an additional message to the lower display might identify the choices available to him. In any case, the eighth lighted pushbutton is agreed upon as the "oops" button, allowing the operator to cancel requests or backtrack through a menu.

We need not detail this much further. The operator presumably presses one of the illuminated buttons (the others are quite inert), causing a message to be sent back to the host. A conversation then progresses between them in which he enters, perhaps, a job number, a finished weight, any defect codes, and so on. At the completion of the exchange, he is given the opportunity to make another report, and if he chooses not to, his badge is returned with a solid "clunk."

This is a fairly pedestrian example of an interchange between the great thinking monster and a plant employee, but is quite typical. A terminal may also play a more dynamic role, however.

Since all the communication logic is already implemented and is, in fact, much more flexible than would be dictated by the requirements of the basic terminal devices, it is possible to add some more classic "industrial control" types of I/O. In particular, sensors of voltage, machine conditions, and temperatures as well as control outputs for processes, interlocks, and other lamps or warning devices, can be added with little difficulty. This permits the terminal, still quite general, to be married to a machine that may not quite justify its own local controller but can still profit from a balanced combinaton of human and computer control. We go into the details of this in Chapter 14, but in the meantime, we can view the terminals as the basic building block of this system.

13.4.3 Time and Attendance Stations

Keeping track of employee time with old-fashioned cards is a bit of a nuisance, although it does, like paper tape, offer the constant reassurance of hard copy. Many workers are as distrustful of "computerized time clocks" as they are of management, for the two are inextricably linked.

Perfectly adequate T&A stations can be created with the basic innards of an IDAC terminal, but with only a four-digit display, a badge reader, and, optionally, pushbuttons for IN and OUT. This last item is uncertain: in traditional factory settings, with little or no overtime and piece work, the buttons are totally superfluous since even a simpleminded program can determine from the times which transactions represent employee arrival and which represent departure. In other cases, someone might show up at 7:30, leave for lunch, leave for dinner, and finally go home at about 9:15—only to show up again three hours later when something hits the fan. All ambiguity can be resolved with the IN/OUT buttons, but it must be remembered that it represents one more potential source of trouble. Few employees are likely to take chances with their paychecks, but there will always be someone who "has it all figured out" and does things that confuse the system.

The four-digit display, of course, receives an update of the correct time every minute. This is one distinct advantage to having the T&A terminals integrated into the network: all potential problems from slightly disagreeing clocks can be avoided, together with the annoyance of employees entering via the slow one and exiting via the fast one to pick up a free minute or so every day.

13.4.4 The Blanchard Control System

In Chapter 2 we discussed at some length the use of a microcomputer to continuously adjust the downfeed rate of a Blanchard grinder's head as a function of the loading perceived, thus providing some control over the rate of diamond wear and the occurrence of machine "crashes."

The system as we described it is perfectly capable of standing alone an doing its job without communication from any entities other than the operator and the Blanchard itself. But as we pointed out in Chapter 9, the incorporation of such processes into intelligent hierarchies introduces a measure of adaptability that would not otherwise exist. In the Blanchard situation, this could involve the subtle alignment of tooling wear rates and scheduled preventive maintenance, or perhaps the adjustment of the feedrate to optimize product throughput on a larger scale — trading off some local efficiency for global considerations with a correspondingly higher cost.

Philosophy aside, the integration of such a system into a communications network is simply a matter of adding the UAR/T and its related hardware, then tacking on a software module that drives it. The interaction with the host is at a relatively high level (target values, reports of completion, etc.), and thus does not impose the kinds of speed requirements that would be associated with real-time control via the communication lines. The control itself is purely local, and the software is designed such that abrupt severance of the link to the host has no effect upon the work in progress.

Refer to Figure 2-8 to refresh your memory of the relationships between the system's major components.

13.4.5 Furnace Controller

Operation of a high-energy furnace capable of firing refractory blocks involves no small measure of danger, both to nearby human beings and to the equipment itself. Dependence upon a communications line, no matter how well implemented, would be an invitation to disaster.

The basic objective of an immersion-arc furnace is the generation of extremely high temperatures in the "bath" of molten ceramic material that will be poured into the molds. This is accomplished by the passage of many kiloamps through an electrode above the bath, with the achieved temperature largely a function of the distance between the electrode tip and the surface of the material. The task of the control system is to maintain the desired temperature by adjusting the electrode position (via a stepper motor) as a function of the thermal differential between input and output cooling water. It also keeps track of the electrical current flow, and adjusts it as well.

That's the basic idea, and it is easy to see how things can go awry. Life- and equipment-threatening conditions flourish around such a furnace, and any control system running the show must be foolproff and highly dependable. It thus makes sense to ensconce a rugged processor nearby, with the host computer serving as a "watchdog" and making high-level decisions about the process. In any case, the local controller should be capable of performing its job with no interruption even in the event of big-system failure, and a 100% redundant backup unit is desirable.

13.4.6 Radio Units

One of the more entertaining aspects of this system is brought about by the customer's need for two terminals in rather unconventional places: a fork-lift truck and a cab crane.

We may quickly discard cables, streetcar-style tracker arms, magnetic fields, and audio. Somewhat more worthy of consideration is infrared, but there are just too many unknowns (although flooding a room with appropriately modulated infrared is gaining favor in office environments for wireless communication with terminals).

No, we must brave the FCC licensing requirements and press ahead with radio.

Some thought given to the problem suggests that the most reliable approach is the use of modems to convert the serial bit streams into audio. These can then be interfaced with the microphone and speaker "ports" of the communications gear to produce a mobile terminal—assuming, of course, that at both ends of the link a similar arrangement is provided. A couple of problems make it nontrivial, such as noise (which might call for beefed-up error detection and correction logic) and the transmit–receive delay of the radio itself (which calls for a bit of external circuitry). But other than that, there are few complicating factors.

Choice of frequency range and modulation type are reasonably critical for the noise considerations. Citizens' Band is out—it would take an hour of repeated attempts to slip through the skip and manage an ungarbled message. A good, quiet frequency can be found (in most parts of the country) in the UHF business band, wherein FM is the standard technique. A number of high-quality units exist which accommodate this band, and we'll choose the Standard 790L, a rugged 6-watt transceiver costing about $600 to $700.

Wait, this is crazy. Why do we want a microcomputer-based terminal on a fork-lift truck (Figure 13-6)?

In the customer's plant, there is place called the mold room. Stored there are hundreds of visually similar but distinctly different molds, used in groups to form sets of custom refractory blocks with fit the furnaces of their customers. When someone orders a set of blocks, he needs all of them, not almost all of them. That's where the problem comes in.

These things are incredibly precise. Even different copies of the same furnace design have irregularities that render them unique, and it is an expensive and time-consuming process to create new molds to replace ones that have been misplaced. The point of all this is that full sets of molds for each customer are in the mold room—but sometimes many months pass between successive occasions for their use and, well, you know how it goes. Things get buried.

Standard procedure was originally to hand a list of mold numbers to the driver of a Clark narrow-aisle "reach truck" and turn him loose in the mold room to find them. On a good day, he succeeds. Can we improve the odds?

How about this: since the molds are stored (up to five deep, front to back) on shelves lining the walls, it is possible to create a three-dimensional matrix in a

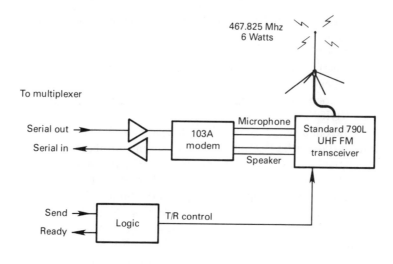

Figure 13-6   A terminal is mounted on a fork-lift truck, powered by an inverter power supply and interfaced with the multiplexer via a radio link and a modem. The result is an extremely useful material handling system which is in constant communication with the plant computer.

computer somewhere which represents the current position of each one. If the computer could direct the actions of the lift-truck driver in his shufflings of the molds, it could not only improve his chances of finding a complete set, but also leave the others in a known configuration. How better to do that than maintain an ongoing dialogue with him via a terminal?

"Take the first three molds from location 15C and move them to location 3A. Got it?"

"Yes"

"Is the one that is now in front at 15C number 566-G12?"

"Yes"

"Good. We want that one. Pick it up and place it on the pallet."

"OK."

Etcetera. The host computer keeps track of all the moves (this is all reminiscent of those little "16-puzzles" with sliding numbered tiles) and leads the operator by the hand through the steps necessary to fill an order.

Of course, once we have a smoothly running mobile computer terminal such as this, all sorts of things become possible. But let's look at the cab crane.

Here, we have basically the same problem, but it is complicated by the inhospitable conditions that confront any human being who attempts to perfom the job by hand. The scenario is as follows. A vast area is filled with cans containing cooling refractory blocks. These things are not only very hot, but they have so much thermal mass that it takes them many days to cool to the point where they can be handled safely. The problem of losing track of part of an order is just as acute, and it is even harder for employees to wander around the area, reading the numbers off the cans. For this reason, a cab crane is suspended above them on girders, capable of being "addressed" in a manner reminiscent of an X-Y plotter.

Of course, this instantly suggests the possibility of adding automatic positioning, or at least some kind of position feedback. That turns out to be conceptually trivial, with shaft-position encoders wedded to appropriate gears in the cab transport mechanism. With this facility, the host computer can again work with a three-dimensional matrix to keep track of the many blocks that enter and leave the cooling area.

13.5 SYSTEM SUMMARY

We have chosen as an applications example a relatively undemanding environment for factory-wide distributed control, the purpose being examination of network behavior without becoming mired in nested twelfth-order control systems. It should be clear that this kind of approach is a flexible basis for many different types of operations, ranging from fairly sleepy warehousing problems to complex continuous processes such as paper milling and plastic production. Higher system speed requirements might justify the use of more multiplexers with fewer channels each, and there is certainly nothing intrinsically magical about the number 16. Well, perhaps there is, but that's a subject for math theorists.

Anyway, this system can be considered a potentially useful framework for a variety of industrial control and data collection systems. Let us now examine the design of the terminals in some detail, seeing how the microcomputers and their associated hardware and software components are assembled.

14

The Data Collection Terminals

14.0 INTRODUCTION

In Chapter 13, we discussed a factory-wide communication network which included a number of terminals, units that were introduced in Section 13.4.2. We now examine the design of these "IDAC" terminals in enough depth to witness the application of interfacing techniques, software philosophy, and awareness of the environmental requirements.

These systems are quite useful as examples, largely because they are not, like the Blanchard controller, so specialized that we must acquaint ourselves with a complex process before even considering the role of the micro. The "system requirements" have been presented in the preceding chapter, allowing us the luxury of considering the design more or less *in vitro*.

There is another advantage to the irreverent dissection of the IDAC terminals: they were designed by the author. This makes available for discussion the informal as well as the formal parts of the design process, and also allows candid analysis of certain weaknesses without offending anybody.

So let's lapse into an anecdotal style, and trace the development of the industrial data collection terminals.

14.1 THE IDAC PROJECT

Once upon a time, in a certain midwestern town known for its Derbies and bourbon, I sidestepped the imminence of employment by starting a tiny, undercapitalized engineering firm called Cybertronics. Its rocky history is hardly appropriate to these pages, so suffice it to say that it suffered most of the ills known to small business, plus a few.

After the company went through its "delusions of grandeur" phase, which included a classy place in an industrial park, wild diversification, and an embezzling bookkeeper, it withdrew to a relatively low-key home environment. Therein, having dispensed with the parts business, the computer store, the surplus shop, and the miscellaneous other diluting attempts at meteoric growth, I was free to spend time on microprocessor system design. Somewhere in the midst of that period, I was approached by the refractory factory and invited to quote on the industrial data collection system.

Having already had some contract involvement with the firm, I was familiar with their environment, and having used microprocessors for various purposes since their inception, I was comfortable with the idea of putting them to work in a high-reliability factory setting.

I think it would be useful now to trace the development of the system from proposal to customer acceptance. Much of the knowledge associated with microcomputer-based design is, as we have noted, heuristic, and an isolated analysis of the various bus interfaces and software modules would serve us only poorly in that regard.

14.1.1 The Proposal

It has to start somewhere. Only in relatively hierarchical engineering departments of rather large companies are you likely to suddenly find yourself confronted with a fully defined specification complete with cost analysis. Although a one-man engineering firm located in a residence is hardly a universally applicable model, it is nevertheless closer in many ways to the typical situation: you are approached by someone who wants a machine and together you work out the details.

In this case, the customer's DP wizard presented me with the problem, but it was not mine to solve single-handedly. Together, we spent a few weeks, off and on, analyzing the communications requirements of various jobs in the plant and sketching possible front panel layouts. All we really had to go on was the obvious need for display, data entry, and employee identification via badges—and, of course, the need for surpassing ruggedness.

CRTs were considered, and rejected because of their likelihood of damage. Standard keyboards were tossed out for that reason, as well as their user complexity. We needed something that could be operated effectively with no training.

For the display, we hit upon two 32-character lines presented upon Burroughs "SSDs"—gas discharge display units requiring only the application of parallel ASCII data and a strobe. At the time, there were only one or two other units of similar design on the market, but today there seems to be quite a plethora of gas discharge, vacuum fluorescent, and liquid crystal displays based on the same idea—an easy-to-use module that requires no logic support or tricky timing.

At the time, the SSDs (Model SSD-0132-0060) were on the order of $300 apiece. This underscored the need for their protection, so we agreed to cover them with ¼-inch Plexiglas plates.

Continuing with the physical devices, we concluded that the best way to handle operator data entry would be via a matrix of heavy-duty industrial-grade lighted pushbuttons. That way, we could use the lit condition of a given key to inform the operator that it was one of his options—unlit buttons would be inert. After much digging around in manufacturers' catalogs, we discovered the Microswitch PW series—desirable not only because of its wide range of contact styles and actuator types, but because the units mount on 1.5-inch centers, making them the smallest rugged oil-tight pushbuttons on the market at the time.

The badge reader was a tough one. Representing the most vulnerable part of the system to both environmental contamination and human abuse, this device was destined to be a weak link. Because of a distrust of the optical units of the day, we went with the Hickok 264A, a mechanical brush-type reader with a solenoid-operated deck. The brush assembly later proved to be a source of trouble from a reliability standpoint and the open design rendered absurd all our other attempts to keep the enclosure sealed, so a subsequent change was made to optical readers. But the design was based on the Hickok, so we'll use it here.

Selection of other components was a process based on different constraints, now that the three points of interface with the operator had been defined. It was intuitively clear, from an admittedly rough understanding of system function, that a minimal processor would be adequate—leading to selection of the Intel card about which we have spoken in the past.

Allowing an amp or so for the yet-undefined interface card that would host the necessary random logic, it was possible at this point to identify the power supply requirements. The V_{cc} adds up to about 5.5 amperes; various MOS devices collectively need a few hundred mils of both $+12$ and -12 volts; the rather archaic 2708 EPROMs (2716s were at this time presenting four- to six-month lead times) need a trickle of -5 volts; and the Burroughs displays each need 30 mA of $+250$ volts. Reasonably certain that future additions would involve tacking on more loads, we decided to be conservative and use some all-American overkill: the power supply chosen was a Xentek XPT-120S, providing 12 amperes of $+5$ and 2 amperes each of $+12$ and -12 volts. The -5 volts was rather trivially generated from the -12 volts using a 7905 three-terminal regulator, and the $+250$ volts for the displays, although available from dc-to-dc converters being advertised, was produced with a little open-frame supply from Power-One.

The astute reader will at this point be leaping up and down, shouting "gotcha!" Yeah, we didn't use a switching power supply, and we paid well for the decision. At the time, however, switchers only became cost-effective on a dollars-per-watt basis at over 200 watts, and multiple-output units were very hard to find. It turned out, of course, that the cooling problems resulting from the use of an inefficient dissipator were severe, and a retrofit was necessary. But we're skipping ahead; we'll have more to say about that problem later.

Somewhere during this embryonic period (during which, I might note, there was yet no contract with the customer—this was still officially speculative), it was necessary to give some thought to the extent of the software design that would be required. After all, I was going to have to come up with a price and delivery quote, and since human time—mine in particular—was the prime commodity, I had to develop some idea of how much of it was going to be involved in the project.

... And it was rectumology time again! It's always tempting to oversimplify the scope of the programming task that lies ahead, perhaps because we think so comfortably at the WHAT end of our much-heralded spectrum that we easily confuse a simple English expression of a problem with the amount of work necessary to code it. "Ah, let's see ... seems that all this thing needs to do is wait for an event—a badge, keystroke, or message—and then either send it over the comm line or shuttle it appropriately to the front panel devices. Gee, do we really need a processor?"

Seems pretty simple, doesn't it? And the hardware—why, that's almost trivial, involving merely the interface of a few devices to the processor's bus. So to come up with a price, I merely took a guess as to the number of hours involved, multiplied by a reasonable rate, added the cost of the hardware, and then tacked on a fudge factor large enough to accommodate the virtually certain absurdity of my projections.

I quoted the terminals at $5,800 each and rather ambitiously presented a proposal promising delivery five months after receipt of order.

Right.

14.1.2 Getting Down to Business

Suddenly the terminal design was no longer something to idly sketch during lulls in my other activities. I received a $28,500 purchase order, which included the first three terminals in the system, the multiplexer, and one radio link.

So. Where to begin? I began sketching front panel layouts, realizing that the process of cutting holes in the enclosure was likely to be somewhere in the critical path, relying as it did upon a local sheet metal shop for panel cutouts and painting. As shown in Figure 14-1, the development process includes three parallel paths, with rather marked interdependence developing as the project progresses.

This drawing is somewhat reminiscent of a PERT chart, but I wasn't that formal about it. Nevertheless, it was clear that certain steps in the system's development had prerequisites, such as the need for functional hardware before the code could be integrated fully and the need for a cabinet before the hardware could be installed. The approach was intuitive, based on the same kind of awareness that prompts one to get purchase orders off to the component vendors right at the beginning of a job.

The question of front panel layout turned out not to be as trivial as I had imagined, although a pleasant arrangement was finally settled upon. One of the early mistakes was selecting an enclosure on the basis of volumetric calculations on the chosen components: although it ended up being a nearly optimal (?) size, it introduced some rather uncomfortable constraints during the panel design process. The final configuration left only 0.1 inch of slop over a 20-inch vertical dimension. This hardly accommodates later vendor changes on panel components without very careful study.

Anyway, Figure 14-2 shows some of the considered approaches, with the one in part d of the drawing ultimately selected as the best. Note that the displays are at the top, obviating the tendency for the operator to visually obscure them with a forearm during data entry. Also, the lighted pushbuttons are in two groups: the eight "alphas" on the left (each of which transmits a message of one byte

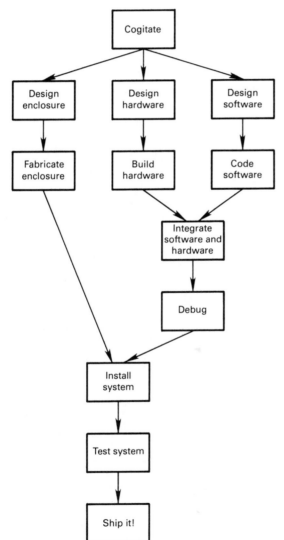

Figure 14-1   The IDAC terminal development includes three paths which ultimately converge at system installation. Careful attention to timing is necessary to avoid long waits for system components that could be developed concurrently with others.

when depressed while lit) and the 10 "numerics" along with CLEAR and SEND on the right (which are used to enter a numeric data string of up to 30 bytes, echoed in the lower SSD and transmitted upon depression of a lit SEND button). The badge reader, only used upon initiation of a transaction, is at the bottom.

Armed with this sketch, it was possible to make an actual-size template, send the 14-gauge steel padlockable NEMA 4 enclosures ordered from the Hoffman Electric Company (Model 20H16CLP) to the local sheet metal shop for hole cutting and painting, and move on to more logical matters.

14.2 IDAC HARDWARE DESIGN

No matter what was eventually to be done with the software, one thing was clear: the basic objective of all the hardware was simply the facilitation of communication between the various I/O devices and the processor. There was no need for random logic in a functional sense, for the 8085 had plenty of spare time to take care of trivia such as switch debouncing and Hollerith-to-ASCII decoding.

The Intel SBC 80/04 processor, as we have noted a time or two, is a capable

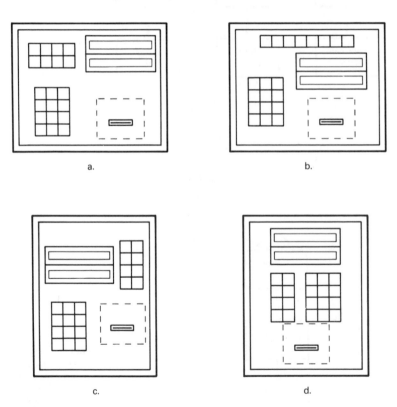

Figure 14-2   An important part of the "cogitation" phase is the choice of a front panel layout: (a), (b), and (c) were discarded in favor of (d).

but limited board. There is no access to the bus, only the availability of three I/O ports which are software defined [one of the first acts upon encountering the initialization routine at location zero is the writing of a byte to a status port (0) that specifies the behavior of ports 1, 2, and 3]. These are created by an 8155 chip, and appear to the outside world on a 50-pin edge connector.

But there was obviously the need to interface more than three devices with the processor, so the ports were used to create a bus structure. In its overall form, it is shown in Figure 14-3.

Each of the devices connected to these buses is the named piece of hardware in association with appropriate interface circuitry. The control port is decoded and used to govern the information flow in the resulting structure: by enabling devices onto the input bus or latching information into them from the output bus, it is the means by which all data transfers in the terminal are accomplished. This, it will be remembered, is the same scheme we used in the multiplexer to handle the 16 UAR/Ts—but this time we will explore it in detail, now that there are some more interesting I/O devices to discuss.

14.2.1 The Control Decoder

Figure 14-4 is a schematic of this key link between the SBC and everything else. Consisting simply of three chips—74LS138 3-to-8-line decoders—the circuit accepts a 5-bit control code from the processor on port 3 and produces active-low commands to the rest of the system's circuitry.

Each 74LS138 decodes, when enabled by the low–low–high combination of pins 4, 5, and 6 respectively, the three bits present on pins 1, 2, and 3. The enabling scheme allows cascading into larger decoders as we have done here, without the bother of external inverters and enabling circuitry. In its present form, this is a 1-of-24 decoder, of which 20 possible states are used.

Take a moment to look at the labels of the 20 generated commands. Each is shown as a mnemonic beginning with the letter "Z" along with a brief English description of the operation. These active-low "Z" commands correspond precisely with the labels used in the software, allowing the convenient use of macros to accomplish most of the I/O operations. A general input macro exists, for example, which simply outputs the appropriate control byte to enable something onto the input bus, reads the input port, and returns the value read in the accumulator. Fetching the UAR/T status byte is thus accomplished by simply saying

INPUT ZSWE

Macros similarly exist for generalized output operations and simple strobes without associated data transfers, such as the pulse that causes a badge to be ejected.

As we progress through Sections 14.2.2 to 14.2.8, you may find it valuable to refer occasionally to Figure 14-4. Since we will leave the SBC at the black-box level (it's only a computer!), the control decoder is as close to the CPU as we'll need to get in our discussion of the IDAC hardware design.

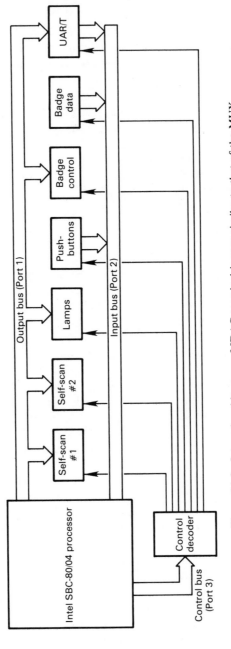

Figure 14-3 Internal architecture of IDAC terminal is very similar to that of the MUX, with a control decoder governing interaction between the logical devices and the two data buses.

334

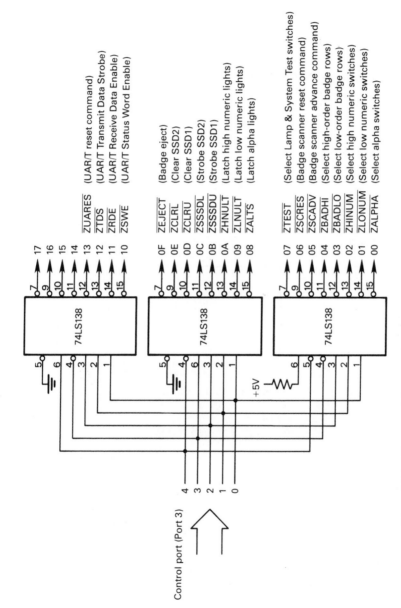

Figure 14-4 The control decoder of the terminal simply consists of three 74LS138s set up as a 6-to-24-line decoder. The outputs constitute the full battery of internal control lines in the machine.

14.2.2 The Self-Scan Displays

Burroughs made things easy for us by rendering the SSDs interfaceable with parallel ASCII. They behave just like latches: all that is necessary to add a character to the display is the presentation of the 6-bit uppercase ASCII subset code on the input data lines and the generation of an active-low strobe. The display module provides some handshaking (acknowledgment), but we have no need for it here.

The circuitry involved with the displays is shown in Figure 14-5. The output bus from the processor card (port 1) is shown passing through the diagram, with its low 6 bits picked off and applied to the two SSDs. Each display receives two signals from the control decoder: a "strobe" which latches the data present at that moment on the data lines, and a "clear" which erases any information presently being displayed.

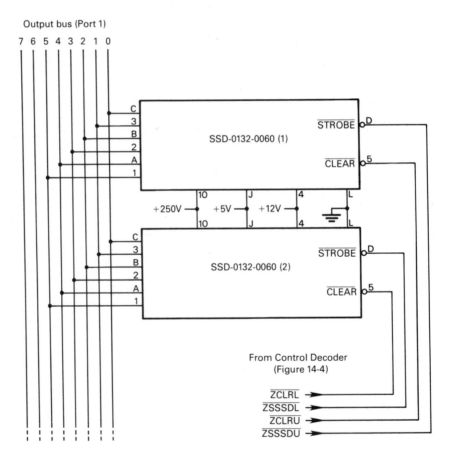

Figure 14-5 The simplest interface of all in the IDAC terminal is the display, with nothing more than the interconnection of data and control lines to the Burroughs SSDs.

The characters are converted to 5 × 7 format by logic in the SSDs and are added to any data currently on the one-line "screen" from the right, moving the rest over. A continuous stream of text transmitted to the display will flow smoothly through it from right to left.

In IDAC operation, it is necessary to treat the displays in two ways. Sometimes the information already in view must be discarded in its entirety and replaced by a new message, and other times it must be possible to simply append new data to whatever is already there. To accommodate this, each display is treated by the software as two logical devices—one with, and the other without, an initial "clear" command.

Note that the four control commands to the displays ($\overline{\text{ZSSSDU}}$, $\overline{\text{ZCLRU}}$, $\overline{\text{ZSSSDL}}$, and $\overline{\text{ZCLRL}}$) can be traced directly to the control decoder of Figure 14-4. That's all there is to the SSD logic, and we will see how the software handles communication with them in a later section.

14.2.3 The Badge Reader

The Hickok Model 264A badge reader similarly simplifies our interface task by incorporating a scanner to present us with a succession of badge column images upon command.

The badge is basically a 22-column Hollerith card, with 12 rows in which punched holes can be made to encode employee ID. In this application we found it unnecessary to allow column data outside the range 0 to 9 since all employee numbers were four- to six-digit numeric values. This decision further simplifies the interface task by eliminating the need to perform a full Hollerith-to-ASCII code conversion.

The task is complicated somewhat by the fact that each column of the badge is presented as a 12-bit word, something that must be broken into two parts to make it successfully through the bus. Consider Figure 14-6.

The reader, being something of a black box from the system standpoint, is shown undetailed here. Everything we need to know about it can be expressed in the relationships among its input and output lines.

Operation begins when somebody inserts a badge, mechanically releasing the spring-loaded deck which carries the brushes and actuating a microswitch in the process. This switch closure results in the creation of a signal called "Ready," which we show connected to the input of a tri-state buffer.

This is as far as things would get were it not for the fact that the terminal's software spends the bulk of its time scanning for conditions, of which one is this ready line. Periodically, the $\overline{\text{ZBADHI}}$ control line will go low, enabling this signal onto bit 4 of the input bus (together with the four high rows of the badge data, but we'll get to that in a moment).

At this point, the software has within its grasp the knowledge that a badge has been inserted, and if it is an inappropriate time for such an occurrence, it can immediately eject it by creating the $\overline{\text{ZEJECT}}$ command. Otherwise, it transmits a message to the host computer via the mux which simply states the condition.

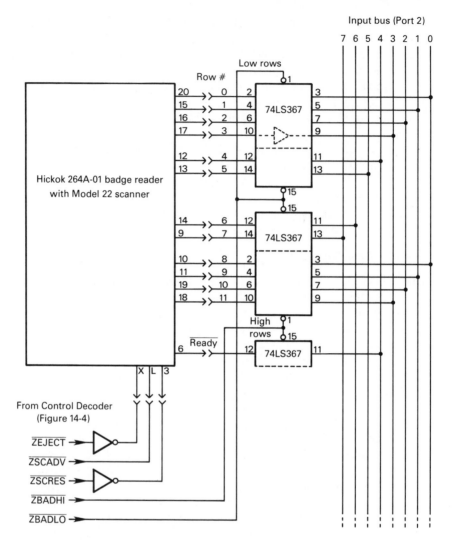

Figure 14-6 The badge reader presents one 12-bit column at a time for each occurrence of a scanner advance ($\overline{\text{ZSCADV}}$) pulse. These data are applied to the input bus via tri-state buffers.

Now, the terminal's processor has some work to do. The badge scanner can be thought of as a 12-bit wide, 22 : 1 multiplexer with its own address counter. Reading an entire badge is accomplished by resetting this counter to column 0 with the $\overline{\text{ZSCRES}}$ command, after which it is advanced column by column with the $\overline{\text{ZSCADV}}$ command. At each state of the scanner's internal counter, the 12 bits corresponding to the current vertical slice of the badge appear on the parallel output lines.

To inhale these for use, the processor need only enable them via the 74LS367 tri-state buffers onto the input bus. This is accomplished in two chunks: the low 8 bits via the command $\overline{ZBADLO}$, and the high 4 (plus the "Ready" again) via $\overline{ZBADHI}$. In each case the processor simply inputs port 2 while the command is active, yielding the 12 bits of interest. As we shall see later, these column data are then translated to ASCII and appended to a message buffer that will be shipped to the host computer in response to the "badge read" command.

14.2.4 The Pushbuttons

The other front panel input device is the array of 20 lighted pushbuttons, arranged as one group of eight and another of 12. These are the operator's primary point of interaction with the machine.

The software considerations, surprisingly, turned out to be nontrivial. A map must be maintained internally representing the lamps which have been lit by system command, allowing unlit buttons to be ignored. Physical debouncing must be performed, with 200 ms of "solid down time" necessary before a depression can be considered as valid, and simultaneous actuations must be ignored.

In Chapter 5 we discussed various means of performing keyboard encoding, and in Figure 5-10 presented the software-scanned matrix that was used in the terminals. The same circuit is shown in more detail in Figure 14-7.

Here, the process of enabling groups of eight switches onto the input bus can be trivially accomplished without the need for tri-state buffers, since the "open" condition of a given switch is already a high-impedance state. But look what happens when one of the control lines goes low ($\overline{ZALPHA}$, for example): the logic "0" now present on the cathodes of the diodes appears on any bus line to which a depressed pushbutton is connected. In a fashion somewhat similar to the badge reader, the processor can obtain a complete image of the 20 pushbuttons (and the test switch, located inside the enclosure) by successively producing the four commands shown and reading the input port while each is active.

The diodes, incidentally, prevent the switches from interfering with other bus activity. If they were not included in the circuit, depression of a button or two could garble other bus activity by causing logic levels to fight one another.

The test switch is a three-position DPDT toggle (center off, spring return from the LAMP TEST position). In that position, bit 7 of the bus is taken low when $\overline{ZTEST}$ is active (allowing the software to place the message "LAMP TEST" on the upper SSD), and the line labeled $\overline{LAMPS}$ is taken to ground. As we will see in the next section, this forces all the lamps on without involving the processor in any way, eliminating all unknowns save the lamp driver circuitry and the bulbs. The other position of the switch, SYSTEM TEST, is nonmomentary, and allows the performance of the series of diagnostics described in Section 11.4.3.

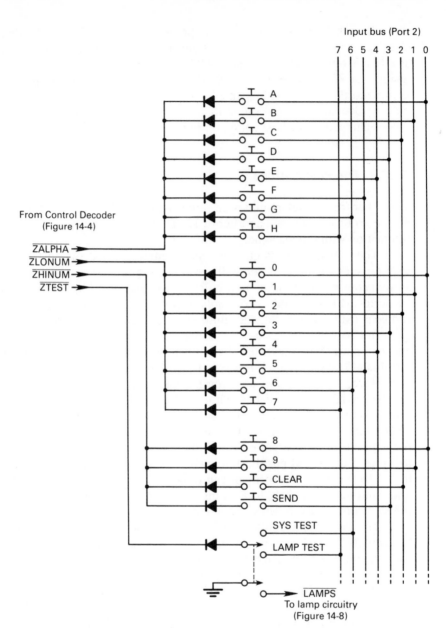

Figure 14-7 Interface with the 20 pushbuttons is accomplished without the aid of additional TTL circuitry, leaving the tasks of debouncing, decoding, and error detection to the program.

14.2.5 The Lamps

Figure 14-8 depicts the circuitry involved with the conversion of bit patterns to photons.

In normal terminal operation, the lamps on the Microswitch pushbuttons are illuminated in direct response to messages arriving from the host which specify device codes 5 and 6 (E and F). In each case, a map is specified in the data field of the message in the form of N's and F's, meaning on and off, respectively. The first device code implies a map of eight alpha light states; the second, of 12 numeric light states.

We don't want to burden the processor (or the programmer!) with the need to continually refresh the lamp states, so we provide five 4-bit latches (74175s) to remember the bit patterns that we output. In the case of the alphas, for example, the processor would place a byte on the output bus representing the desired pattern, then create via the STROBE macro the command $\overline{\text{ZALTS}}$. This is inverted, creating a leading positive edge (that proves, by the way, to be totally superfluous—the 74175s are actually quad-D flip flops, and would be just as happy to latch the data on the trailing edge of a low-going pulse), and applied to the latches. This results in a "snapshot" of the output bus which appears, inverted, on the latch outputs.

The information remains in the latches until updated by a later command or until cleared by the system reset line, shown connected to pin 1 of each of the five chips.

At this point, the lamp driver circuitry takes over, which has already been described in detail in the text associated with Figure 5-13. Note the use of the LAMPS signal generated by the test switch.

Note also that the +5 volts common to the light bulbs could be sourced via a current-sensing circuit, allowing the terminal to perform its very own lamp test by illuminating them one at a time and checking for the appropriate 50 mils or so.

14.2.6 The UAR/T

The communications logic comprises the only hardware subsystem that involves both the input and output buses. As shown in Figure 14-9, the UAR/T is controlled by four signals from the decoder, $\overline{\text{ZUARES}}, \overline{\text{ZTDS}}, \overline{\text{ZSWE}},$ and $\overline{\text{ZRDE}}$. The first of these is simply a reset, which establishes known conditions upon system startup, and the others have already been described in detail.

UAR/Ts require transmit and receive clock signals which are 16 times the desired baud rate. In this design, the terminals interact with the mux at 1200 baud (or 120 characters/second), thus creating the need for a 19.2-kHz clock. This could have been provided by a dedicated crystal oscillator, but I cheated: by bringing a 983,040-hertz clock signal up from the SBC 80/04 card on a shielded wire, a reasonably close approximation could be derived through division by 51. The circuit that accomplishes that, consisting of two 74177 binary counters and a 7474, is shown in the lower half of Figure 14-9.

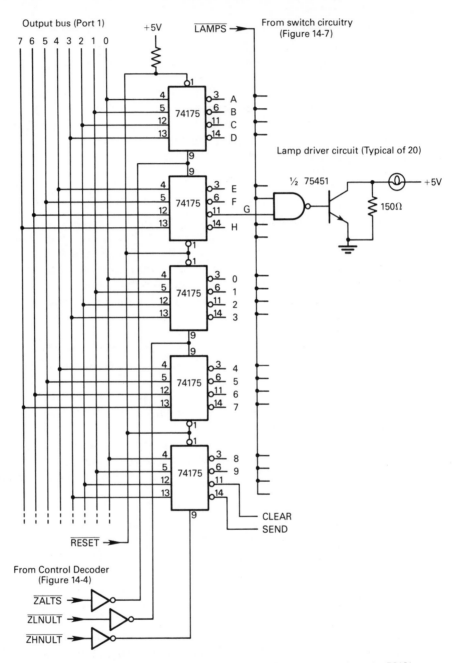

Figure 14-8 The lamps associated with the pushbuttons are driven by 75451s connected to TTL latches. Changing the configuration of lit bulbs involves writing a binary image to the appropriate latches via one of the three control lines.

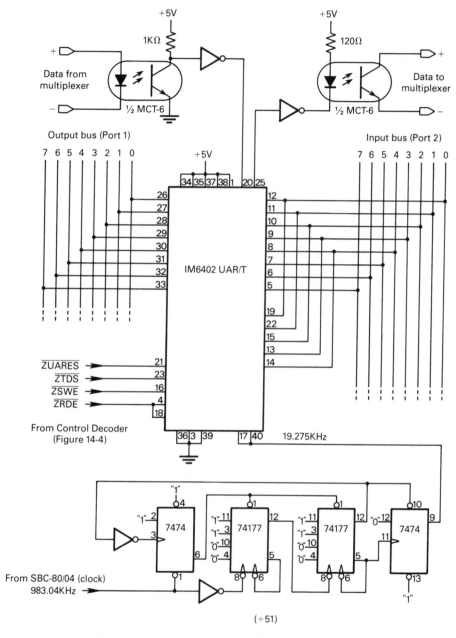

Figure 14-9 The communications logic in the IDAC terminal consists essentially of a UAR/T tied to the I/O buses and controlled by the four signals shown at the left. The current loop interface is at the top, and the baud-rate clock is generated by the three-chip circuit at the bottom (which divides a high-frequency signal from the CPU board by 51).

I must admit with some embarrassment that this was not always the way the clock was generated. Initially, in both the multiplexer and the terminals, the UAR/T baud-rate clock was produced with a 555 timer, a capacitor, a couple of resistors, and a trimpot. It worked quite solidly on the bench, but once the terminals were installed in widely varying thermal environments, ranging from direct summer sunlight to snowy doorways, problems began to occur. The customer became a bit tired of running around with a jeweler's screwdriver diddling trimpots, and complained. The resulting circuit completely solved the problem, although it does have the disadvantage of requiring a jumper on the SBC which connects the clock signal to an unused connector pin.

Anyway, the rest of the UAR/T circuit should look quite familiar by now — we dealt with both it and the associated current loop interfaces back in Chapter 5.

We have just used up all of the control signals generated by the circuit of Figure 14-4.

14.2.7 Adding Other Devices

As has been suggested by the simplicity of device interfacing, grafting other hardware onto this system is not particularly difficult. Somewhat typical of such enhancement might be the addition of hardware to allow the sensing of certain interlocks and the application of control voltages to a process if the host determines that conditions are appropriate. Indeed, just about anything in the "control" milieu can be tacked onto this architecture — although the effort begins to become counterproductive as the highly variable communication delays and susceptibility to host system problems suggest the need for more localized control.

The addition of, say, an eight-channel analog front end to the terminal is no more forbidding than any of the devices we have discussed. The three-chip control decoder has four spare lines, and the addition of another 74LS138 provides another eight. The software, of course, would require a bit of patching, but as we shall soon see, it is structured enough that the only additions would be embodied in stand-alone modules which have no effect upon the overall design. Such a change should hardly be considered a "field mod," of course, but it is comforting to know that the engineering effort expended in the initial creation of the terminal can be repaid in part by the savings on subsequent, similar jobs.

14.2.8 Miscellaneous IDAC Circuitry

Before looking into the software considerations, we should spend a moment on the less glamorous parts of the terminal's hardware design.

There is more to providing power than bringing in ac, running it through a transformer, rectifier, regulator, and filter, then cabling it off to all corners of the system. From the moment the power line enters the enclosure, we must concern ourselves with noise and various other manifestations of reality.

The first device to encounter a 120-volt ac difference of potential is an

R-Tron line filter (an RNF-1L3), which was shown in schematic form back in Figure 6-2. This removes most of the high-frequency garbage that is a constant companion of virtually every electromotive phenomenon in a busy factory, making it somewhat less offensive to edge-sensitive components looking for a volt or two of input excursion.

This slightly cleaner ac is then fused, switched, and made available on a barrier strip to which are screwed three MOVs whose function is the rapid absorption of voltage transients above a safe level. From here, it is cabled (via twisted pair) to the two open-frame power supplies and the badge reader's ejection solenoid. At one time, it was also cabled to the fan.

The fan, you say? After all this talk about hostile environments, this guy cuts ventilation holes in the box and pulls air through it with a fan?

Hey, how do you think I learned about the hostile environment?

Indeed, the IDAC terminals started their careers with 30% efficient linear power supplies which convert the differential between regulated and unregulated voltages into heat. It quickly became obvious that the box was going to cook itself in short order, so to speak, if something wasn't done to get rid of some of that waste heat. Hence the fan. And the filters. And the unending hassle of clogged pieces of foam that became so after perhaps four hours of use in a particularly dirty part of the plant. Maintenance people started leaving them off, and the boxes filled up with dirt.

Much thought was given to the problem, with all sorts of esoteric schemes considered for moving the heat outside: Peltier effect devices, heat pipes, Hilsch vortex tubes, air conditioners, fluid exchange systems, and heat sinks.

And, of course, switching power supplies. It turns out that their vastly increased efficiency of at least 70% makes a rather significant difference in the scope of the problem—effectively, ahem, eliminating it in this case.

The fans hit the junkbox.

Anyway, once the power has been massaged into something a little more palatable to logic circuitry, it is cabled about as needed. The +5-volt supply provides a remote sensing feature which allows the regulation point to be determined by the voltage at the other end of the accumulated IR drops, and that is used to keep noise immunity high. The various dc voltages find their way to the front panel (where most of the action is) via a separate cable from the logic lines, there to terminate in another barrier strip from which those hardware entities needing power can drink at will. A bit of remote filtering is provided at this point as sort of a high-level decoupling to consume any particularly obnoxious I_{cc} spikes before they have a chance to propagate throughout the box.

And, of course, the +250 for the SSDs is insulated, since it stings a bit.

The only other item in the "miscellaneous hardware" category is yet another barrier strip, this one providing all necessary interface with the outside world (via a MIL-spec connector on top of the cabinet). In particular, it carries the four current-loop lines, a ground for the communications cable shield, and a READY line for handshaking with a radio controller (this whole thing can be mounted on the fork-lift truck, plugged into the modem/radio combination, powered with an appropriately scaled inverter power supply, and driven off into the murky darkness of whatever warehousing operation might profit from such a bizarre-looking manifestation of microprocessor technology).

14.3 IDAC SOFTWARE DESIGN

The highest ROM address used in the IDAC terminals is 06AB hex—or decimal 1707 bytes of code. This is resident in a pair of 2708 EPROMs plugged into the SBC 80/04 board.

Contained therein is a program, written with the 8085 subset of the Z-80 assembler, that handles all terminal operations via the bus structure we have just discussed. I will not assail you with a complete listing, but there is some value in a discussion of the various modules and their functions.

14.3.1 Initialization

If you were to look at an assembly listing of the IDAC software, you would note that the first page is given over to a module named INITIALIZATION. This is where execution begins on system reset; this is location zero. Various things happen which prepare the system for all subsequent activities.

First, the stack pointer is set to 4000H, one plus the top of the RAM space (3F00-3FFF). As subroutines are called and data are pushed, the top 20 or so locations will be used for the stack.

Then, all of RAM is cleared via a four-instruction loop. This is a brute-force method of initializing flags and counters—generally allowing execution to begin with a clean slate. The only value that has to start as something other than zero is something called TPOINT, which is a 16-bit pointer to a keypad buffer. This is initialized to the buffer's first location.

Following this, the ports are defined. As we mentioned earlier in the chapter, the I/O structure has to be specified by an output instruction to the 8155—in this case, with a parameter of 0DH. This sets up port 1 as output, port 2 as input, port 3 as output, and disables the interrupts that can be associated with ports 1 and 2. The high 2 bits of this byte control the system timer, and we'll be dealing with those momentarily.

The program then initializes the front panel devices, writing 0's to all of the lamp latches, delivering CLEAR commands to both SSDs, and performing a badge ejection.

The last operation performed by this module before it passes control to the executive is the setup of the system timer to interrupt the processor every 5 ms. This is used in the determination of pushbutton downtime, and also renders the executive synchronous to simplify the addition of other software—a primitive form of time slicing.

Establishment of this interval is performed by writing the hex value 26C to ports 4 and 5—the 14-bit counter which divides the 122.88-kHz crystal clock to determine the repetition rate of the output. There is also a pair of bits which allow specification of timer mode—single pulse output, square-wave output, automatically repeating pulses, and so on. The latter is selected. (The value 26C, or 620 decimal, yields a 5.047-ms interval because the basic pulse length of the clock input is 8.14 μs). Once this is done, the timer is started by another output to port 0, and the initialization is complete.

14.3.2 The Executive

The system then spends the bulk of its time waiting for something to do. Take a long look at Figure 14-10.

Upon arrival from the initialization module, the system immediately checks the UAR/T status word, either processing a received character or continuing with the transmission of a message if one of the two important bits is set. We will see shortly what is implied by this. But if nothing is happening on the communications front, the system enables interrupts and checks for an occurrence of the 5-ms clock (which will be detected no matter when it took place, since the interrupts are latched). If it's not there, interrupts are promptly disabled again, and the program returns to the UAR/T checks.

If there has been a pulse, however, a whole series of tests is performed. First, a byte called QUEUE is examined to see if any messages need to be transmitted to the host. In keeping with the structured concept, individual device handlers are not allowed to speak to the host themselves—they just set a flag indicating to a higher-level program that they have assimilated a message, and would somebody please do something about it? Here, the discovery that a QUEUE bit is set will result in a message transmission, as long as one is not already in progress (we are interested only in the top level right now—there is a lot of logic in some of these blocks).

Second, a flag called KIKFLG is checked to see if something called a "kick delay" is in progress. This was an afterthought, and resulted from the observation that holding a badge in the reader against the system's will resulted in frantic buzzing of the solenoid with the eventual blowing of the fuse. The kick delay is about 1 s, and must elapse between the command to eject and the actual firing of the solenoid. Here, the discovery that the delay is in progress simply results in the subtraction of 5 ms from the time remaining and, if it reaches zero, the actual ejection of the badge.

Third, the microswitch indicating that a badge has been inserted is checked, as described in Section 14.2.3. If it discovers that one is there and, in fact, that it was inserted since the previous scan, a routine (GOTBAD) is invoked to deal with it. This, as you may recall, involves sending a message to the host noting the occurrence.

Fourth, the test switch is interrogated via the circuitry of Figure 14-7, with

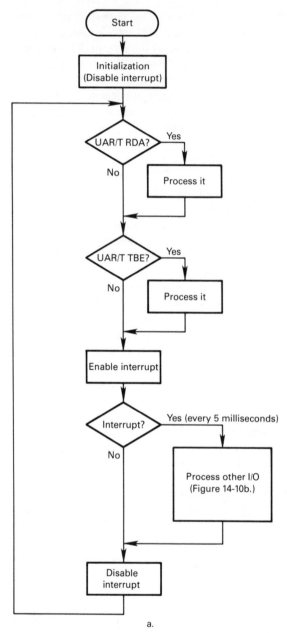

a.

Figure 14-10 IDAC terminal software is simply a series of tests for conditions that require attention, together with modular blocks of code that perform whatever processing is required for each. The interrupt used is simply a handy way to implement a real-time reference for switch debouncing, and does not suggest the existence of high-priority external conditions.

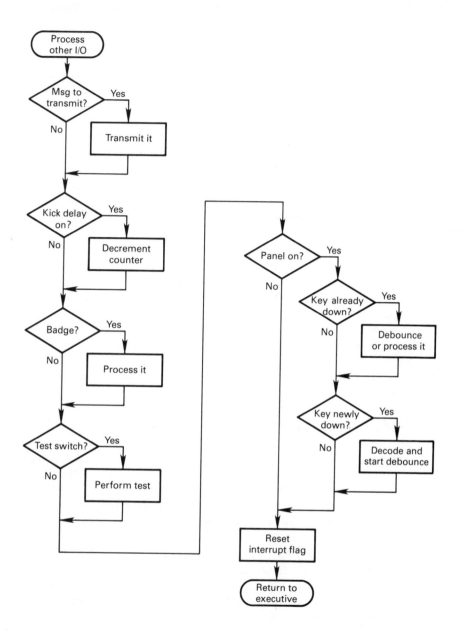

b.

anything other than the center-off position resulting in the execution of a test routine.

Fifth, the PANST flag is checked to see if the keyboard is enabled. One of the "devices" accessible to the system-level programmer is called "panel control." This allows the terminal to be rendered inoperative without having to kill power or somehow handle it at the mux or host level: if the terminal is directed to be off, the message "TERMINAL OFF" appears in the top SSD, any badge inserted will be promptly ejected, and all pushbutton depressions will be ignored. The only way to bring the unit out of this autistic state is to send it a "panel on" command. Anyway, it is obviously superfluous to go through the pushbutton tests if the machine is disabled, so the logic provides for an early termination of the interrupt sequence if this is the case.

If it is not the case, however, a flag is checked to see if any pushbutton is currently depressed and, like the kick delay, a counter's value is appropriately updated if so. After 200 ms of uninterrupted downtime, the switch is considered valid and acted upon in an appropriate manner.

If a button is newly depressed, it is decoded and the debouncing process is begun.

At the completion of this series of tests, the interrupt flag is reset, new interrupts are disabled, and control returns once again to the UAR/T logic to complete the loop.

Now let's look at the contents of the boxes that were referenced so vaguely in Figure 14-10.

14.3.3 Incoming Data Handling

The logic that falls into this category is fairly extensive, given the fact that it includes not only communications overhead but also the performance of whatever action is directed by the message contents. It might thus be somewhat overwhelming to charge into this at the flowchart level. Instead, let's try to picture the events associated with the arrival of a single character from the host computer.

The first thing that must be done, once the executive has discovered that the byte has been received and is currently sitting in the UAR/T's holding register, is to inhale it so it can be put to use. The command is ZRDE, and in a trice, the byte is in the accumulator. It is promptly AND'ed with 7FH to mask off the high-order bit, then is saved in register B.

Are we awaiting a checksum? There would be nothing in the nature of the received byte to tell us whether it happens to be one, so we must check a flag which would have been set upon receipt of a carriage return. If it is set, we leap off to a section of code called CHECK, wherein the byte is compared to the checksum internally accumulated during the receipt of the message, resulting in the transmission of either an ACK or a NAK back to the multiplexer as appropriate.

If we are not awaiting a checksum, we can check the byte to see if it is the backslash character that implies the start of a message, as shown in Figure 14-11. If it is, then we go execute a body of code that sets a pointer to the start of the received message buffer, initializes the internal checksum, and sets a "receive in progress" flag.

If the received byte is neither a checksum nor the start of a new transmission, perhaps it is an ACK or NAK corresponding to a message that was recently transmitted by the terminal. Another flag is checked to see if we are indeed waiting for acknowledgment, and if so, the byte is compared to the values corresponding to the two possible responses. If it is an ACK, the flag is cleared and the whole business forgotten; if it is a NAK, the message "TRANSMIT ERROR" is displayed on the upper SSD and another attempt is made to send the same message.

If it was none of these things, it just might be garbage. The "receive in progress" flag is checked at this point, and if it is not set, the new byte is ignored. If it is set, there are two possibilities.

It could be anything but a carriage return, in which case it is added to the receive message buffer at the pointer location (whereupon the pointer is incremented), or it could be a carriage return, in which case the "awaiting checksum" flag is set. In either case, the byte is added into the locally accumulating checksum, the message buffer is checked for overflow, and control is returned to the executive.

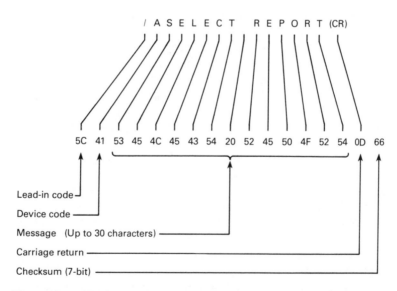

Figure 14-11 The format of a typical message includes a preamble (with a lead-in code and terminal ID), the message body, a carriage return, and a checksum for error detection.

But how do we act upon a received message? Naturally, we cannot be certain that it is good until the checksum is determined to be correct, so together with transmitting an ACK to the multiplexer as noted above, control is passed to a location in the program known as ANALYZ.

14.3.4 Received Message Analysis

At this point, the receive message buffer (RECBUF) can be assumed to contain a complete message to one of the terminal's devices. Now what?

The first thing we have to do is look at the device code (the second byte in the message) and determine therefrom just how the data field should be handled. It would hardly do to display lamp maps on an SSD. The first few instructions of ANALYZ successively compare device codes to the one in the message, jumping off to appropriate routines as indicated. (This, incidentally, is an example of the DO CASE structure.)

The first of these is the one that handles device code 0—the panel on/off command. In a routine appropriately labeled ONOFF, the data field of the received message is checked for "N" or "F" (with an abandonment of the task if it is neither) and the subroutines PANON or PANOFF executed accordingly. In either case, a message is displayed on the top SSD indicating the status of the terminal, and the flag PANST, which we used to permute the scan in the executive, is loaded with either a 1 or a 0. In the event that the "panel off" command was issued, the lamps are all extinguished and the lower SSD is cleared as well, rendering the terminal apparently inert but for the "TERMINAL OFF" message in the top display. When this is complete, control returns to the executive.

The next possibility is a message to the self-scan displays. Recall that we set them up as four devices, each display being accessible with or without a preliminary CLEAR command. The first part of the code labeled SELFSC, then, determines on the basis of the device code which of the possibilities will be realized. At this point, the specified display either does or does not get cleared, and the message's data field is passed to a display utility for the dirty work of getting it out to the screen.

It's not too dirty, actually—just a simple loop that is invoked with a pointer aiming at the start of a message (terminated conveniently by a carriage return) and a byte in register B indicating which of the two displays is the lucky one. Note that a display can be cleared by the applications software in the host computer through the simple expedient of sending a blank message field to a device whose code implies a preliminary clear.

The next possibilities are the lamps, with device code E implying a new configuration of the eight alphas and code F doing the same for the numerics. Recall that the data fields of these messages are "maps" consisting of N's and F's for on and off, respectively. When the software (at ALIGHT or NULITE) receives these, it converts the sequence of ASCII characters to either an 8- or a 12-bit word, then simply outputs it to the circuitry in Figure 14-8. The binary maps are saved in RAM, because they will be used to screen subsequent pushbutton depressions.

The only other possibility presently associated with analysis of a received message is a command to the badge reader, which will be either "E" for eject, or "R" for read. Control is passed as indicated to the routines KICK and GIVEIT, the latter being the setting of a bit in the QUEUE byte (since the reader logic actually reads and translates a badge upon insertion—it just doesn't tell the host about it until asked).

All that we have just described in the preceding two sections belongs in the little box labeled "Process it" associated with the "RDA?" test in Figure 14-10.

14.3.5 Data Transmission

If the news from the UAR/T status byte was that a character had been fully transmitted, leaving the buffer empty, a different set of actions must be undertaken. We arrive at a block labeled BYTDUN (some of these labels are a bit obscure, but the field is limited to six bytes—although others can be tacked on with no ill effect: BYTEDONE would be seen as BYTEDO).

Here, the first thing that takes place is the test of a flag which indicates whether or not it is time to send the checksum, suggesting that the byte whose transmission spawned the "transmitter buffer empty" condition was a carriage return. If so, of course, the checksum is sent; if not, another flag is checked to see if that was what happened the last time. If so, the "transmit in progress" flag (TIP) is cleared, and the "awaiting acknowledgment" flag is set.

If none of these boundary conditions exist, it can be assumed that we are somewhere in the midst of a message transmission. The communications pointer is fetched, the referenced byte from the message buffer obtained, the pointer incremented, the byte added into the locally accumulating checksum, and the character transmitted via a little subroutine called COMOUT. And there things rest until 8.3 ms later.

It should be noted that this whole process starts when the logic we will discuss in Section 14.3.6 initiates a message transmission by sending the leading backslash and setting TIP. This is all that must be actively done to transmit a message, since the rest is more or less automatically driven by the transmit buffer empty condition associated with each byte.

14.3.6 Transmit Initialization

The first thing that is checked by the executive upon occurrence of the 5-ms interrupt is the QUEUE byte: if it is not zero, it is safe to assume that some module has assembled a message that is ready for transmission to the host. The block of code that is invoked thereupon is represented as "Xmit it" in Figure 14-10.

Immediately upon arrival in this module, the program checks TIP to see if a message is already under way. Things might get a bit garbled if a new one were begun at this time, so it simply ignores the condition and returns to the executive. Otherwise, however, it sets TIP and disassembles the QUEUE byte to see which device initiated the need to communicate.

At present, there are only four possibilities: an alpha pushbutton (invoking the routine SNDALF), a numeric string terminated by the SEND key (SNDNUM), the 22-byte badge image (SNDBDG), or the "B" indicating badge insertion (SNDBEE).

And from this point there's not much to it: the appropriate routine establishes a communication buffer containing the "\", the device code corresponding to the type of message, and the data. It then passes control to a block of code common to all of them which simply masks off the proper bit in QUEUE, begins checksum accumulation, and transmits the first character. From that point on, transmission progresses as described in the preceding section.

14.3.7 Badge Reader Logic

In our flowchart of Figure 14-10, we show two events connected with the badge reader: detection of an ejection delay in progress and insertion of a new badge. The first case is trivial, consisting of a few lines of code that fetch and decrement the delay counter, restoring it and exiting if it does not become zero in the process. If it does, it issues an eject command to the reader via the $\overline{\text{ZEJECT}}$ control line.

The second case is somewhat less trivial. If the executive detects the actuated condition of the badge sensing switch, a module called BREAD (for Badge READ—what else?) is invoked which immediately checks a flag to see if the condition is an old one (since the switch stays closed the entire time that the reader is holding a badge). If it is, it is simply ignored; otherwise, a translation routine is executed which fills the badge buffer (BADBUF) with ASCII numeric values derived from the hole patterns sensed by the reader. Any byte found to be non-numeric simply results in the letter "X."

At the conclusion of this process, the reader's "badge ready" QUEUE bit is set, and control is returned to the executive.

14.3.8 Test Routines

Proceeding downward through the series of tests in the executive flowchart, we come next to an interrogation of the test switch. There are two active positions in which it can be.

The first is LAMP TEST and, as we have described, the true "testing" that takes place here is accomplished by the hardware itself. There is, however, a brief routine associated with it which first displays the message on the upper SSD, then waits for the switch to be released before clearing it.

The other one, SYSTEM TEST, begins by displaying the message, "SELF-TEST: LPB ECHO." This implies the first of the test functions, which is simply the illumination of any key that is pressed. This is obviously trivial, but allows a very revealing test of the switches, the input logic, the processor, the output logic, the lamp drivers, and the lamps themselves. If lamp "G" fails to light on depression, for example, but works OK in LAMP TEST mode, then a serviceman knows the problem is somewhat more profound than an open filament. Subsequent poking around with other devices using the same bus bit should allow reasonably tight localization of the difficulty.

The software that accomplishes this is just a short loop which inhales the switch bytes, complements them to establish proper logic polarity, and promptly outputs them to the lights.

This test routine also allows the badge reader's operation to be observed: an inserted badge will have its contents displayed on the lower SSD, then will be promptly ejected. Most of this is pirated from the badge reader logic (by faking an actual badge insertion, then repairing the modified QUEUE byte, tacking a carriage return onto BADBUF, and displaying it via the SSD utility).

The communications loopback test involves the multiplexer, and requires its cooperation. It would be pleasant to make this a purely local function, but testing all components right out to the opto level involves various clumsy schemes which are only marginally justifiable. Since this aspect of the program was an afterthought anyway, it was adjudged desirable to accomplish it without any hardware changes.

As written, it simply transmits a message to the mux preceded by an identifying special character (a ^), directing the processor there to send the same message back. It then compares them and displays an error on the upper SSD if they are not the same. To avoid torturing the mux with endless loopback requests, this is done only once for each entry into the system test routine.

14.3.9 The Pushbutton Handler

In the Figure 14-10 flowchart, we arrive next at the pushbutton tests (which are avoided if the "panel off" condition exists). There are two:

The first one checks to see if a flag called KEY is set, indicating that someone depressed a pushbutton sometime in the recent past and as of 5 ms ago, had not released it. It begins by calling a routine called SINGLE, which returns the CPU's ZERO flag reset if more than one button is currently down. This results in execution of WIPOUT, which kills the KEY flag and returns to the executive.

Assuming that one and only one key is depressed, the program then performs a few tests to ascertain that it is the same one that initially resulted in the setting of the KEY flag. If not, it performs WIPOUT; otherwise. ...

It continues the debouncing process by decrementing a timer called BOUNCE which was originally set to 40 upon initial detection of the key (this corresponds to 200 ms; although 400 was the ISA specification, it was just too slow). Note that the combination of this logic effectively kills bounce, for it sees any bounce of the button before 200 ms elapses as the equivalent of a release, thus executing WIPOUT.

If the debounce period is successfully undergone, the logic determines whether it is alpha or numeric. If alpha, then the byte representing the switch closure (the input induced by the $\overline{\text{ZALPHA}}$ control signal) is translated to an ASCII character in the range A to H and placed in a buffer for transmission to the host. Then all the alpha lamps are extinguished (indicating to the operator that he has no more choices—at least until the host sees fit to present him with some), and the appropriate QUEUE bit is set to initiate transmission.

If the key is discovered to be numeric, a flag is interrogated to see if it is the first one that has been depressed. If so, the lower SSD is cleared to allow subsequent display of the numeric string accumulating in the buffer. The program then checks for the two special cases: CLEAR, which clears the key buffer and the display, and SEND which initiates transmission of the entire string. If it is neither, the key is translated to an ASCII numeric value and added to the buffer, making sure in the process that it does not overrun the available space (if it does, it can be assumed that someone is playing, and the whole thing is cleared).

When a numeric string is sent, by the way, all the numeric lamps are extinguished. This protocol with the lamps allows easy direction of operator attention and action.

After all this, it is reasonably obvious what takes place if the executive detects, not a KEY flag, but a newly depressed pushbutton. It checks immediately for multiple depressions, then compares the switch map to the lamp map which represents the currently illuminated bulbs. If the switch is not associated with an illuminated lamp, it is ignored completely. Otherwise, the KEY flag is set and the process described above begun with an appropriately initialized BOUNCE counter.

14.4 IDAC INTEGRATION

At last. Somewhere in this book, we had to plunge headlong into the guts of a system and see how the logic that we have been obliquely referencing actually works. A gritty discussion of IDAC software is hardly casual reading, but hopefully it provides a reasonably tangible image of the steps a processor must undertake to accomplish the trivial WHAT-level task so blithely characterized in one sentence at the end of Section 14.1.1.

We have seen how it works, but how did we manage to convert the original, flawed conceptualization of this into a solid and correctly functioning system? As Figure 12-1 rather humorously suggests, writing a piece of code, assembling and burning it into ROM, then plugging it into the target hardware is a process fraught with much struggle. The IDAC software was developed with the aid of the RAM-port, described in Section 12.2.4.

Construction of hardware and initial coding proceeded more or less in parallel (which is quite a trick, given the fact that I worked on the project alone), culminating in a great unknown mass of modules and a great unknown agglomeration of wires and chips. At this point, the initial smoke test was undergone with everything sensitive unplugged, then when it seemed safe, power was applied to the CPU, the interface card, and the front panel devices. It was still quite untestable, of course, with the exception of some more or less isolated pieces of logic (such as the baud-rate generator) which were independent of the software.

The RAMport was plugged into the ROM0 socket, which at this point was jumpered to believe that the device was a 2716 EPROM. (This was an endless pain, by the way: a 2708 accidentally plugged into a 2716 socket doesn't even last long enough to appear on the bus; the +5-volt level on pin 21 irrevocably and permanently makes all addresses appear to contain FF.)

Anyway, with the SBC's program memory now resident in BEHEMOTH's address space, it was possible to begin executing test routines. Piece by piece, the hardware was determined to be correct (or fixed if it wasn't). First with a scope, then with the front panel devices, data were followed through the bus structure until I could be reasonably certain that it all worked.

Then came the code.

Figure 14-12 shows the configuration of the system during the software debugging process. This is the same as the corresponding drawing of Chapter 12, with the addition of the communications line linking the host with the terminal. Since all the terminal's software is in one way or another involved with communications, it was necessary from the very inception of debugging to provide a means of simulating the host computer.

In practice, this amounted to moving the most recent version of the code into the RAMport's address space, placing it into the external mode, removing the reset from the SBC, and observing the terminal's behavior while executing from the console a program called COMSIM—the communications simulator. Messages could be typed on the keyboard and replies seen on the screen, allowing all aspects of the IDAC code to be explored from the host system's viewpoint.

Finally, after many days (weeks?) of effort, the umbilicus was removed, and ROMs plugged in to replace it. It was working. Producing other terminals then became a relatively simple task—with the software replicated by the system and provided in the form of a pair of chips, labeled "IDAC Rev. 36."

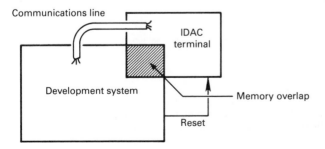

Figure 14-12 Software and hardware development of the IDAC terminals were aided with the use of the RAMport, together with a communications line to allow simulation of the host computer.

14.5 FINAL ASSEMBLY AND TEST

The mechanical configuration of the terminal's hardware components is represented by the sketch of Figure 14-13. The general intent is the optimization of the various requirements presented by front panel layout, heat distribution, noise, cable requirements, and aesthetic appeal.

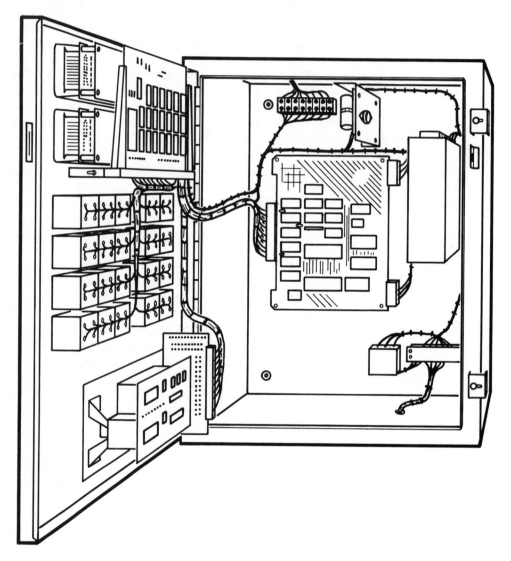

Figure 14-13 Internal view of the finished terminal. All logic not embodied in the software itself is contained on the small board mounted behind the two displays at the top, and is interconnected with everything else via the wiring harness. In normal operation, the sealed enclosure is padlocked shut.

The interface card is mounted on a bracket fashioned from aluminum angle and channel, and is kept from swaying back and forth by a pair of card guides. Its edge connector is the focal point of a massive wiring harness, consisting of various colors and sizes of tightly bundled stranded wire, interconnecting virtually everything in the enclosure. The dc power has its own private harness (partially for noise purposes, and partially because a short circuit that melts a V_{cc} wire or two would be even more depressing if it fuses a few dozen signal lines as well).

The components located in the back of the enclosure—power supplies, ac distribution hardware, the CPU board, and the comm line barrier strip—are fastened to a panel which is itself bolted onto four threaded studs welded into the back of the cabinet (they come with it). This enabled the more mechanical aspects of the assembly to be performed in a relatively convenient fashion.

Once the hardware and software were cajoled into productive cohabitation, it was time to do some testing. First tentatively, then with a measure of gusto, things were done to the machine which under normal circumstances would have set the designer's teeth on edge. The machine's ability to withstand rapid power cycles, elevated temperatures, a Jacob's ladder, comm line reversal, sharp mechanical transients, degradation of ac line voltage, human abuse, and foul language were all tested in turn, followed by a full week's burn-in during which a BASIC program running in the development system kept up a barrage of communication to the lights and displays.

It was then shipped. Installation and early field trials were not without tribulations, but the terminal (and its brethren) soon formed the basis for the plant-wide network that we described in Chapter 13.

All things considered, the project came together rather smoothly. A few months of use in the factory helped refine the design somewhat, encouraging various changes that we have noted throughout the discussion.

Ultimately, the units were placed under a maintenance contract, which, considering their reliability, was probably the most profitable part of the whole job.

15

A Potpourri
of Applications

15.0 INTRODUCTION

We have bandied about philosophies at levels ranging from pushbutton interfacing to natural language understanding. We have irreverently poked about inside bus structures, considered the special requirements of industrial environments, and examined some software tools. We traced the system development process, even exposing a few fiascos that ignored an important tenet or two. And we explored a real system in rather lengthy detail.

We should do one last thing—devote a chapter to a celebration of possibilities, of things done and things yet to be done.

Microprocessors are so all-pervasive that it is absurd to try to deal with them as a collective "field." I suppose we could talk about "pure micros" just as people talk about "pure math," but the real beauty of these beasts is their generality and applicability to almost every engineering and scientific discipline. Although we have indeed discussed real applications quite a bit, we have hardly begun to suggest the widely divergent areas in which these delightfully flexible tools can be applied.

Let us then move toward that long-lusted-after "THE END" with a spirited overview of microcomputer applications that have at least a vaguely industrial character.

15.1 WORKING IN THE FREQUENCY DOMAIN

There are many phenomena out there that are only clumsily approached with the classic, intuitive methods. If you were interested in observing the frequency of a flute note, for example (and were not blessed with perfect pitch), you could always convert it into a waveform with a microphone, amplifier, and oscilloscope. You could then determine its frequency by counting the number of graticule lines from one zero crossing to the next, multiplying by the time/division switch setting, and taking the reciprocal. Or you could apply the signal to a frequency counter.

There are ways to automate these processes with a small computer, but they quickly become rather useless when the input waveform is not a pleasantly sinusoidal flute note but is instead a complex, real-time phenomenon that is visually indistinguishable from random noise. If you were building bridges, for example, and wished to avoid the embarrassment of having a structure fall down when subjected to certain combinations of wind and truck traffic, you might be strongly interested in the frequencies at which the bridge resonates. This information would be useful in the "fine tuning" of the design to avoid such cataclysmic and abrupt ends to one's civil engineering career.

But how might this information be obtained? It is a reasonably simple matter to subject the structure to an impulse (which, by definition, contains all frequencies) and monitor various points with accelerometers to determine resonant modes. The problem is that the signals observed at the sensors look somewhat like garbage. But they are not.

The information actually desired here is the frequency spectrum—a plot of the amplitudes of the various sine waves which, in concert, yield the final signal. In the bridge example, this is complicated by the fact that the spectra are somewhat time-variant, but in essence, such a plot derived from an accelerometer at some arbitrary time after the impulse would represent the resonant modes experienced by that particular element of the structure.

The process that needs to be performed here is comparable philosophically to the operation of a classic spectrum analyzer, which in one fashion or another sweeps a high-Q filter through the frequency range of interest and produces a plot of the output amplitudes yielded by the application of the filter to the input signal.

But there's a faster and more accurate method that lends itself to all sorts of analytical applications: the *Fourier transform*. Given a time-domain waveform (presumably derived by sampling an analog input channel), the application of the transform produces a corresponding frequency spectrum which can rather trivially yield information as to the relative phase and amplitudes of the various components that make up the original signal. Most commonly used is the *fast Fourier transform* (FFT), which is an approximation of the continuous transform that takes advantage of certain symmetries by using a power-of-2 number of samples (or a small integral multiple thereof).

The general effect of this domain transformation is shown in Figure 15-1. In parts a and b, two time-domain waveforms are shown which are each formed of the same three frequency components—a fundamental and its first two odd harmonics (1000, 3000, and 5000 hertz, for example). The two harmonics are at the same amplitude, which is half that of the fundamental. The difference between

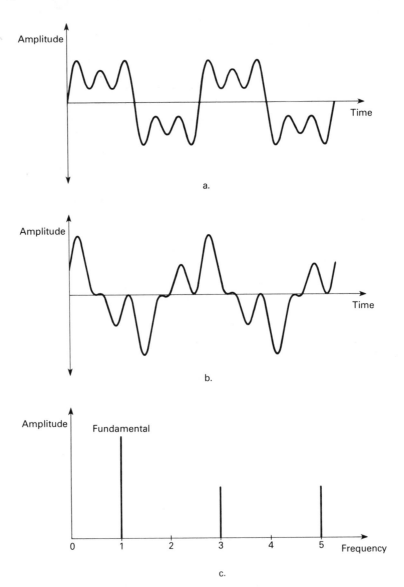

Figure 15-1 Signals (a) and (b) each consist of a fundamental and its first two odd harmonics at half the fundamental's level. The difference lies in the relative phase, something not revealed by the simple spectrum in (c).

the signals is due to the fact that in the first case, the three components all start out in phase, but in the second case, the fundamental lags by 90 degrees. This clearly has a profound effect on the output.

If a "pure" transformation from time to frequency domain were applied to the signals of parts a and b, the result would be as shown in part c. It can be seen that this type of plot would far more readily satisfy one's curiosity about the frequency components found within the signals than would the signals themselves—for it is a direct representation of amplitude vs. frequency. Note, however, that phase information is not represented. In some applications, it is superfluous, but in structural analysis and things like that it matters quite a bit. Fortunately, the FFT algorithms that exist to serve us provide a "phase plot" as well.

Unfortunately, real FFTs don't produce nice, clean-cut spectra like the one in Figure 15-1c. To see why, let's take a look at Figure 15-2.

The waveform of interest (part a) is a continuous cosine wave. Its "scope trace" is on the left, and corresponds to its pure spectrum shown on the right—a pair of infinitely narrow peaks centered around the zero frequency. This is meaningless in the real world, of course, for there is no negative frequency. It can be considered an artifact of the transform process.

This is very nice but isn't at all realistic. One cannot operate on a signal over all of time, for at some point it is necessary to stop looking at the data and produce some usable result. This spawns the need for a "sampling window" which defines the period during which the input waveform is discretely sampled (by an A/D converter, or whatever) and stored in memory. Unfortunately, the window itself has a spectrum and has to be considered right along with the signal of interest, because the actual data provided to the FFT algorithm is the point-by-point multiplication of the window and the cosine wave.

To understand the effect of this upon the result, we have to look at the spectrum of the window alone, shown in the right half of Figure 15-2b. It is a $(sin\ x)/x$ curve centered about the origin.

In the next drawing, the windowed cosine wave is transformed. Note the effect on the spectrum! Suddenly, there is some uncertainty concerning the narrowness of the spectral peaks corresponding to the positive and negative values of the signal's frequency. Instead of having all the energy in one vertical line, the plot implies that there is a rather bumpy distribution of the energy around the point of interest. But we know this to be untrue. The phenomenon is called *leakage*, and is the result of this multiplication between the time-domain signals: their convolution in the frequency domain. This unavoidable problem is a direct result of analyzing a finite, rather than an infinite, data record.

The severity of these confusing side lobes can be ameliorated somewhat by shaping the sides of the window in certain ways, accepting the concomitant attenuation. Depending upon the type of information that is desired from the spectrum, which may include considerable additional processing (transfer functions, Nyquist plots, etc.), different special windows can be applied to the data sampling. Some of the better known are Hamming, Hanning, Kaiser–Bessel, and exponential. They have the general effect of reducing the leakage associated with the FFT process.

There is one other effect that has to be noted in the same context as the windowing phenomenon. Just as the data record derived from the original signal is not infinitely long, the space between the samples is not infinitely small. The sampling rate selected for a signal-processing task has a couple of important implications.

First, it must be greater than twice the value of the highest-frequency component of interest in the signal. Failure to accomplish this results in a phenomenon known as *aliasing*, in which a sinusoidal component above one-half the sampling frequency appears upon transformation to be much lower than it actually is. This can be readily demonstrated by drawing a sine wave and a set of too-widely spaced sample points, as has been done in Figure 15-3. Connecting the dots yields a phantom sine wave that is obviously much lower in frequency and which would certainly have disastrous effects upon any signal processing task dependent upon it.

This minimum acceptable sampling rate, by the way, is called the *Nyquist frequency*.

Second, this concept has a rather interesting effect upon another aspect of the Fourier transform itself. Think about the series of values that would comprise the data record in memory if a sine wave of precisely one-half the Nyquist frequency were sampled. The information describing the shape of the wave would be lost; the record would consist of simply a series of impulses. Yet the existence of a sine component at that frequency would be unambiguously indicated by the computed spectrum.

This suggests that the sample "clock" itself, just an infinite train of impulses, possesses a spectrum comparable to a sine wave of half its frequency (despite the implication of all frequencies in the term "impulses"). This is shown in Figure 15-2d, actually a third function that has to be multiplied in the time domain by the signal of interest.

The convolution of these three functions is shown in Figure 15-2d and e. Compared to the ideal spectrum in part a, the reality of part e is a bit depressing. But such are the sacrifices one makes when subjecting the continuous and infinite functions of the real world to the discrete and finite representations of a computer. The result of an FFT is a set of spectral windows in contiguous memory addresses, representing the convolved transformations of the signal of interest, the sampling window, and the sampling rate itself. But the utility of the operation is unquestioned.

Consider, for example, the problems associated with maintaining large turbines such as those found in power-generation facilities or jet aircraft. The effects of failure vary greatly between those two applications areas, but they have one thing in common: *catastrophe*. There is thus a brisk market for vibration analysis equipment that not only points out a dangerous imbalance, but detects tiny abnor-

Time domain   Frequency domain

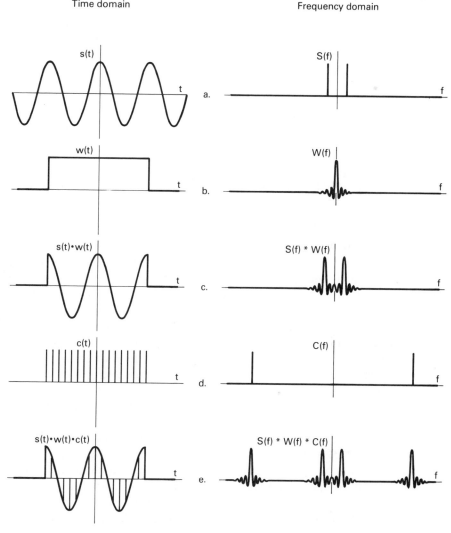

Figure 15-2   The fast Fourier transform (FFT), although extremely useful in the analysis of time-domain phenomena, introduces a number of artifacts that must be taken into account in a design (see the text). (Based on Figure 7 of "A Guided Tour of the Fast Fourier Transform" by G. D. Bergland, *IEEE Spectrum*, vol. 6, pp 41–52, July 1969, © 1969 IEEE.)

malities, detectable as frequency-domain glitches, which may result in later failure.

And then there is the automobile business. If you were attempting to build a comfortable car, you would like to know how it feels to the driver under all road conditions. Does the complex mechanical structure comprised of tires, shock ab-

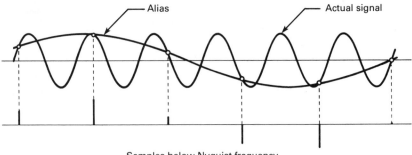

Figure 15-3   Sampling a signal at or below twice the highest frequency of interest can yield "aliasing" errors appearing as phantom spectral components of lower frequency.

sorbers, chassis, isolation mounts, and seats couple unpleasantly high amplitudes of vibration or displacement to the human being? With techniques of finite-element modeling and modal analysis, the perceptible effects of vehicular bounce, pitch, torsion, and the like can be determined before customer complaints begin to pour in. It's hardly a trivial set of calculations, but it is even less so without computers and frequency-domain analysis.

There is much more besides. Image recognition, about which we spoke in Chapter 9, depends heavily upon the FFT, as do many methods of speech processing. Digital filtering finds applications in any field wherein analog filters might be of use, and the computer simulation of analog waveform processing circuitry is a common engineering tool. Other digital signal-processing techniques apply to Doppler radar systems, analysis of biomedical signals, seismic exploration, and so on.

This body of frequency-domain techniques is a potent tool, and now that numerous MSI and LSI devices exist to turn many of the once-formidable logical cranks, it can be added to your design repertoire without a hard-core plunge into impossibly complex number crunching. In fact, 1024-point transforms can even be performed by S-100-sized boards with relatively low-key software support, offering speeds that allow real-time analysis of audio spectra and most mechanical phenomena.

15.2 SHARING THE TASK

It is not at all uncommon for the requirements of a system to call for some distribution of processing resources. It may be that a system that could single-handedly perform the job would be too expensive (either in dollars or programming time), or the nature of the task may be such that multiprocessing is the only intelligent way to go about it.

As microprocessors continue to offer more and more "horsepower" per dollar, it becomes ever more reasonable to apply them in a distributed fashion that would, not too long ago, have been absurd. This is clearly appropriate in the AI world, where the brain is an architectural model, but is becoming so in other fields as well. We already spoke of one case—the distributed industrial system.

But that system does not really exemplify the idea. Although it does provide task distribution by using a number of processors, they are loosely coupled and are unable to share the load of a given problem except in a relatively abstract sense.

The insertion/withdrawal force tester (IWT) of which we have spoken at various points (refer to Figure 5-17) is a good example of somewhat more tightly coupled processors. Even though one of them, the 9511 floating-point processor, is specialized and interfaced as a support device, it has to be considered as a computation site in its own right. By relieving the 8085 of the need to deal with repetitive and lengthy mathematical operations while simultaneously worrying about machine timing, measurement, and control, it actually lowers the design cost of the system. Much more thought would have otherwise been necessary (and, possibly, a more robust processor).

There are a lot of ways to accomplish this load-sharing idea, depending upon the nature of the application. Arithmetic processing capability is available in the form of a "coprocessor" which inhales the same instruction stream seen by the host—something that may not always be appropriate. A number of slave CPUs can be placed on an accommodating bus (such as Intel's Multibus), or they can be arranged in an I/O connected network.

The important thing here is not the detailed nature of the interconnection but the idea of subdividing the task. In the human brain, there is a rough pyramidal structure that loosely maps onto the morphology of the system, extending from the high-level, conceptual activity of the cortex through the "system housekeeping" functions of the *diencephalon*, down further through the I/O processors of the *medulla* and *pons*, at last to the bit-level control functions of the spinal cord. Throughout this complex hierarchy, there exists massive parallelism which enables such wonders as the real-time assimilation of a complex battery of continuous sensory input. The evolution of the brain emphasized the complexities of survival and adaptation, eschewing the orthogonal capability of linear mathematical operation. In contrast, we use as our basic computing tools devices that evolved in a climate of data processing—of mathematics. It has always been possible to crank a little more speed out of a machine and make it appear "smart," but now that applications complexity has grown beyond the level at which that approach works, it is necessary to reconsider the problem. The design of the brain has withstood the test of time.

Why not borrow from it?

15.3 SHARING THE PROCESSOR

A totally opposite veiwpoint to the one expressed above is appropriate in many a situation. Consider a production batching operation.

Typically, this might consist of a terminal under control of the "batch man," a printer that produces tickets, and some control logic that interfaces with the batching equipment itself—valves, gates, scales, moisture meters, conductivity and temperature sensors, hydraulic actuators, and so on. Although a number of different tasks must be performed, many of them simultaneous yet asynchronous with respect to one another, there is probably not a massive processor load associated with any one.

It therefore makes little sense to distribute CPUs throughout the system if the total processing time is substantially less than that which could be provided by one, used intelligently. It is to this end that the techniques of multitasking (time slicing, etc.) are useful.

This, of course, is simply time sharing—that classically "big system" operating mode that allows one computer to use a large number of humans. But microcomputers aren't the sluggards they once were, and even a single 8085 or Z-80 can run circles around some of the formerly respectable machines in the airconditioned computer rooms of yesteryear. And the new super-zoomers ... well, let us just say that the definition of "minicomputer" has been forced to change recently under the onslaught of highly capable integrated systems.

At the heart of a multitasking system is a real-time executive that "context switches" as necessary to service the independent tasks, presenting to each a virtual CPU that, as far as the task is concerned, belongs to it alone.

This has all sorts of applications. As we mentioned, batching systems might be amenable to this approach, as might other processes that require a fair amount of operator, printer, and/or host system interaction. The limitation arises when one task becomes so demanding that it monopolizes the system (effectively terminating lower-priority jobs) or when there are so many tasks that the overhead associated with switching between them unreasonably burdens the system.

15.4 ROBOTICS

Taking another major conceptual leap (after all, this chapter *is* called a "potpourri"), let's look briefly at some applications areas that fall within the field of automata.

There is an undeniable, economically motivated shift toward robotics in manufacturing environments. This, although it appears revolutionary to certain panicky special-interest groups, is but a logical extension of factory automation which began over a century ago. Albeit primitive by comparison, numerically controlled lathes, automatic presses, conveyor systems, and wirewrapping machines are all "robots" in the sense that they apply a measure of logical control and feedback to a manipulative task. The difference manifested by the ones that make people nervous is adaptability—something that has traditionally been a uniquely human attribute.

Here we have a bonanza in microcomputer applications. Even a single-arm manipulator presents the need for task training, monitoring of stresses in joints, control of servos and steppers, visual and kinesthetic feedback, overload protection, and a level of planning that accomplishes motions in three-dimensional space with something approaching economy. Just as the human system profits from the use of a processing hierarchy which provides dedicated controllers at the muscular level, so can a robot design gain from the use of multiple processors that free the goal-direction logic to dedicate itself to its complex task. With the unlimited range of application requirements presented by industry, there will be "micro" design opportunities here for years to come.

This is closely involved with the creation of sensory systems that purport to mimic those of the human being. We discussed in Chapter 9 some of the problems that vastly complicate vision—but it is something we are going to have to figure out sooner or later. The utility of a general-purpose robot is largely related to its ability to garner meaningful information about the world and act on it in real time. Systems that look about with a TV camera and then sit inert for five minutes furiously applying numerical feature-extraction algorithms to what they saw are interesting for a while but not very useful. Systems that require their visual input to consist of one and only one mechanical component (with which it is already familiar), oriented in any way on a uniform, featureless background, are perhaps useful in highly reduced situations, but they can hardly handle the complex range of sensory data routinely assimilated by even the least skilled human machine operator. Picking up a bolt, for example, is trivial when the part is resting on a light table and presenting a sharp profile to a camera, but surprisingly complex when it is oriented randomly in a large bin full of similar bolts.

It should be noted that work under way in the research environment is always considerably more advanced than the applied systems earning their keep in a factory. It takes a while to make the conversion. But even in the ivy-covered halls of the "ideal" world—perhaps especially there—a keen awareness of the remarkable distance we have yet to progress pervades all efforts. It is a field that will be rife with entrepreneurial fervor for quite some time.

15.5 MISCELLANEOUS APPLICATIONS

This is one of those grandiose titles that cannot help but evoke a chuckle in anyone even vestigially aware of the micro's ubiquity. It is included here as a parting gesture—something that may spark an idea or two, or at least render a hopelessly sketchy image of some of the widely divergent tasks to which these tools have been applied. To wit:

A 6800-based unit monitors the process of rolling steel, at rates up to 6000 feet per minute.

The World Trade Center saves energy with microprocessor control of lighting throughout the 110-story building, with human override of default patterns performed through Touch-Tone phones.

Ac motor efficiency and life are improved with adjustable frequency control techniques which allow precise synchronization between motors, decreased energy requirements, and reduction of electrical and mechanical surges through "soft starting."

An 8085-based multitasking system controls the process of batching concrete, with automatic queueing of jobs and accommodation of different levels of moisture in the aggregate components.

The transient response of a linear system's transfer function can be modeled with a micro, then plotted to observe step response and other nonobvious characteristics.

The penetration rate of a hydrostatically driven blasthole drill is automatically optimized by monitoring hydraulic feed pressure, air pressure in the drill blowout system, and current consumption of the drill's rotational drive motor.

Micros in Danish fuel delivery trucks meter the oil and print customer tickets.

A 16-bit processor controls a robot spot-welder, controlling rotational motion to a resolution of 1% of a degree and translational motion to 0.001 inch over a 36-inch span.

Smart digital panel meters with plug-in keyboards handle localized closed-loop control tasks.

An SBC displays the thermal profile of an oven used for the continuous processing of thermoplastic resins.

The sewer systems of many cities are routinely monitored for flow rate and depth by CMOS processors equipped with ultrasonic sensors.

Aluminum foil is wound on institutional-sized rolls using low-cost microprocessor-based controllers.

A major appliance manufacturer uses plant communication and data-base systems for quality control of refrigerator components, automated test of dishwasher timers, and computer-aided assembly of washing machine pumps.

This is all thoroughly insane, of course. Any attempt, even if admittedly halfhearted, to catalog a representative sampling of microprocessor applications is akin in its absurdity to an equivalent attempt to relate the pervasiveness of the wheel—or the transistor. There was, perhaps, a time when micros were rightly considered esoteric, but that was before the combined effects of customer demand, human ingenuity, and economic motivation brought about their metamorphosis from complex and confusing devices to powerful and well-supported systems.

Oh, they have their problems: people. The ones who can apply an intuitive grasp of the technology are few and expensive. No application can be undertaken without a software effort, and that activity is yet so imbued with black magic that another device generation will probably pass before it stabilizes into a staid, and not terribly entertaining, engineering profession. But there will always be a "cutting edge"—always that quiver of excitement in the latest issues of the trade journals heralding memory densities and integrated processing power that were once but idle musings over the coffee cups of visionary tinkerers. There will always be magic in the technology of knowledge, the teaching of machines, and the seemingly endless compression of processing power into tiny spaces.

And there will always be, as long as the economy survives, an industrial marketplace thirsting for the nectar of high technology.

The gap between the established and the newly possible grows no smaller. It's a wide-open field.

Index